Biological Interactions with Surface Charge in Biomaterials

RSC Nanoscience & Nanotechnology

Series Editors:
Professor Paul O'Brien, *University of Manchester, UK*
Professor Sir Harry Kroto FRS, *University of Sussex, UK*
Professor Ralph Nuzzo, *University of Illinois at Urbana-Champaign, USA*

Titles in the Series:
 1: Nanotubes and Nanowires
 2: Fullerenes: Principles and Applications
 3: Nanocharacterisation
 4: Atom Resolved Surface Reactions: Nanocatalysis
 5: Biomimetic Nanoceramics in Clinical Use: From Materials to Applications
 6: Nanofluidics: Nanoscience and Nanotechnology
 7: Bionanodesign: Following Nature's Touch
 8: Nano-Society: Pushing the Boundaries of Technology
 9: Polymer-based Nanostructures: Medical Applications
10: Metallic and Molecular Interactions in Nanometer Layers, Pores and
    Particles: New Findings at the Yoctolitre Level
11: Nanocasting: A Versatile Strategy for Creating Nanostructured Porous
    Materials
12: Titanate and Titania Nanotubes: Synthesis, Properties and Applications
13: Raman Spectroscopy, Fullerenes and Nanotechnology
14: Nanotechnologies in Food
15: Unravelling Single Cell Genomics: Micro and Nanotools
16: Polymer Nanocomposites by Emulsion and Suspension
17: Phage Nanobiotechnology
18: Nanotubes and Nanowires: 2nd Edition
19: Nanostructured Catalysts: Transition Metal Oxides
20: Fullerenes: Principles and Applications, 2nd Edition
21: Biological Interactions with Surface Charge in Biomaterials

How to obtain future titles on publication:
A standing order plan is available for this series. A standing order will bring
delivery of each new volume immediately on publication.

For further information please contact:
Book Sales Department, Royal Society of Chemistry, Thomas Graham House,
Science Park, Milton Road, Cambridge, CB4 0WF, UK
Telephone: +44 (0)1223 420066, Fax: +44 (0)1223 420247, Email: books@rsc.org
Visit our website at http://www.rsc.org/Shop/Books/

Biological Interactions with Surface Charge in Biomaterials

Edited by

Syed A. M. Tofail
Materials and Surface Science Institute, University of Limerick, Ireland

RSC Publishing

RSC Nanoscience & Nanotechnology No. 21

ISBN: 978-1-84973-185-0
ISSN: 1757-7136

A catalogue record for this book is available from the British Library

Published by The Royal Society of Chemistry,
Thomas Graham House, Science Park, Milton Road,
Cambridge CB4 0WF, UK

Registered Charity Number 207890

For further information see our web site at www.rsc.org

Preface

Similar to many other objects, tools and devices that we use on a day-to-day basis, electrical charge can be present at the surface of medical devices, many of which are insulators. Even for the devices or materials that are not insulators semi-conducting and insulating properties of the top surface may still produce surface charge.

Electrostatic interactions between a solid surface and a biological species have long been considered to be a contributing factor to the adhesion of biological species on the given solid surface. Despite this, electrical behaviour and surface charge are not currently considered important in the design of mainstream biomedical devices. There are problems such as encrustation or plaque formation that often compromise the long-term performances of indwelling medical devices. Knowledge of the surface charge in biomaterials is important as it can provide means to minimize such deposits at the device surfaces. On the other hand, bone growth has long been known as responsive to electrical stimulation, and information on surface charge creation in bone or dental implants can foster healing and improve bone-implant bonding.

A number of metal oxide nanoparticles show high photocatalytic activity, which originates from the ability of these materials to generate surface charge that breaks water to form antimicrobial reactive oxidative species. These reactive oxidative species are very effective against Gram-positive bacteria such as the MRSA superbug. An understanding of the generation of surface charge and its associated photocatalytic property can lead to novel methods of exploiting photocatalytic nanoparticles more responsibly and more effectively in applications in which conventional methods have proven ineffective, *e.g.* in combatting hospital-acquired infections.

The origin of this book lies in many of its contributing authors' engagement in research and education in the field of manipulating surface charge in biomaterials through electrical, chemical and electro-chemical methods to mediate

RSC Nanoscience & Nanotechnology No. 21
Biological Interactions with Surface Charge in Biomaterials
Edited by Syed A. M. Tofail
© Royal Society of Chemistry 2012
Published by the Royal Society of Chemistry, www.rsc.org

biological reactions to a desired end. A consortium, Bioelectric Surface, funded by the European Commission under the FP-7 NMP (Nanosciences, Nano-technologies, Materials and new Production Technologies) programme, undertook a systematic study of electrical modifications of biomaterials to understand, control and exploit biological reactions. It was soon realized that information related to surface charge and its biological implications had been scattered over numerous specialist contributions in a range of scientific disciplines such as nanoscience and nano-biointeractions, nanomedicine, electrostatics, surface modifications, protein immobilization, polymer and ceramic processing, biomaterials, photocatalysis, clinical microbiology, biofilms, MRSA, textile processing, stent manufacture, restenosis, plaque resistance, encrustation, bone growth and so on. It was felt important that specialists from these varied disciplines require the learning the languages of each other to contribute effectively in this field.

This book is thus an attempt to bring more generalization into the concept of surface charge in biomaterials so that a researcher or a biomedical device engineer can find it useful to familiarize with the interplay of surface charge and its biological role that may influence the life and performance of a medical device. It provides non-specialist overviews of a number of concepts relevant to surface charge, its creation, control and measurements (Part I). The book then provides an overview of basic biological interactions such as protein and cellular interactions on solid surfaces (Parts II and III), before discussing some relevant biomedical applications in which manipulation of surface charge have been demonstrated or can be potentially exploited (Part IV).

The book benefits from the contributions not only from experts in different specialised fields such as nanomanipulation, protein immobilization or surface charge measurements, but also from practitioners such as clinicians and device engineers from industry. I would like to express my gratitude to all of them. The subject matter of this book is highly interdisciplinary and goes well beyond my own area of expertise. I would also like to thank Dr. Tewfik Soulimane, a structural biologist, who has played a critical role in planning the outline of this book.

I would like to thank Dr Merlin Fox, Commissioning Editor of the Royal Society for Chemistry, for inviting me to edit this book and for his patience and flexibility during the various stages of preparation and publication of this book. Finally, I would like to thank my wife, Sanjida, for her support and my daughters, Yasna and Iman, for being so lovely and accommodating during the preparation of this book.

Contents

RSC Nanoscience & Nanotechnology No. 21
Biological Interactions with Surface Charge in Biomaterials
Edited by Syed A. M. Tofail
© Royal Society of Chemistry 2012
Published by the Royal Society of Chemistry, www.rsc.org

Part III   Cellular Interactions with Abiotic Surfaces

**Chapter 8   Interactions of Bone-forming Cells with Electrostatic Charge
                at Biomaterials' Surfaces   107**
U. Hempel, C. Wolf-Brandstetter and D. Scharnweber

**Chapter 9   Interactions of Biofilm-forming Bacteria with Abiotic
                Surfaces   122**
S. Robin, T. Soulimane and S. Lavelle

.

Chapter 10 Endothelial Cells and Smooth Muscle Cells: Interactions at Biomaterials' Surfaces 136

M. Wawrzyńska, B. Sobieszczańska, D. Biały and J. Arkowski

Part I
Electrostatic Charge on Biomaterials' Surfaces

CHAPTER 1

Electrical Modifications of Biomaterials' Surfaces: Beyond Hydrophobicity and Hydrophilicity

S. A. M. TOFAIL* AND A. A. GANDHI

University of Limerick, Materials and Surface Science Institute, National Technology Park, Limerick, Ireland

1.1 Introduction

When a biomaterial is placed inside the body, a biological response is triggered almost instantaneously at the top few nanometres of the biomaterial. It is commonly understood that electrical properties such as local electrostatic charge distribution at biomaterials' surfaces play an important role in defining biological interactions *e.g.* protein adsorption and cell adhesion.[1,2,3] Protein adsorption is the initial event that takes place within the first few milliseconds at the biomaterial surface.[4] The adsorbed proteins interact with selected cell membrane protein receptors. The accessibility of cell adhesive domains (such as various specific amino acids of adsorbed vitronectin, fibronectin and laminin) may either enhance or inhibit subsequent cell attachments and proliferation.

Biological membranes such as those found in cell membranes are subjected to an electrical field gradient in excess of 10 mV m^{-1}.[5] In other words, a cell with a

RSC Nanoscience & Nanotechnology No. 21
Biological Interactions with Surface Charge in Biomaterials
Edited by Syed A. M. Tofail
© Royal Society of Chemistry 2012
Published by the Royal Society of Chemistry, www.rsc.org

thickness of few nanometres is subjected to an electromotive force of few millivolts. In a physiological environment, this substantial electric potential will essentially polarize its surrounding body fluid to form an electrical double layer. Similar double layers are formed around proteins and other biomolecules and biomedical device surfaces *in vivo*. Interactions of biomolecules with biomedical device surfaces thus can be seen as interactions between these double layers, which are further influenced by ionic interactions, chemotaxis and biochemical factors. The thickness of these double layers is dependent on the magnitude of the surface charge and usually lies in the nanometer range.

The adsorption of biological species is promoted primarily by two types of physical forces, electrostatic attractions and hydrophobic interactions. Most inorganic surfaces are negatively charged at normal pH, but usually they also contain hydrophobic domains. There can also be spatial variations of polarity of surface charge (Figure 1.1). A protein usually has positive, negative and hydrophobic patches thus enabling a protein to be attracted to most surfaces, on the one hand, by electrostatic attractions between positive patches and negatively charged groups on the surface, and, on the other hand, by hydrophobic interactions between hydrophobic domains of the protein and the surface.

For control and utilization of the adhesion and growth of biological species, *e.g.* proteins and cells, adhesion mechanisms on biomaterials used in implantable devices need to be properly understood. Biomaterials that are currently used in such devices, *e.g.* in cardiovascular and urinary stents and coatings in hip prosthesis, do not specifically address this interfacial phenomenon in device designs. For devices that remain in the body for medium to long term, biological interactions can cause encrustation, plaque formation and aseptic loosening in these device surfaces. These problems contribute to a patient's trauma and even increase the risk of death. A detailed knowledge of such interactions will not only help to produce a desired biological response but also

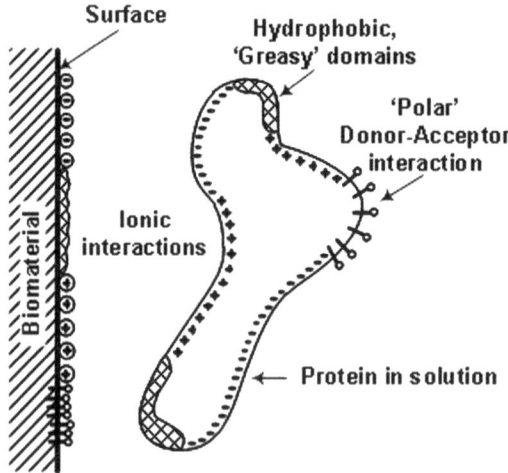

Figure 1.1   Protein adsorption on Biomaterials' surfaces (adapted from ref. 12).

pre-screen many inappropriate designs of biomedical devices long before any expensive animal or potentially risky clinical trials. Electrical modification of biomaterials can thus serve dual purposes: it can help in the understanding of biological interactions and, in turn, provide knowledge to engineer surface charge in biomaterials for improved medical devices.

1.2 Characteristics of Biomaterials' Surfaces and their Modifications

Relevant surface characteristics of biomaterials can be listed as follows:

- Contact angle (hydrophobicity/hydrophilicity)
- Electrical Charge
- Topography and roughness
- Porosity (pore diameter, interconnectedness, open and closed pores, pore orientation) and surface area
- Mechanical compliance/stiffness/hardness
- Water content (dry/wet)
- Chemistry (elements present, contamination overlayer, the presence of polar/non-polar group, dangling bonds, acid–base characteristics)
- Crystallinity, surface relaxation and reconstruction.

These characteristics are greatly affected and often unintentionally modified by: packaging and exposure to environment (*e.g.* hydrocarbon contaminants), handling (*e.g.* contamination, alteration of topography), storage (*e.g.* residual stresses can result in dimensional changes) and sterilization process (temperature, water absorption, duration of heating, the use of ionizing radiation, the use of oxidizing gases and duration of degassing).

At present, biomedical devices are designed primarily for functional uses while maintaining the inertness of the device so that, when used, they do not create any adverse effect in the biological system. Biomaterials' surfaces form an interface between two entirely different systems: biological and non-biological. The surface properties of the biomedical device play a critical role in defining the biological response.[6] Present day surface engineering of biomedical devices is focused on modifying the topography, surface chemistry, wettability, water stability and water-binding properties of the biomaterial surface.[7]

In order to develop the next generation of biomaterials, emphasis must be placed on bioactivity rather than the material's bio-inertness. To achieve this end, there has been an increased interest in the modification of the surfaces chemically [*e.g.* by using plasma, surfactants or functional groups, or drugs), mechanically (*e.g.* by roughening the surface) and biologically (*e.g.* by modifying surface with ribonucleic acid (RNA)].[8] While a plethora of information is now available[9] that describes the effects of such modifications on biological interactions, the interdependent nature of such modifications poses a problem in determining the exact role of different processes in a biological environment.

Protein interactions with polymers and other biomaterials used in medical devices, particularly cardiovascular and blood-contacting surfaces, have been studied for a long time.[10,11,12] Table 1.1 provides a list of variables critical in designing biomaterials' surfaces for a specific application and their corresponding protein properties.[12] The importance of electrical properties of biomaterials' surfaces to trigger a specific protein response can be readily seen in Table 1.1. What is absent from current day biomedical device design is an adequate consideration of the surface charge, its distribution and the electrical double layer potentials, which are interrelated and critical in defining biological interactions (Table 1.1).

Surface engineering of biomaterials to produce a desired topography, morphology, chemistry or hydrophilicity/hydrophobicity can have important implications for surface charge to the extent that the beneficial effects of the engineering can be compromised. This, in turn, results in unreliable performance of biomedical devices *in vivo*. Another important issue is the effect of sterilization on surface charge and the performance of implantable devices. As it will be discussed in the following sections, the sterilization method can itself create surface charge in devices. Unfortunately, there are inadequate investigations and data available on the effect of sterilization. The current limitations of medical device design can be summarized as below:

- A focus on biofunctionality and bio-inertness
- Bio/non-bio interactions at the surface are largely ignored
- Surface charge at the device surface is not a design criterion
- Hydrophobic, hydrophilic and neutral patches on the same surface make control of biological response extremely difficult
- Quantitative insight to biological reactions at the surface is absent
- Performance of bio-inert surfaces is unreliable and contributes to patients' trauma and risk of death
- the effect of sterilization on device performance is not adequately known.

Table 1.1   Biomaterial and protein surface properties that determine protein interaction at the biomaterial surface. Properties in italics are not considered in current biomedical device design. Adapted from ref. [12].

Biomaterial surface properties	Protein surface properties
Topography	Isoelectric point (point of zero surface charge)
Surface chemistry	Net charge and charge distribution
Surface charge and distribution	Conformation
Wettability (hydrophilicity/ hydrophobicity)	Placement and nature of hydrophobic patches
Water stability	Binding of low molecular weight species
Water binding	Charge transfer groups or regions
Electrical double layer potential	Bound water and electrical double layer

1.3 Electrical Modifications of Biomaterials' Surfaces

Electrical modifications of Biomaterials' surfaces depend on the ability of certain biomaterials to possess, store or generate surface charge under suitable conditions. Metallic biomaterials are good conductors of electricity and do not store charge. Thin metal oxide layers grown naturally or artificially on these metals usually possess semiconducting properties. These layers, if grown sufficiently thick, can work as an insulator, which can be polarized electrically. Examples will be anodically grown oxides of Ti or Ta on Ti or Ta metals or alloys. Alternatively, a sufficiently thick layer of the metal oxide deposited by chemical or physical means can also be polarized. Common polymeric and ceramic biomaterials are insulators and can be used as coatings or devices and can be electrically modified by electrical polarization.

When insulators are polarized, and if they can store this polarization, they are called 'electrets': a material that has a permanent macroscopic electric field at its surface.[13] In reality, the term 'permanent' in the definition may be misleading as electret polarization is subject to decay due to thermal, internal and environmental reasons but this decay time can be sufficiently long. This has made possible wide applications of electrets in electronics and xerography. 'Dipolar electrets' are overall electrically neutral but possess a macroscopic electrical dipole moment, which gives rise to the 'vectorial' effect (discussed in Chapter 17). Slow cooling of an insulator from a high temperature in the presence of a high electric field produces a dipolar electret due to the dipoles being 'frozen' with a net orientation along the direction of the applied field. This is normally referred to as the 'contact poling' process. This gives rise to the dipolar charge originating from the alignment of dipoles in dipolar materials containing dipolar molecules, and in ferroelectric materials. 'Space charge electrets', on the other hand, possess a net macroscopic electrostatic charge that results from the addition of charge to the surface and bulk of a material by bombarding it with an electron beam, ion beam or corona, contacting directly with a charged electrode, or transferring ions to or from the material being 'poled'.

Electrical charges can be created and stored in an electret in the form of dipolar or real charges. Real charges, when they occur at the surface, are called 'surface charge'. When they occur in the bulk, they are called 'space charge'. Most insulating materials, especially organic materials, are space charge electrets. Even in a dipolar electret, space charges can form during the 'poling' process due to trapped charges in defects or grain boundaries from the surface to deep inside the bulk of the insulator. This is why the term 'surface charge' can often be ambiguous, as the method of creating such surface charge may also induce dipolar charges and, more occasionally, space charges.

It is also imperative to discuss the polarity of the surface after electrical modification. Surface charges or space charges with polarity similar to the adjacent forming electrodes are called 'homocharges' and those with opposite polarity are called 'heterocharges'. Homocharges may occur as a result of charge carrier injections from an electrode or charge source (electron beam, ion beam or corona), followed by the capture of injected carriers in traps near the place of charge injections.[14] While this sounds simple, the spatial distribution of such homocharges can be quite complex and, in the case of poling with the help of electron beam, can also give rise to significant amount of 'heterocharges' due to the dynamic balance between primary electron, secondary emission electrons and the positive holes created at the surface region due to electron emission. Electret domains created by electron beam can show both positive and negative charges at the surface and sub-surface zone.

A simple way of modifying surfaces electrically is to apply a coating which is known to be highly electropositive or electronegative. In the cases where a minimum chemical change at the surface is desired, artificial polarization is employed to produce an ordering of molecular dipoles or an uncompensated surface charge.[15] The generation of surface charge will ideally involve very little or no change in the chemistry, topography or morphology of the surface, as such a modification of the surface, in turn, can alter the surface charge characteristics. Various methods can be employed in this pursuit, such as:

(i) Thermoelectrical poling with solid contact
(ii) Liquid-contact poling
(iii) Corona discharge poling
(iv) The electron beam method
(v) Electromagnetic radiation poling.

Conventional thermoelectrical poling requires metallic coating on the two opposite surfaces of an insulator and may not be suitable for poling biomaterials. Yamashita *et al.*[16] has eliminated this problem by simply sandwiching a hydroxyapatite ceramic pellet between two platinum electrodes, which were removed after poling was complete.[17] The nature of the contact between the electrode and the insulator can have important implications for the polarity of the charge, as a small air gap can give rise to surface monocharges in addition to the dipolar charges. After the removal of the applied field, the total polarization gradually decreases to a quasi-steady level, which is mainly the remaining 'frozen' dipolar polarization associated with the difference in relative permittivity between the ambient and the poling temperature.

To our knowledge, liquid-contact poling has not been reported for poling biomaterials. In this method one of the electrodes is a conductive liquid such as ethanol or water.[18] Similar poling can be achieved by using a solid material that is liquid and conductive at the poling temperature. In this method the control of charge density is easier and the lateral charge distribution is uniform. Corona discharge poling is described in Chapter 4 and will not be described here. Electron beam poling is usually used to produce real charges stored in the bulk of the material. A 20 keV field has been used in a traditional scanning electron microscope to polarize local micron-scale areas of hydroxyapatite thin films.[19] Hydroxyapatite ceramic pellets have also been poled with lower energy electron beams ($<$1keV).[20] In a thermal equilibrium state a very stable distribution of negative charges can form. The electron beam poling method has the advantages of an easily controllable density, location and lateral distribution of the injected negative charge. Ion beam implantation or ion beam poling, while similar in principle, has the disadvantages of damaging the material due to the larger size of the ions.

Electromagnetic radiations such as Gamma ray, X-ray, ultraviolet (UV) or visible light can also produce electrets under an external electric field, which is usually applied through transparent conductive electrodes. Under an electric field, photogenerated carriers will be separated and will move to the electrodes of opposite polarities. These carriers may be trapped near the electrodes to create a space charge polarization. After removal of the radiation and the electric field, the charge distribution enters into a quasi-equilibrium state and decays. These electrets are less stable than the ones described earlier. The radiation method has rarely been used for polarizing biomaterials.

Similar to nanomagnets, 'nanoelectrets' can also be produced by exploiting the methods described above. Discrete nanodomains can be created at the surface of such electrets to obtain a surface charge at the nanoscale level and can offer the advantage of studying biological interactions with electrostatic charge at high lateral resolutions. The principles of electrical modification to obtain discrete electrostatic domains with high lateral resolution in the nano, submicron and micron scale can be listed as follows:

- direct manipulation of the surface by applying a dc electric field through a conductive tip (Figure 1.2) in a scanning probe microscope (SPM). This can be accomplished within the settings of an electrostatic force, a piezoresponse force or a Kelvin force microscopy. The nanoresolutions of SPM in lateral dimensions facilitate the achievement of discrete domains of preferentially charged electrostatic domains.
- inserting a pre-patterned conductive grid in the passage of ions from a corona source to the Biomaterials' surfaces

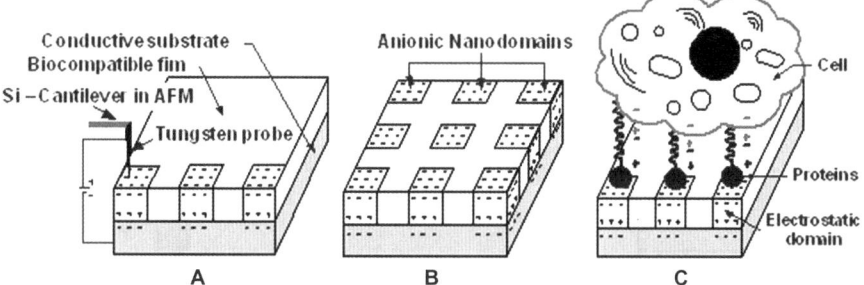

Figure 1.2   Schematic representations of electrical modification of planar Biomaterials' surfaces by using a conductive probe in scanning probe microscope (SPM): (A) the creation of electrostatic surface charge; (B) the creation of an array of nanodomains of electrostatic charge; (C) biological interactions with such nanodomains (not to scale).

- excitation of the space charge due to photon irradiation in photocatalytic materials
- electron beam lithography, focused electron beam writing or direct poling by an electron beam raster
- direct writing by a focused ion beam
- contact poling through a conductive grid
- low- and high-pressure plasma passed through a pre-patterned mask
- contact poling by arrays of nanowires/nanocolumns/nanodots (Figure 1.3)
- non-contact poling by electron beam through a pre-patterned mask (Figure 1.4)

Creating nanodomains by using a conductive probe in SPM can create such nanoelectret regions (nanodomains heretofore) at regular intervals with positive, negative or mixed polarities. An advantage of such technique is that the magnitude of the electric potential can be measured immediately after the creation in a second scan by the same probe by measuring electrostatic force or

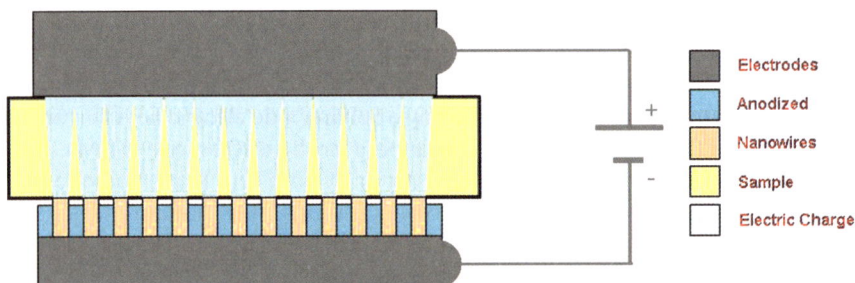

Figure 1.3   A schematic diagram of contact poling using anodized Al-and Ni-filled wires as an electrode.

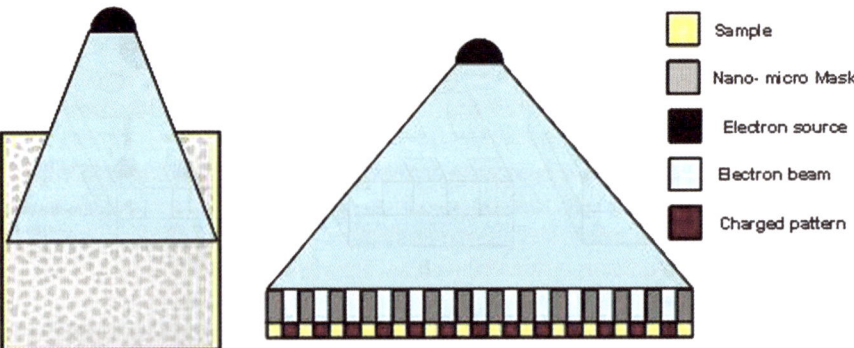

Figure 1.4   A schematic diagram of non-contact poling method using pre-patterned masks with micro/nano features.

Kelvin force (surface charge measurements by these techniques are described in Chapter 3). Sample heating can be accomplished by using a miniature heater on which the electret sample will rest during polarization. The creation of nano-domains by SPM is painfully slow and restricts its use in industrial applications. Non-contact methods such as corona discharge or electron beam poling methods can be employed for the creation of large area nanodomains in the shortest possible time.

1.4   Effects of Electrical Modifications: Beyond Hydrophobicity and Hydrophilicity

Electrical modification of orthopedic materials has attracted considerable attention due to the following findings:

- macroscopic polarization of hydroxyapatite (HA) increased bone like-apatite growth by a factor of six on the negative surface while no growth was observed on the positive surface,[16]
- similar selectivity of polarized hydroxyapatite has also been demonstrated in bacterial adhesion[21] but the selectivity was lost during the *in vivo* process,[22]
- controlled nucleation and growth of crystals from organic templates and in a number of biomineralization systems suggest that biominerals nucleate on a surface that exposes repetitive patterns of anionic groups.[23]

The *exact* role of electrical modification in defining biological interactions, *e.g.* adhesion of biomolecules and cells is still unclear. The original report[16] of polarizing hydroxyapatite claimed a high selectivity in the mineralization of calcium phosphates from a high-ionic-strength saline solution depending on the polarity of the charged hydroxyapatite electret. Although this remarkable selectivity has not so far been independently verified for hydroxyapatite, there is a working hypothesis that negative polarization in endoprosthetic devices has a beneficial effect in osteogenesis.[24] For example, *in vivo* and clinical studies on corona poled Ta-oxide (150–450 nm thick) and Teflon films (30–40 μm) on Ta have demonstrated that these electrets were able to stimulate physiological processes even after a protracted duration in conducting liquids.[25] A greater mineralization of cartilage tissues on the negatively charged surface was observed, while the positively charged implants exerted less effect on miner-alization, similar to an unpolarized specimen.

The Derjaguin–Landau–Verwey–Overbeek (DLVO) theory and the ther-modynamic method are the two approaches that are used in understanding the adhesion of biological species on solid surfaces. These theories, as it is described in Chapters 9 and 11, have undergone some modifications with the intention to describe biological adhesion. The DLVO theory, similar to the interaction of a colloidal particle with a solid surface, presupposes that the total interaction between a cell, for example, and a solid surface is a summation of the van der

Waals and electrostatic (mainly Coulombic) interactions. An insulator particle in an aqueous solution attracts counter ions against its surface charge to form a surrounding electrical double layer. As cells and natural biomaterials surfaces are usually *negatively* charged, a repulsive electrostatic energy is caused by the overlapping double layers of cells and the solid surface. This repulsive energy decreases as the ionic strength of the aqueous solution, due to shielding of the surface charges, by ions increases.[26] It is thus conceivable that within a conducting medium, such as the body fluid, the effect of the electric field emanating from an electret will be short range due to the screening effect. It is currently not certain how significant this electric field will be. The difference between the zeta-potential of a poled and an unpoled electret at physiological concentrations is usually too small to account for any significant influence on biological behaviour.

It has been long known that electrostatic contributions in wetting phenomena are substantial.[27] Conventional electro-wetting theories describe the changes in the cosine of the contact angle due to an external electric field as a quadratic function of the applied voltage.[28] This means that, despite the negative or positive polarity of the capacitive charge at the surface of the insulator, an increased surface potential will make the surface hydrophilic. Indeed, contact poling of hydroxyapatite ceramic exhibited a relatively higher *hydrophilicity* than an unpoled ceramic although the difference between the negatively and positively polarized surfaces was negligible.[29] Paradoxically, hydroxyapatite ceramics poled by low-energy electron beams have shown an increasing hydrophobicity as the surface potential increased due to poling.[30] This effect was annihilated by annealing the poled ceramic, which made it extremely hydrophilic.

Another important issue that needs attention is the sterilization process. The nature of sterilization, such as autoclaving, or exposure to gamma or UV radiation, an electron beam or an oxidizing agent such as ethylene oxide, may all should have implications for the charged electret. The exact nature of this effect is currently unknown. It can be safely stated that the observed *in vitro* and *in vivo* beneficial effects of electrical modifications of biomaterials is a combination of an intended polarization from deliberate modification and an unintended modification (enhancement or deterioration) due to sterilization.

1.5   Conclusions

Various methods of electrical modification of biomaterials have been discussed in this chapter along with their advantages and disadvantages. Electrical modifications to impart electrical charge in biomaterials have been reported to have beneficial effects, in particular in osteogenesis. Higher rates of calcification and cellular activities have been reported on the negatively polarized Biomaterials' surface. This general observation cannot be reconciled from a purely hydrophobicity/hydrophilicity of the surface as the current theory do not suggest any influence of polarity of the surface on wetting behaviour. More fundamental studies are required to obtain a proper understanding of the electrical modification on the reported selectivity in the stimulation of mineralization.

References

1. P. C. Zhang, A. M. Keleshian and F. Sachs, *Nature*, 2001, **413**, 428.
2. Y. Wang, S. Gao, W-H. Ye, H. S. Yoon and Y-Y. Yang, *Nat. Mater.*, 2006, **5**, 791.
3. C. C. Streimer, T. R. Gaborski, J. L. McGrath and P. M. Fauchet, *Nature*, 2007, **445**, 749.
4. L. Vroman and E. F. Leonard, *Ann. N.Y. Acad. Sci.*, 1977, **283**, 1.
5. R. Pethig, in *Electronic Conduction and Mechanoelectrical Transduction in Biological Materials*, ed. B. Lipinski, Marcel Dekker, New York, 1982.
6. D. F. Williams (ed.), *Medical and Dental Materials*, Willey-VCH, Weinheim, 2005.
7. H. Liu and T. J. Webster, *Biomaterials*, 2007, **28**, 354.
8. J. Michael, R. Beutner, U. Hempel, D. Scharnweber, H. Worch and B. Schwenzer, *J. Biomed. Mater. Res. B Appl. Biomater.*, 2007, **80**, 146.
9. D. M. Brunnette, *Titanium in Medicine: Material Science, Surface Science, Engineering, Biological Responses and Medical Applications*, Springer, Berlin, 2001.
10. R. E. Baier (ed.), *Adv. Chem. Ser.*, 1975, **145**; S. L. Cooper and N. A. Peppas, *Adv. Chem. Ser.*, 1982, **199**; M. A. Lopes, F. J. Monteiro, J. D. Santos, A. P. Serro and B. Saramago, *J. Biomed. Mater. Res.*, 1999, **45**, 370.
11. M. A. Lopes, F. J. Monteiro, J. D. Santos, A. P. Serro and B. Saramago, *J. Biomed. Mater. Res.*, 1999, **45**, 370; D. C. Miller, Nanostructured polymers for vascular grafts, PhD thesis, Purdue University, 2006.
12. J. D. Andrade (ed.), *Surface and Interfacial Aspects of Biomedical Polymers*, Plenum Press, New York, vol. 2, 1985.
13. L. S. McCarty and G. M. Whitesides, *Angew. Chem. Int. Ed.*, 2008, **47**, 2188–2207.
14. K. C. Kao, *Dielectric Phenomena in Solids: With Emphasis on Physical Concepts of Electronic Processes*, Elsevier, Amsterdam, 2004.
15. S. B. Lang, *Modern Bioelectricity*, Marcel Dekker, New York, 1988.
16. K. Yamashita, N. Oikawa and T. Umegaki, *Chem. Mater.*, 1996, **8**, 2697.
17. K. Yamashita, N. Oikawa and T. Umegaki, *Chem. Mater.*, 1996, **8**, 2697.
18. R. W. Chudleigh, *J. Appl. Phys.*, 1976, **47**, 4475.
19. T. Plecenik, S. A. M. Tofail, M. Gregor, M. Zahoran, M. Truchly, F. Laffir, T. Roch, P. Durina, M. Vargova, G. Plesch, P. Kus and A. Plecenik, *Appl. Phys. Lett.*, 2011, **98**, 113701.
20. D. Aronov, G. Rosenmana, A. Karlov and A. Shashkin, *Appl. Phys. Lett.*, 2006, **88**, 163902.

21. M. Ueshima, S. Tanaka, S. Nakamura and K. Yamashita, *J. Biomed. Mater. Res.*, 2002, **60**, 578.
22. S. Itoh, S. Nakamura, T. Kobayashi, K. Shinomiya, K. Yamashita and S. Itoh, *Calcif. Tissue Int.*, 2006, **78**, 133.
23. S. Aryal, K. C. R. Bahadur, S. R. Bhattarai, P. Prabu and H. Y. Kim, *J. Mater. Chem.*, 2006, **16**, 4642.
24. L. S. Pinchuk, V. I. Nikolaev, E. A. Tsvetkova and V. A. Goldade (ed.) *Tribology and Biophysics of Artificial Joints*, Elsevier, Amsterdam, 2006.
25. M. S. Morgunov, V. P. Homutov and I. M. Sokolova, *Proceedings of 8th International Symposium on Electrets*, Paris, France, 1994, pp. 863–868.
26. K. Hori and S. Matsumoto, *Biochem. Eng. J.*, 2010, **48**, 424–434.
27. L. F. G. Fokkink and J. Ralston, *Colloids Surf.*, 1989, **36**, 69–76.
28. M. Paneru, C. Priest, R. Sedev and J. Ralston, *J. Am. Chem. Soc.*, 2010, **132**, 8301–8308.
29. M. Nakamura, A. Nagai, T. Hentunen, J. Salonen, Y. Sekijima, T. Okura, K. Hashimoto, Y. Toda, H. Monma and K. Yamashita, *Appl. Mater. Interfaces*, 2010, **1**, 2181–2189.
30. D. Aronov, R. Rosen, E. Z. Ron and G. Rosenman, *Process Biochem*, 2006, **41**, 2367–2372.

CHAPTER 2

Photocatalytic Effects in Doped and Undoped Titania

M. KOPACZYŃSKA,*[a] M. VARGOVÁ,[b] K. WYSOCKA-KRÓL,[a] G. PLESCH[b] AND H. PODBIELSKA[a]

[a] Institute of Biomedical Engineering and Instrumentation, Wrocław University of Technology, Wybrzeże Wyspiańskiego 27, 50-370 Wrocław, Poland; [b] Department of Inorganic Chemistry, Faculty of Natural Sciences, Comenius University, 842 15 Bratislava, Slovakia

2.1 Introduction

Heterogeneous photocatalysis is an intensively developing field in environmental engineering. The preparation of environmentally friendly novel nanostructures, with the ability of selective photocatalytic degradation of organic and non-organic pollutants or water purification, is one of the major interests of researchers.[1] The use of semiconductor photocatalysts in chemical and biological applications arises from their environmental friendliness: they do not transfer any toxic contaminants during chemical precipitation and they do not produce any hazardous waste.[2] Many studies have focused on the crystalline forms of titania, namely anatase and rutile.[3]

Among the many candidates for semiconductor photocatalysts, titanium dioxide (titania) is widely known because of its self-cleaning properties, easy accessibility and capability for environmental purification.[4] Moreover, TiO_2 offers efficient photoactivity,[5] high photostability and relatively lower cost.[6] The photocatalytic activity of nanocrystalline TiO_2 materials has been studied for a long time since the report of the Fujishima and Honda effect in 1972.[7]

RSC Nanoscience & Nanotechnology No. 21
Biological Interactions with Surface Charge in Biomaterials
Edited by Syed A. M. Tofail
Published by the Royal Society of Chemistry, www.rsc.org

Since then titania has been widely used as a photocatalyst to generate charge carriers in, for example, photocatalytic water splitting, hydrogen sensing, antibacterial activity,[8–10] cancer treatment,[11] and reductive and oxidative process initiation upon exposure to ultraviolet (UV) light.[12] When UV light illuminates titania surfaces, two types of photochemical reactions can take place: a photo-induced redox reaction of adsorbed reactants, and a photo-induced hydrophilic transition of TiO_2.

Two general strategies have been developed to increase the photocatalytic activity of TiO_2 for visible light irradiation: the use of an organic dye as a photosensitizer or doping titania with metallic or non-metallic elements. The application of an organic dye that absorbs visible light has worked perfectly under conditions where oxygen is excluded and the degradation of the dye is decreased in the electrolyte solution by the efficient quenching of the dye oxidation state.[13] Whereas, the doping of TiO_2 material with metallic or non-metallic elements promotes the titania photoresponse to the visible spectra.[14] TiO_2 has been doped with various metals, including V, Cr, Mn, Cu, Fe, Ni, Pt, Ag, Au and rare earth metals, or with non-metallic elements, such as carbon, nitrogen, sulfur and other non-metallic elements, to introduce visible light absorption in titania.

Improving the spectral sensitivity of photocatalysts may significantly enhance the photoprocesses by absorbing distinctly more light. The design of more selective photocatalysts that can be activated by visible light and can be applied in medical diagnosis or therapy (in cancer treatment, implantology or as an antibacterial material) or novel photocatalysts for the production of energy resources (from biomass or from water splitting) are among the major goals of application of doped and undoped titania photocatalysts in biological and chemical applications.

2.2 Surface Charge Generation and Photocatalysis in Nanoscale Titania

Titania nanomaterials have an electronic band gap at approximately 3.0 eV and a very high absorption in the UV region. A titania photocatalyst absorbs photons with energy equal or higher than the band gap of TiO_2. Electrons are excited from the valence band (VB) to the conduction band (CB), leading to creation of electron–hole pairs. These charged carriers can recombine, radioactively or non-radioactively, or may migrate and become trapped at the surface and react with the chemicals (with electron donors or acceptors) adsorbed on the surface to decompose these chemicals.[15–18]

The fundamental photocatalysis process is expressed in eqn (1):

$$TiO_2 + hv \rightarrow TiO_2(h^+ + e^-) \tag{1}$$

where h^+ is the positively charged hole, e^- is the electron, h is the Planck's constant and v is the frequency of the incoming photon.

The efficiency of the heterogonous photocatalyst TiO_2 depends on its physicochemical properties such as: crystalline phase/structure and the concentration of surface defects in the crystal structure;[19,20] specific surface area and pore size;[21] the different degree of hydroxylation of the surfaces (density of hydroxyl groups), and the points of zero charge (PZC) of the photocatalysts;[22,23] size and agglomeration of catalyst particles, surface acidity, and the number and nature of trap sites (both in and at the surface).[24]

TiO_2 exists in three different crystalline phases: anatase, rutile and brookite. Anatase and rutile have been studied for their photocatalytic activity and both structures of TiO_2 have similar application as photocatalytic materials, but anatase showed a higher photocatalytic activity.[25] Anatase structures have a large specific surface area and a higher degree of hydroxylation of the surface than rutile structures. Some experiments showed that a mixture of anatase (70–75%) and rutile (30–25%) is more active than a single phase structure.[26,27] The high temperature needed to obtain rutile crystallites allows the agglomeration of particles and lowers the activity of the specific surface area. Amorphous TiO_2 is not photocatalytically active.

Particle size is a very important parameter in heterogeneous catalysis because of the influence of electron–hole recombination[28] and the influence on the efficiency of a catalyst due to specific surface area. Small particle diameter increases the band-gap energy and can enhance the redox potential of photogenerated electrons and holes, *i.e.*, smaller particles will be more photoactive than larger ones so they can enhance the catalytic activity further. A titania catalyst with a high specific surface area has many active centres to absorb reagents. Hydroxyl groups on the surface usually trap holes thus decreases the degree of recombination of excited electron–hole pairs. Too large a specific surface area indicates a presence of structural defects and thus increases the rate of recombination.

In addition to photocatalytic splitting of water, photocatalytisis has seen to result in a very high wettability called 'superhydrophilicity'. Some photocatalytic materials acquire superhydrophilic properties after UV illumination, which causes the formation of electrons and holes, but they react in a different way. If the surface is illuminated by UV light, the contact angle of water becomes smaller (a contact angle close to zero indicates that water spreads perfectly on the surface). The TiO_2 photocatalyst decomposes hydrophobic molecules present on the surface of material. The application of a titania photocatalyst as a superhydrophilic material is a recent interest for many environmental applications.

2.3 The Shift in Photocatalytic Band Gap: Rare Earth-Doped Titania

Recent studies have demonstrated that doping TiO_2 with metal ions of d or f electronic configuration can slow down or reduce the rate of electron–hole (e^-/h^+) recombination significantly. Furthermore, doping with metal ions can extend the photoresponse of TiO_2 into the visible region and inhibit anatase–rutile phase transformation.[29–31] Metal ions are incorporated into the TiO$_2$ lattice so that impurity energy levels are formed in the band gap of TiO$_2$. Such doping process also creates complexation centres on the TiO$_2$ surface, thus enhancing its photoactivity. However, there exists an optimum of doped metal ion concentration, above which the photocatalytic activity decreases due to the increase in electron–hole recombination.

The main objective of doping is to induce a red shift, *i.e.*, a decrease in the band gap or the introduction of intra-band gap states, which results in more visible light absorption. In fact a number of dopants mentioned in Section 2.1 are added to cause this 'red shift', of which metal ion doping is popular. The interest in modifying TiO$_2$ with rare earth metals from the lanthanide series arose due to the distinctive chemical, physical and electronic features of rare earth metals.[32] Lanthanide ions are able to form complexes with various Lewis bases through the interaction between the f-orbitals of the lanthanide metal and the functional groups of Lewis bases, thus providing an effective absorbability of organic pollutants on the TiO$_2$ surface.[33] Lanthanide ions are able to effectively trap conductive band electrons when they are confined to the TiO$_2$ semiconductor's surface.[34] The special electronic structure $4f^x5d^y$ of lanthanide ions could lead to different optical properties and dissimilar catalytic properties, and the redox coupling of Ln^{n+}/Ln$^{(n-1)+}$ is able to form the labile oxygen vacancies with relatively high mobility of bulk oxygen species.[35]

For lanthanide-doped TiO$_2$, the 4f level plays an important role. This energy state takes part in interfacial charge transfer and elimination of electron–hole recombination by acting as efficient electron scavenger.[36] Since the ionic radii of lanthanides are of a much larger size than that of Ti^{4+}, it is difficult for these rare earth ions to enter into the lattice of TiO$_2$ and replace the Ti^{4+} ion.[37] Some authors suggest that the lanthanides form oxide species dispersed on the surface of TiO$_2$. However, partial replacement of Ti^{4+} ions with Ce^{4+} ions in the TiO$_2$ lattice resulting in appearance of the new crystal phase of cerium titanate has also been reported.[38] Likewise, Wu *et al.*[39] suggested that La^{3+} ions can be introduced into TiO$_2$, which resulted in the increase of TiO$_2$ lattice parameters and cell volume and the formation of structural defects. However, lanthanide ion doping leads to formation of Ti–O–Ln structures, as many researchers have reported.[40] This results in the appearance of a new electronic state in the middle of the TiO$_2$ band gap. This introduces structural defects in TiO$_2$ lattice and changes the band-gap energy. Thus, the excitation energy is expanded from UV to visible light. Electrons excited from VB by absorbing visible-light photons could be captured by the inter-band trap site. The formation of two sub-energy levels (defect level and 4f level) in Ln–TiO$_2$ may be a key reason to eliminate the recombination of electron–hole pairs and to enhance the photocatalytic activity.

Based on above-mentioned beneficial properties of Ln ions, their modification into TiO$_2$ photocatalysts showed a significant improvement effect in the degradation of organic dyes. Moreover, such modifications can lead to the red shift of absorption edge of TiO$_2$ improving visible light activity. Enhanced photocatalytic activities were observed at certain doping content of different rare earth metal ions (La, Ce, Er, Pr, Gd, Nd and Sm).

A large red shift in Gd-doped TiO_2 can extend the absorption range to the visible region and enhance activity. Such a red shift may have originated due to a better crystallinity, or due to the appearance of a new electronic state within the band gap, which is conducive to change transfer transitions between the CB or VB of TiO_2 and the f electrons of rare earth ions. Half-filled orbitals of Gd were also reported to explain the enhanced photoactivity of Gd-doped TiO_2. Malato *et al.*[29] reported the highest efficiency of Gd-doped TiO_2 due to the higher ability to transfer charge carriers to the interface, while El-Bahy *et al.*[40] attributed the enhanced photoactivity to lower band-gap energy.

Liu *et al.*[41] demonstrated that 2,3-dichlorophenol (2,3-DCP) can be degraded under visible light irradiation with Ce^{3+}–TiO_2 hydrosol. The proposed mechanisms include the Ce 4f level function as the inter-band photocatalysis mechanism. In addition, the formation of the charge-transfer surface complex of TiO_2 and 2,3-DCP with visible light response contributed to the degradation of 2,3-DCP. Enhanced activity of Ce-doped TiO_2 under both UV and visible light has been reported in other studies and the higher activity was mainly attributed to the charge-transfer transition between earth ion *f* electrons and the TiO_2 conduction or valence band and the efficient separation of photo-generated electrons and holes.[33] Ce–TiO_2 samples also exhibit red shifts of the absorption edge and a significant enhancement of light absorption at a wavelength of 400–500 nm.

Increased dye degradation and mineralization ability under both UV and visible light was exhibited by Eu-doped TiO_2 hydrosols.[42] Eu^{3+}–TiO_2 did not show any spectrum absorption in the visible range and only absorbed in the UV range due to the change-transfer process in the bulk TiO_2. This means that the catalysts cannot drive the direct photocatalysis under visible light irradiation and Vis/X-3B/Eu^{3+}–TiO_2 hydrosol system only underwent a sensitized photocatalysis reaction. An important red shift for Eu-doped TiO_2 was reported in many studies. The authors suggest that the absorption red shift could be attributed to the formation of energy levels associated with the defects or impurities.[43] The increased photoresponse of these catalysts to visible light and thus enhanced photocatalytic activity were observed.

Doping of the precipitated TiO_2 powder by Eu(III) ions can indeed increase its photocatalytic activity. A nearly 30% increase in the rate of mineralization was seen in Eu^{+3}-doped Ti^{+4} in comparison to undoped Ti in one of our investigation of photoactivity through the measurement of phenol photo-mineralization under irradiation of a metal-halogenide lamp, which has similar radiation characteristics to solar light. A molar ratio of 0.005 was used for Eu to Ti (Figure 2.1).

2.4   Probing Photocatalysis

Photocatalytic activity is usually evaluated by monitoring a change in selected parameters. The most often used parameter is the concentration of the original pollutant. A commonly used method of measuring the original pollutant is UV-visible spectroscopy. Monitoring of the degradation of dyes particularly

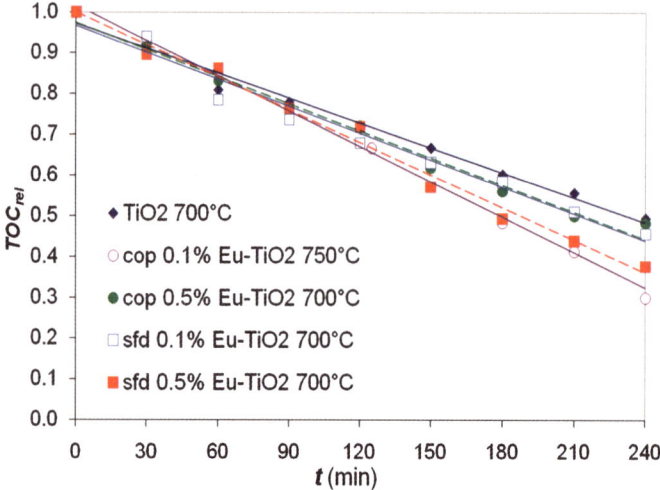

Figure 2.1 Time dependence of (TOC$_{rel}$) of Eu-doped TiO$_2$ samples with various Eu/Ti ratio (expressed in mol%) prepared by co-precipitation of Eu and Ti oxide (cop) and surface doping of freshly prepared TiO$_2$ before annealing (sfd).

employs this technique. Gas chromatography for following the concentration of phenol was reported by de la Cruz Romero *et al.*[41] High-performance liquid chromatography (HPLC) equipped with UV detection was used for the determination of the azo dye X-3B and 2,3-DCP. The use of liquid chromatography with UV detection (LC-UV) for the determination of the concentration of initial pollutants as the most reliable and versatile technique for the analysis of organic pollutants dissolved in water.

The degradation of pollutants is associated with the formation of reaction mixtures containing one or several contaminants and their degradation products. Therefore, the analysis of these reaction mixtures requires analytical methods that enable the separation and identification of the large number of compounds, with very different chemical properties present in a wide interval of concentrations.[44] Very sophisticated equipment and techniques, particularly in the field of gas chromatography coupled to mass spectrometry (GC-MS) and liquid chromatography coupled to mass spectrometry (LC-MS), have been developed to enable the efficient determination of organic compounds. GC-MS has doubtlessly been the most applied technique, since it offers important advantages in the analysis of complex samples and its use is relatively simple and is therefore established experience in the field. LC-MS is progressively gaining acceptance, since it presents several advantages over gas chromatography, some of which are very important for identification of reaction intermediates in advanced oxidation processes.

Another basic parameter used for evaluation of photocatalysis is mineralization. Since hydroxyl radicals react non-selectively, numerous intermediates are formed *en route* to complete mineralization at different concentrations.

Therefore, a complete mass balance of the photocatalytic degradation processes is imperative, because the photocatalytic treatment must destroy not only the initial contaminant, but any other organic compound as well. The determination of the total organic carbon (TOC) and/or the measurement of the chemical oxygen demand (COD) of the irradiated solution are generally used for monitoring the mineralization. It is appropriate to analyse the COD when TOC determination is unavailable, since it is a cheaper method. However, the information that it provides on mineralization of the compounds is not as reliable as TOC, because COD analyses are usually not accurate for pollutant concentrations in the order of mg L^{-1} and because oxidizable inorganic species cannot be distinguished from the organic pollutants. Comparison of the degradation monitored by the concentration of the initial pollutants and the mineralization process reported in studies reveals the importance of the TOC technique.[41,42]

Electron paramagnetic resonance (EPR) is a valuable tool for the study of photogenerated charge carriers in TiO_2, which play a key role in the photo-induced oxidative heterogeneous processes on its surface. The key factor for the photoactivity of TiO_2 is the process of the formation of electron and hole pairs upon photoexcitation with light showing higher energy as the band gap E_g.[45] Also, for photoactivity the recombination of electron–hole pairs, their interfacial transfer and reaction with adsorbed species on the surface of the nanocrystals play a crucial role. Photogenerated holes are trapped by the surface hydroxyl ions forming reactive hydroxyl radicals in aqueous media or initiate the direct oxidation of the adsorbed inorganic or organic substrates. In oxygenated systems the photogenerated electron reacts with oxygen producing superoxide anion radicals, which may act as oxidizing agents or as an additional source of hydroxyl radicals *via* the subsequent formation of hydrogen peroxide. Consequently, the irradiation of TiO_2 particles in aqueous media leads to formation of short-lived and highly reactive oxygen species on the surface ($\cdot OH$, $\cdot OOH/O\cdot$, O_2^-, $^1O_2$). They have the ability to oxidize a large variety of organic and inorganic substances. The oxidation ability can lead also to distinct disinfection and antimicrobial properties. Due to the short life time, the continuous wave (cw) EPR investigation of trapped paramagnetic species and adsorbed radicals upon irradiation of oxygen-saturated TiO_2 aqueous suspensions at ambient temperature, at which the photocatalytic reactions proceed, is difficult;[46] typically, low temperatures are necessary.[47] The EPR lines of photo-induced holes trapped at O^- lattice sites and electrons trapped at Ti^{3+} and O_2^- sites have been observed at 10 K.[48]

The photoproduced reactive and short-lived radicals may be conveniently studied at ambient temperature by EPR spectroscopy employing the spin-trapping technique.[49] It involves the interaction of short-lived radicals with a diamagnetic EPR-silent compound (spin-trap) to produce a stable EPR-active free radical product. Mostly the spin trapping agents DMPO (5,5-dimethyl-1-pyrroline *N*-oxide), TMPO (3,3,5,5-tetramethyl-1-pyrroline *N*-oxide) and POBN [α-(4-pyridyl-1-oxide)-*N-tert*-butylnitrone] have been used.

On the other hand the short-lived photo-induced free radicals can be detected by investigation the process of the termination of semi-stable free radicals, *e.g.*, TEMPOL (4-hydroxy-2,2,6,6-tetramethylpiperidine *N*-oxyl), DPPH (1,1-diphenyl-2-picrylhydrazyl), ABTS^{-+} [2,2-azino-bis(3-ethylbenzothiazoline-6-sulfonic acid) diammonium salt], added into the photoactive TiO$_2$-containing system. The EPR spectrum of the semi-stable free radicals is gradually eliminated during irradiation, and this system represents a simple technique for the evaluation of TiO$_2$ photoactivity.

We compared the photoactivity characterized by measurement of the TOC decrease and by the EPR study of the rate of building spin-trapped radicals. Anatase powder prepared by precipitation reaction was studied. Good photocatalytic activity in photomineralization of phenol measured by TOC under UVA light was observed. The photoactivity was decreased when a metal halogenide visible light lamp, which emits only a small part of light in the UVA region, was used. In both cases the highest activity showed the powders annealed at 700 °C. The results of photoactivity measurements performed by EPR spin-trapping study of the photogenerated radicals using a DMPO spin-trap proved that the maximal rate of building of photogenerated radicals and the maximum photoactivity was, in the case of the titania powder, annealed at 700 °C, which is in accordance with the results obtained from TOC measurements. The formation of photo-induced charges at various wavelengths using various cut-off filters was studied as well. The data showed that the best activities were observed using a Pyrex glass filter (transparent for $\lambda > 300$ nm) and for cut-off filters that are transparent for irradiation at $\lambda > 365$ nm. For the $\lambda > 400$ nm cut-off filter, weak photoactivity could be observed, but the photoactivity under $\lambda > 436$ nm was not significant. Since the energy gap of the obtained anatase lies in the UVA region (3.2 eV), one may conclude that the small photoactivity observed in the anatase used in the present study under has been caused by the absorption of a portion of UVA radiation from the spectra of a metal halogenide visible light lamp.

2.5   Conclusions

The precipitated TiO$_2$ powder doped by Eu^{+3} ions showed an increase in the photocatalytic activity. The photoactivity properties were examined by the TOC method exploiting phenol photomineralization under the irradiation of a metal halogenide lamp with the radiation characteristics of solar light. Samples with molar ratios of Eu/Ti 0.005 and 0.001 were prepared and examined. In the samples with the molar ratio 0.005, the rate of TOC successfully decreased compared to the non-doped TiO$_2$ by nearly 30%.

The TOC study was also performed with Ag-doped TiO$_2$ amorphic nano-powders that were prepared by sol-gel route. The results showed that this powder acts as an effective photocatalyst in mineralization of phenol, and a decrease in the TOC value with irradiation time was observed. However, the presence of metallic silver in the anatase form of TiO$_2$ leads to slight decrease in the photocatalytic activity in comparison to the non-doped sample. This effect of Ag-doped titania on the photocatalytic activity can be related to the fact that the deposited silver is present on the surface in form of metallic particles with a size of approximately 33 nm. The study of the photocatalytic properties of the anatase form prepared by precipitation reaction indicated good photocatalytic activity in photomineralization of phenol under UVA light and a decreased photoactivity under illumination with the metal halogenide visible light lamp.

References

1. T. Ochiai, K. Nakata, T. Murakami, A. Fujishima, Y. Yao, D. A. Tryk and Y. Kubota, *Water Res.*, 2010, **44**(3), 904.
2. N. H. Ince and I. G. Apikyan, *Water Res.*, 2000, **34**, 4169.
3. I. Sopyan, M. Watanabe, S. Marasawa, K. Hashimoto and A. Fujishima, *Chem. Lett.*, 1996, 69–70.
4. A. Fujishima, T. N. Rao and D. A. Tryk, *J. Photochem. Photobiol. C., Photochem., Reviews*, 2000, **1**, 1.
5. H. Yu, H. Irie and K. Hashimoto, *J. Am. Chem. Soc.*, 2010, **132**(20), 6898.
6. K. Hashimoto, H. Irie and A. Fujishima, *Jpn. J. Appl. Phys.*, 2005, **44**(12), 8269.
7. A. Fujishima and K. Honda, *Nature*, 1972, **238**, 37.
8. X. Pan, I. Medina-Ramirez, R. Mernaugh and J. Liu, *Colloids Surf., B Biointerfaces*, 2010, **77**(1), 82.
9. F. R. Marciano, D. A. Lima-Oliveira, N. S. Da-Silva, A. V. Diniz, E. J. Corat and V. J. Trava-Airoldi, *J. Colloid Interface Sci.*, 2009, **340**(1), 87.
10. H. Koseki, K. Shiraishi, T. Asahara, T. Tsurumoto, H. Shindo, K. Baba, H. Taoda and N. Terasaki, *Biomed. Res.*, 2009, **30**(3), 189.
11. T. Lopez, E. Ortiz, M. Alvarez, J. Navarrete, J. A. Odriozola, F. Martinez-Ortega, E. A. Páez-Mozo, P. Escobar, K. A. Espinoza and I. A. Rivero, *Nanomedicine*, 2010, **6**(6), 777.
12. H. Kim, S. Lee, Y. Han and J. Park, *J. Mater. Sci.*, 2005, **40**, 5295.
13. A. R. Khataee, M. Fathinia, S. Abera and M. Zarei, *J. Hazard. Mater.*, 2010, **181**(1–3), 886.
14. Y. L. Wei, K. W. Chen and H. P. Wang, *J. Nanosci. Nanotechnol.*, 2010, **10**(8), 5456.
15. A. Mills and S. Le Hunte, *J. Photochem. Photobiol., A*, 1997, **108**, 1.
16. O. Carp, C. L. Huisman and A. Reller, *Prog. Solid State Chem.*, 2004 **32**, 33.
17. J. T. Yates Jr, *Surf. Sci.*, 2009, **603**, 1605.
18. A. Fujishima, X. Zhang and D. A. Tryk, *Surf. Sci. Rep.*, 2008, **63**, 515.
19. Z. Zhang, C-C. Wang, R. Zakaria and J. Y. Ying, *J. Phys. Chem. B*, 1998, **102**, 10871.
20. C. Su, B-Y. Hong and C-M. Tseng, *Catal. Today*, 2004, **96**, 119.

21. T. Alapi, P. Sipos, I. Ilisy, G. Wittmann, Z. Ambrus, I. Kiricsi, K. Mogyorosi and A. Dombi, *Appl. Catal., A*, 2006, **303**, 1.
22. A. Di Paola, E. García-López, S. Ikeda, G. Marcì, B. Ohtani and L. Palmisano, *Catal. Today*, 2002, **75**, 87.
23. A. Di Paola, G. Marcì, L. Palmisano, M. Schiavello, K. Uosaki, S. Ikeda and B. Ohtani, *J. Phys. Chem. B*, 2002, **106**, 637.
24. N. Serpone, *J. Photochem. Photobiol. A*, 1997, **104**, 1.
25. A. L. Linsbigler, G. Lu and J. T. Yates Jr, *Chem. Rev.*, 1995, **95**, 735.
26. D. S. Muggli and L. Ding, *Appl. Catal. B*, 2001, **32**, 181.
27. T. Ohno, K. Sarukawa, K. Tokieda and M. Matsumura, *J. Catal.*, 2001, **203**(1), 82.
28. Z. B. Zhang, C. C. Wang, R. Zakaria and J. Y. Ying, *J. Phys. Chem. B*, 1998, **102**(52), 10871.
29. S. Malato, P. Fernández-Ibáñez, M. I. Maldonado, J. Blanco and W. Gernjak, *Catal. Today*, 2009, **147**, 1.
30. F. B. Li, X. Z. Li, M. F. Hou, K. W. Cheah and W. C. H. Choy, *Appl. Catal., A*, 2005, **285**, 181.
31. Y. Zhang, H. Zhang, Y. Xu and Y. Wang, *J. Solid State Chem.*, 2004, **177**, 3490.
32. F. Han, V. S. R. Kambala, M. Srinivasan, D. Rajarathnam and R. Naidu, *Appl. Catal. A*, 2009, **359**, 25.
33. K. T. Ranjit, I. Willner, S. H. Bossmann and A. M. Braun, *Environ. Sci. Technol.*, 2001, **35**, 1544.
34. S. Matsuo, N. Sakaguchi, K. Yamada, T. Matsuo and H. Wakita, *Appl. Surf. Sci.*, 2004, **228**, 233.
35. B. M. Reddy, P. M. Sreekanth, E. P. Reddy, Y. Yamada, Q. Xu, H. Sakurai and T. Kobayashi, *J. Phys. Chem. B*, 2002, **106**, 5695.
36. T-x. Liu, X-z. Li and F-b. Li, *Chem. Eng. J.*, 2010, **157**, 475.
37. H-H. Wu, L-X. Deng, S-R. Wang, B-L. Zhu, W-P. Huang, S-H. Wu and S-M. Zhang, *J. Disp. Sci. Technol.*, 2010, **31**, 1311.
38. Y. Xie and C. Yuan, *Appl. Catal. B*, 2003, **46**, 251.
39. X. Wu, X. Ding, W. Qin, W. He and Z. Jiang, *J. Hazard. Mater.*, 2006, **137**, 192.
40. Z. M. El-Bahy, A. A. Ismail and R. M. Mohamed, *J. Hazard. Mater.*, 2009, **166**, 138.
41. D. de la Cruz Romero, G. Torres Torres, J. C. Arévalo, R. Gomez and A. Aguilar-Elguezabal, *J. Sol-Gel Sci. Technol.*, 2010, **56**, 219.
42. Y. Xie and C. Yuan, *J. Mater. Sci.*, 2005, **40**, 6375.
43. F. Vasiliu, L. Diamandescu, D. Macovei, C. M. Teodorescu, D. Tarabasanu-Mihaila, A. M. Vlaicu and V. Parvulescu, *Top. Catal.*, 2009, **52**, 544.
44. C. Sirtori, A. Zapata, S. Malato, W. Gernjak, A. R. Fernández-Alba and A. Agüera, *Photochem. Photobiol. Sci.*, 2009, **8**, 644.
45. D. C. Hurum, A. G. Agrios, S. E. Crist, K. A. Gray, T. Rajh and M. C. Thurnauer, *J. Electron Spectrosc. Relat. Phenom.*, 2006, **150**(2–3), 155.

46. K. L. Yeung, S. T. Yau, A. J. Maira, J. M. Coronado, J. Soria and P. L. Yue, *J. Catal.*, 2003, **219**, 107.
47. G. Li, S. Ciston, Z. V. Saponjic, L. Chen, N. M. Dimitrijevic, T. Rajh and K. A. Gray, *J. Catal.*, 2008, **253**, 105.
48. R. Scotti, I. R. Bellobono, C. Canevali, C. Cannas, M. Catti, M. D'Arienzo, A. Musinu, S. Polizzi, M. Sommariva, A. Testino and F. Morazzoni, *Chem. Mater.*, 2008, **20**, 4051.
49. C. Adán, A. Bahamonde, M. Fernández-García and A. Martínez-Arias, *Appl. Catal., B*, 2007, **72**, 11.

Surface Charge Measurements on Biomaterials in Dry and Wet Conditions

M. GREGOR,[a,b] T. PLECENIK,[a] A. PLECENIK,[a]
C. WOLF-BRANDSTETTER,[c] D. SCHARNWEBER[c] AND
S. A. M. TOFAIL[b]

[a] Comenius University, Department of Experimental Physics, Mlynska Dolina F2, Bratislava, Slovakia; [b] University of Limerick, Materials and Surface Science Institute, Limerick, Ireland; [c] Technical University of Dresden, Max Bergmann Center of Biomaterials, Dresden, Germany

3.1 Introduction

Electrical modifications of the surface of biomaterials can play a significant role in manipulating the interactions of biological species such as proteins and cells on such surfaces. Methods for the charging and polarization of such materials, as well as the measurement of the induced charges, are thus important. Based on the characteristics of the material used, the surface, space or dipolar charge can be induced by various methods such as contact poling, electric or corona discharge and electron or ion beam irradiation. The induced charge can be measured in both dry and wet conditions by several techniques. In dry conditions, the charge measurement methods are generally based on direct measurement of depolarization or the discharging current of the sample, or indirect measurement of charge induced on capacitive-coupled probe.

RSC Nanoscience & Nanotechnology No. 21
Biological Interactions with Surface Charge in Biomaterials
Edited by Syed A. M. Tofail
© Royal Society of Chemistry 2012
Published by the Royal Society of Chemistry, www.rsc.org

The thermally stimulated depolarization current (TSDC) measurement technique has been used especially for electret biomaterials. It is an important tool for the measurement of relaxation phenomena in various polarized or charged materials. The relaxation (*e.g.* depolarization or discharging) is usually achieved by increasing the sample temperature and the resulting induced current is directly measured. It can provide valuable information about the creation, storage and relaxations of charges in bulk material and its surface. On the basis of such information, it is possible to study the influence of the surface and volume charge of biomaterials on mediating interactions of biological species at the Biomaterials' surfaces.

The main disadvantage of TSDC method is that the charge or polarization is destroyed by the measurement. For this reason, measurement methods based on charge induction on the measurement probe by the electric field produced by the charges in the examined material are often used. Among others, the induction probe, field mill and Kelvin probe are the most used techniques and will be discussed here. Methods such as laser intensity modulation method (LIMM) will be discussed in the following Chapter.

Particular attention is given to scanning probe microscopy (SPM) methods, which can be used to analyse the properties of materials with a lateral resolution at the nanometre scale. For the characterization of electrostatic properties, electrostatic force microscopy (EFM) and Kelvin probe force microscopy (KPFM) are the most often used techniques. The SPM methods usually do not need special treatment of the samples and do not require any special conditions. In addition, techniques such as the Kelvin probe are capable of working in a liquid medium, which is favorable for the measurement of biological objects in wet conditions. This technique will be discussed in relation to more conventional streaming potential measurement of surface charge.

3.2   Surface Charge Measurement Principle

It is well known that electrical charge can be created and stored in dielectric materials, which includes a majority of biomaterials and coatings made of ceramic and polymeric materials. Thin metal oxide coatings on metallic implants are usually semiconducting, but, when they are thickened sufficiently, typically over 100–500 nm, they too can be used to store charge. Depending on the method of charge induction or injection, two different major types exist: real charge and dipolar charge. The real charge can be further divided into surface charge (located on the surface of the material) and space charge (located in the volume of the material). The dipolar charge (polarization) can be induced in dipolar materials containing dipolar molecules or in ferroelectric materials.[1]

The real charge may appear due to the charge carrier injection (*e.g.* from an attached electrode, an electric or corona discharge, electron beam irradiation, *etc.*) to the dielectric material, followed by the capturing of the carriers in surface or volume traps of various origin.

There are many ways to form electrical charges in dielectric materials, for example, by applying a strong electric field across the material which may cause injection of charge carriers or polarization, or both, by electron beam irradiation and corona charging on the surface. The corona charging of biomaterials is discussed in detail in Chapter 4. Some charges can be produced by heating of dielectric material in the strong electric and/or magnetic field and then cooling slowly to an ambient temperature. Some dielectric materials can form electric charges simply by a thermal heating without an electric and magnetic field or by applying mechanical pressure.[2]

A common polarization method is contact poling, which is performed by applying an electrical field (E) for time, t, at temperature, T. In the case of ferroelectric materials, this temperature can be below or above the Curie temperature T_c, at which a ferro-to-paraelectric transition takes place. As, for ferroelectric materials, the polarizability is highest at a temperature just below T_c, the poling is most effective if the sample is heated just above T_c and then is cooled back down to room temperature while an electric field is still applied (assuming T_c is above the room temperature).[3]

The stored charge can be measured directly, for example, by TSDC as discussed in Section 3.3. Non-contact and non-destructive methods based on the measurement of the electric fields created by the charges are also applicable. Such measurements provide the way to detect the density of charge at the surface and in the bulk volume.

Electrical fields may be measured by the charge induced on a charge-sensing probe of an instrument capable of measuring charge either directly or indirectly by measuring the current or voltage. These measurements are usually performed by three kinds of measuring methods: induction probes,[4] field mills[5] and electrostatic probes.[6] All these measuring techniques are based on capacitive coupling of a conductive probe carrying the surface charge density to the inducing electrostatic field, E, as shown in eqn (1):

$$\delta = \varepsilon_0 E \qquad (1)$$

where E is the electric field (V m^{-1}), δ is the probe surface charge density (C m^{-2}) and ε_0 is the permittivity of free space (8.854×10^{-12} F m^{-1}).

The electrostatic induction methods measure the charge distribution indirectly. The charge induced on the probe is measured by integration of the current flowing to or from the probe. No discharge or ionization effects take place between the measuring probe and the charge bearing surface. Before measurement, the charge on the probe has to be cancelled and a zero voltage has to be given to the probe (nulling method).

The field mill is similar to the induction probe, except that the DC signal is converted into an AC signal either by alternately shielding and unshielding a conductive plate exposed to the local field or by oscillating the probe (top plate of the parallel-plate capacitor). The electric field mill consists of one or two electrodes, which are periodically exposed to the external electric field. The result is that an alternating signal is generated. The field mill with a stationary reading electrode and a rotating shutter is the most common commercially used type.[5]

The electrostatic (also called Kelvin) probe is one of the most popular devices for charge measurements. It is based on a feedback-null surface potential monitor cancelling the electrical field between the charged surface and the probe by applying a DC voltage to the probe. This technique was proposed by Lord Kelvin.[7] This device can be understood as a simple capacitor consisting of two flat and parallel conductive plates. Its capacitance C is then given by eqn (2):

$$C = Q/U \qquad (2)$$

where Q is an electric charge accumulated by the capacitor and U is the voltage between the electrodes of the capacitor. The capacitance depends on the surface area of the planes A, their distance D and the material between them and can be expressed in eqn (3):

$$C = (\varepsilon \varepsilon_0 A)/D \qquad (3)$$

where ε is relative electric permittivity of the material between the electrodes and ε_0 is the electric permittivity of vacuum. The charge on the tested surface can be calculated from eqns (2) and (3):

$$Q = U[(\varepsilon \varepsilon_0 A)/D] \qquad (4)$$

In 1932 William Zisman of Harvard University introduced a new method, the vibrating Kelvin probe.[8] The probe vibrates in the direction perpendicular to the tested surface. The current flows from and to the probe changes proportionally to the frequency and amplitude of that vibration. When D_1 is the amplitude of the sinusoidal vibrations and D_0 is initial distance between the tested surface and the probe, the capacitance, C, of such a system can be written as eqn (5):

$$C = \frac{\varepsilon \varepsilon_0 A}{D_0 + D_1 \sin(\omega t)} \qquad (5)$$

and the probe current can be expressed as eqn (6):

$$I = U\frac{dC}{dt} = U\frac{d}{dt}\left(\frac{\varepsilon \varepsilon_0 A}{D_0 + D_1 \sin(\omega t)}\right) = -U\varepsilon \varepsilon_0 A\frac{D_1 \omega \cos(\omega t)}{[D_0 + D_1 \sin(\omega t)]^2} \qquad (6)$$

To nullify this current, the voltage between the tested surface and the probe has to be brought to zero. In KPFM techniques, this method is often modified and the alternating voltage is applied to the probe instead of mechanical oscillations. Such Kelvin probe techniques will be described in the following sections.

3.3 Scanning Probe Methods of Surface Charge Measurements

SPM is a relatively new technique which is generally based on interactions of an ultra-sharp tip (probe) with the sample surface. Depending on the respective technique, atomic forces, electrostatic or magnetic forces, tunneling current or other interactions between the SPM tip and the sample surface are measured and recorded as a function of the tip position.

Although the SPM was primarily developed for surface topography characterization, it is now widely used to measure many other surface properties. Among others, EFM techniques allow the study of a surface or volume-trapped charges, polarization, surface potential or work function distribution at a sub-micrometre scale.[9,10]

3.3.1 Principles of EFM: Two-Pass Techniques

In general, EFM techniques are based on the measurement of the electrostatic interaction between the SPM tip and the sample. The force between the tip and the sample is a sum of the charge – (induced) charge interaction and the force given by the tip-sample capacitance, as in eqn (7):

$$F = \frac{Q_S Q_T}{4\pi\varepsilon_0 z^2} + \frac{\partial C}{\partial z}\frac{V^2}{2} \tag{7}$$

where Q_S and Q_T is the charge of the sample and the tip, respectively, C is the total tip-sample capacitance, V is the total potential difference between the tip and the sample and z is the tip-sample separation. In basic EFM techniques this force is measured directly by sensing the cantilever deflection or amplitude/phase changes in dynamic regimes. This is in contrast to KPFM, which is based on the compensation of the potential difference V by applying a DC voltage on the tip. In both cases, conductive SPM tips (usually standard Si tips with conductive coatings) are used. KPFM will be described in Section 3.3.2.

To eliminate the influence of the short-range atomic forces, capillary forces and other non-electrostatic short-range interactions, EFM measurements are most often done in a so-called two-pass regime. In this case, every line of the scan is scanned two times. In the first scan, the surface topography is obtained by an atomic force microscopy (AFM) regime. The SPM tip is then lifted to several tens of nanometres. The information about the surface topography is then used to keep this separation constant during the second scan, when the long-range electrostatic forces are measured. The two-pass regime is also used in magnetic force microscopy (MFM) and, generally, most SPM techniques in which long-range forces are measured.

In the basic EFM setup, dynamic non-contact mode is usually used for the measurement of the electrostatic forces. After the surface topography is obtained (first pass), the conductive SPM tip oscillates at its resonant frequency several tens of nanometres above the surface (second pass). Force gradient of the electrostatic forces between the tip and the sample results in the resonant frequency shift, which is usually detected by measurement of the phase or amplitude shift and recorded as a function of the tip position. If the tip is grounded (zero-biased), charges on the sample induce image charges of the same value but opposite sign on the tip. The tip–sample interaction is thus always attractive and the sign of the charge cannot be distinguished. For biased tips, the force depends on the tip bias and the sign of the charge can be distinguished.

3.3.2   KPFM

KPFM also works with a two-pass regime, but in the second pass the SPM tip does not oscillate and is biased by both DC and AC voltage. In this case, the force, $F(z)$ between the tip and the sample can be written as eqn (8):[9]

$$F(z)=\frac{1}{2}\frac{\partial C}{\partial z}\left[(V_{Tdc}-V_S)^2+2(V_{Tdc}-V_S)V_{Tac}\sin(\omega t)+\frac{1}{2}V_{Tac}^2(1-\cos(2\omega t))\right] \quad (8)$$

where C is the total tip-sample capacitance, z is the tip–sample distance, V_{Tdc} and V_{Tac} are the DC and AC voltage on the tip, respectively, and V_S is the surface potential of the sample. It is easy to see that the resulting force can be divided into three components: the static (DC), first harmonic (ω) and second harmonic (2ω) component. The first harmonic component, which is directly proportional to the difference between the sample surface potential and the tip DC voltage, is extracted by a lock-in technique. A feedback loop is then used to nullify the first harmonic component of the force by adjusting the DC voltage on the SPM tip to the value of the surface potential of the sample (Figure 3.1).

This DC voltage is recorded as a function of the tip position, and the lateral distribution of the sample surface potential is obtained. In the case of

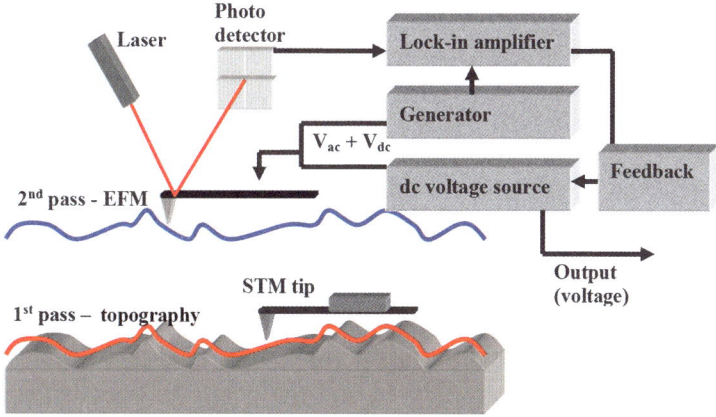

Figure 3.1   Schematic diagram of a basic KPFM.

conductive samples, the obtained surface potential reflects the distribution of the work function on the samples' surface. If the tip is calibrated against material with a known work function, an absolute value of the work function of the measured surface can be determined. For this purpose, tips with a well-defined work function can be prepared.[11] In dielectric samples, the surface potential distribution is mostly influenced by trapped charges and/or polarization of the material, although some other factors, *e.g.* surface contamination or chemical changes, may also have significant influence.

As the measurement is based on nullifying the electrostatic force, the signal obtained is much less sensitive to topographic features and changes of the tip–sample capacitance, as in the case of the basic EFM. This technique thus allows high lateral (tens of nanometres) and potential ($\sim$mV) resolution.[11–13]

In conventional EFM or KPFM, the cantilever is driven either mechanically (basic EFM: force gradient imaging) or electrostatically (KPFM: voltage modulation). In the enhanced KPFM technique, a combination of both mechanical and electrostatic drive is used. The topography and surface potential data are thus obtained within one scan. As the tip does not have to be lifted when a KPFM signal is obtained, better lateral resolution is obtained with enhanced KPFM than with conventional KPFM.

The cantilever is mechanically driven slightly above its resonant frequency ω_c and electrostatically driven at frequency ω, which is much lower or much higher than ω_c. Two lock-in amplifiers are then used for the detection of cantilever oscillations at frequencies ω_c and ω. The first signal (at frequency ω_c) is used to obtain surface topography and maintain constant tip–sample separation, while the second signal (at frequency ω) is nullified by applying a DC voltage on the tip via a feedback loop, similarly as in classical KPFM.

3.4 TSDC Technique

The TSDC technique is widely used for investigations of dipolar and DC relaxation phenomena by applying heat after polarization or charge injection. The heat-induced TSDC current occurs due to the relaxation of a previously oriented molecular, fragment or bond, dipoles and the release of trapped charges as a protons, ions and electrons. TSDC measurements have been widely used for many systems such as polymers,[14] semiconductors,[15] dielectric ceramics,[16] metal oxides,[17] hydrogels[18] and organics.[19] This technique can be used to study effects occurring at both the surface and the bulk.

A common method of polarization of dielectric materials is by the application of an electrostatic field. If the charge or polarization is frozen in the sample by cooling down to room temperature under the field to prevent depolarization by thermal energy, the sample is different from the equilibrium order observed without the application of the field. If the sample is heated, the individual dipoles and their clusters are relaxed and the polarization is removed. The corresponding change in charge distribution can be measured as a depolarization current. In addition, the DC relaxation current caused by the release of trapped charge carriers is induced. This current is collected as a function of temperature.

TSDC measurement is a sensitive method, allowing measurement of very low depolarization currents (down to 10^{-17}A). Polarization usually occurs due to application of the temperature T_p and the electric field E_p ($E_p = 10^1$–$10^4\,\mathrm{kV\,m^{-1}}$) for a time t_p which is sufficient to orient different entities of the material (Figure 3.2). By continuously decreasing the temperature down to T_0 (usually room temperature), the oriented dipoles will be frozen. At the temperature T_0, the external electrostatic field is switched off. To remove a free space charge, the sample is short-circuited for time t_0. After that, the sample is linearly heated and the depolarization current is recorded as a function of temperature with a highly sensitive electrometer.

For polymers, current peaks are usually observed around the glass-transition temperature, at which polarization is completely removed and all trapped charges are released. The position of the peak, its shape and size, magnitude of polarization, activation energy (E) and relaxation time (τ) can be obtained from the depolarization curve. The depolarization current is given by eqn (9):

$$I(T) = \frac{S_{el}\Pi_0}{\tau_0}\exp\left(-\frac{E}{k_BT}\right)\exp\left(-\frac{1}{\beta\tau_0}\int_{T_0}^{T}\exp\left(-\frac{E}{k_BT}\right)dT\right) \qquad (9)$$

from which the activation energy of depolarization can be calculated.[20] In eqn (9) S_{el} is the surface area of the electrodes, which are used for application of an electric field to the dielectric material, Π_0 is the frozen polarization, k_B is the Boltzmann constant, τ_0 is the relaxation time, T_0 is the initial temperature of depolarization, T_i is the temperature of the i-th TSDC maximum and β is the heating rate. If a material contains several traps of different energy or if the activation energy of depolarization varies within the sample, then a

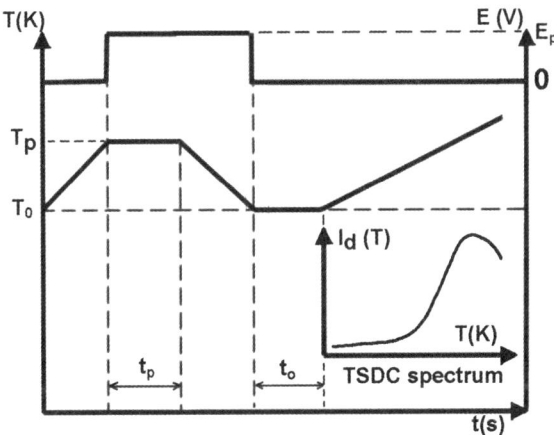

Figure 3.2   Principle of the TSDC method.

corresponding number of current peaks should be observed during the heating. The equation for a TSDC current spectrum with several peaks can be written as eqn (10):

$$I(T) = S_{el}\Pi_0 \int_{E_{min}}^{E_{max}} \sum_{i=1}^{N} w_0^i \exp\left\{ -\frac{E}{k_B T} - \frac{w_0^i k_B T^3}{E\beta} \exp\left(-\frac{E}{k_B T}\right)\right.$$
$$\left. + \frac{w_0^i k_B T}{E\beta} \exp\left(-\frac{E}{k_B T}\right)\right\} f(E)dE \qquad (10)$$

where w^i is the transition probability of the dipoles, τ is the relaxation time and β is the heating rate.

$$w_0^i = \frac{\beta}{k_B T_i^3}(E + k_B T_i)\exp\left(\frac{E}{k_B T_i}\right) \qquad (11)$$

In eqns (9) and (10), Π_0 is equal to electrical charge, Q, produced during depolarization, which can be calculated by integration of the TSDC spectra in eqn (12):

$$Q = \frac{1}{\beta}\int I(T)dT \qquad (12)$$

The TSDC technique is widely used for investigation of relaxation effects on polymers, disperse and porous metal oxides, composites, liquid crystals, solid solutions, *etc.*[20] TSDC studies of biomaterials [polyurethane, hydroxyapatite (HA), *etc.*] and bio-objects (proteins, DNA, cells, tissue, bone, collagen, *etc.*) have also been carried out.[21]

The polarization of HA [$Ca_{10}(PO_4)_6(OH)_2$], which is used in orthopedic and dental implant coatings and bone graft substitutions, has been carrried out by the TSDC method.[22] It has been shown that the electrically polarized HA enhances the bioactivity and the osteoconductivity. Negatively poled HA surfaces showed consideribly higher osteoconductivity and mineralization compared to uncharged surfaces.[23] This is now termed as vectorial effect and discussed further in Chapter 17.

In HA, the polarization is driven by rearrangement of protons at the OH^- ions in apatite columnar channels. As the activation energy of the protons is relatively high, the HA polarization is stable at room temperature.[24] HA has the potential to store high charge density, which lies within a few tens of nano-Couloumbs to a few milli-Coulombs over a square centimetre area (10^{-8}–10^{-3} C cm^{-2}). This high charge storage ability is believed to give rise to a significantly higher biocompatibility compared to uncharged HA.

The typical TSDC spectrum of the HA ceramics is shown in Figure 3.3. The HA disk with a diameter of 1 cm and a thickness of 2 mm was polarized in a DC field of 2 kV cm^{-1} at 375 °C for 1 h with heating and cooling rate of 5 °C min^{-1}. To avoid dipole relaxation during the cooling, the electrical field was maintained until the HA was cooled down to room temperature. Before the

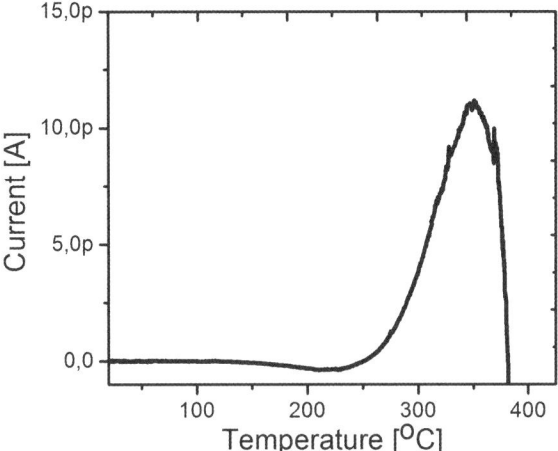

Figure 3.3 The TSDC spectrum of HA disk (2 mm thick, 10 mm diameter) polarized in a DC field of 2 kV cm^{-1} at 375 °C for 1 h.[25]

TSDC measurements, the HA was short-circuited to remove free charge on the surface. The TSDC spectra were collected with a heating rate of 1.6 °C min^{-1}. The current started to increase at 250 °C and reached a maximum at 355 °C and then decreased down at 380 °C. The storage charge calculated from the TSDC curve using equation (3.12) was established to be 1.4×10^{-8} C.[25] This polarization charge can be increased by increasing of the poling temperature or by increasing the electrical field during the poling.[24]

Despite being widely used, the TSDC method suffers from two major drawbreaks: (a) it is a destructive method; and (b) due to the contribution of trapped and dipolar charges from the deep region of the bulk, the amount of charge actually present at the surface cannot be unambiguously determined. To provide such information, LIMM is a better method and will be discussed in Chapter 4. Another limitation of the TSDC method is that it is currently a macroscopic technique, although microelectrodes and high charge amplification can make TSDC measurements possible in the micron scale. The whole TSDC experiment can be very time consuming, especially for ceramic samples for which the TSDC relaxations usually occur at a temperature of a few hundred °C.

3.5 Principles and Applications of Streaming Potential Measurements

Among the principles of measurement described in this chapter, the measurement of streaming potential and streaming current belong to the methods which have an electrolyte, *i.e.* wet conditions, as a prerequisite. This makes this method particularly relevant to implant biomaterials, which will remain in an essentially wet environment during their usage time inside the body. Similar to TSDC, this method provides macroscopic measurements only.

An explanation of the measuring principle, originating from the work of Bikerman[26] on the presence of mobile ions in the diffuse part of an electrochemical double layer, starts with a structure model of this double layer. These models distinguish between a stationary and a mobile layer, both being separated by a shear plane (Figure 3.4). The zeta potential, ζ, is defined as the potential difference between the solid substrate surface and the potential in this shear plane.

Using an external (shear) force parallel to the solid substrate–liquid electrolyte interface will result in a relative movement between the above-mentioned stationary and mobile part of the electrochemical double layer and thus a charge separation which allows for an experimental access to the zeta potential.

Thus streaming potential and streaming current measurements enable a comprehensive characterization of electrosurface phenomena, such as charge formation, surface dissociation phenomena, structural features of self-assembled monolayers, adsorbed proteins *etc.*, at nonconducting solid materials in contact with aqueous electrolytes. Limitations are given for physicochemical inhomogeneous surfaces including roughness, porosity, brush-like coatings *etc.*

In the following, due to the focus on substrate materials and geometries for biomedical implant applications, the contribution will concentrate on investigations with standard electrokinetic substrates, *i.e.* samples having a smooth, non-porous and chemical homogeneous surface.

For this type of samples the measurement of the zeta potential is usually performed using a slit cell forming a rectangular streaming channel built up by two parallel identical substrate surfaces. A typical geometry is shown in Figure 3.5. Reference electrodes for the measurement of the parameters

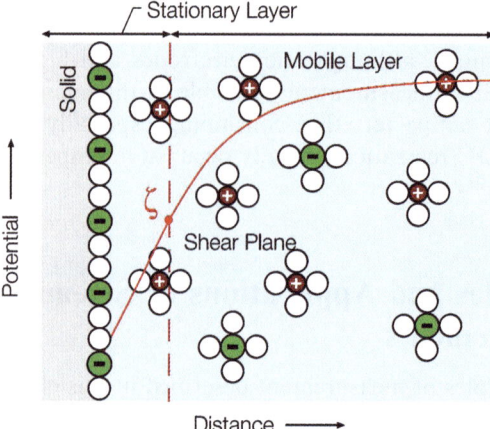

Figure 3.4   Simplified model of the electrochemical double layer.

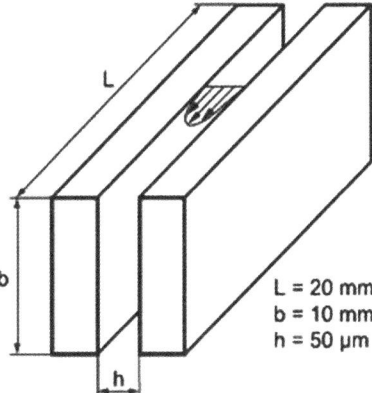

Figure 3.5 Typical geometry of an electrokinetic slit cell. Cell height (here shown as 50 μm) is usually variable.

streaming potential, U_S, and streaming current, I_S, have to be located at the inlet and outlet of the slit cell. An electrolyte pumping system allows for the adjustment of defined pressure differences over the slit cell. This is usually realized by simultaneous computer-controlled adjustment and measurement of the concentration and pH value of the electrolyte.

From experiments performed in this setup the zeta potential can be calculated from streaming potential and streaming current data following the Smoluchowski equations:[28]

$$\zeta(U_S) = \frac{\eta K_B}{\varepsilon_0 \varepsilon_r} \frac{dU_S}{dp}\bigg|_{Du<<1} \tag{13}$$

$$\zeta(I_S) = \frac{\eta L}{\varepsilon_0 \varepsilon_r bh} \frac{dI_S}{dp} \tag{14}$$

where η is the dynamic viscosity of the aqueous electrolyte, K_B is the specific electrical conductivity of the aqueous electrolyte, ε_0 is the permittivity of the vacuum, ε_r is the dielectric constant of the aqueous electrolyte, and P is the pressure difference across the measuring channel.

The boundary condition for eqn (13) with the dimensionless Dukhin number, $Du = K^\sigma/hK_B << 1$ corresponds to high electrolyte concentrations and large slit cell heights.

Typical zeta potential curves as function of the parameter electrolyte pH and concentration, as derived from measurements of the streaming potential and current respectively, are given in Figure 3.6.

For $\zeta = 0$, the data deliver directly the isoelectric point (IEP) of the substrate surface as a parameter to characterize the brutto surface charge of the substrate as a function of the pH. For a more detailed and sophisticated analysis of

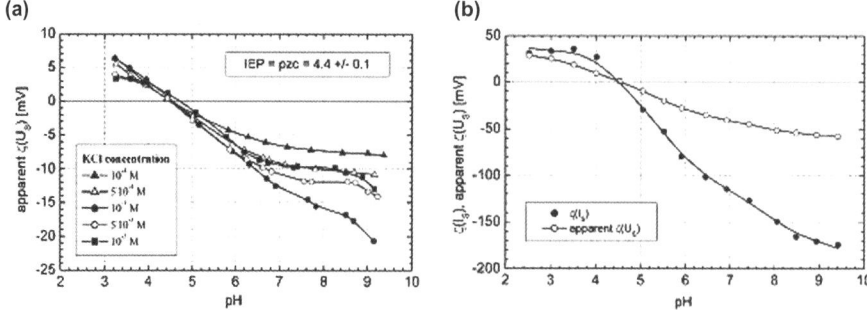

Figure 3.6 Typical shape of zeta potential curves for TiO_2 derived from streaming potential (a) and current data (b) as a function of the electrolyte pH.

streaming potential and current measurements, the reader is referred to Hunter,[29] Lyklema[30] and Zimmermann *et al.*[31]

The surface conductivity parameter K^σ in the Dukhin number mentioned above is given by eqn (15):

$$K^\sigma = F \sum_i |z_i| \int_0^\infty [c_i(x) - c_i(\infty)] u_i(x) dx \qquad (15)$$

where F is the Faraday constant, I is the index of the different ionic species in the aqueous electrolyte, z is the valency of the ions, $c(x)$ is the ion concentration perpendicular to the surface, $c(\infty)$ is the ion concentration of the bulk electrolyte, u is the mobility of the ionic species and x is the distance from the surface.

Thus the surface conductivity gives an additional valuable independent parameter that reflects the ion accumulation at the sample–electrolyte interface.[28] Because K^σ is not related to any hydrodynamic shearing process in the interface, it is not affected by properties influencing the electrolyte flow in the interface region.

Streaming potential and current measurements have been applied in a variety of cases to characterize materials' surfaces relevant for biomedical applications. An overview of polymeric materials used in the biomedical area is given by Werner *et al.*[32] Recently, Zimmermann *et al.*[31] studied poly(L-glutamic acid) chains grafted to glass carriers by a combination of zeta potential (from streaming current measurements), surface conductivity and Dukhin number analysis. In agreement with the presence of carboxylic acid groups and their function as the major charge forming surface chemistry, the IEP at pH $= 2.7 \pm 0.1$ was determined. Furthermore, a strong increase in the surface conductivity when changing from pH 6.0 to 9.0 was detected, which is explained by charge-induced alterations of the structure of the poly(L-glutamic acid) chains. The change from the α-helical relatively compact conformation of the uncharged chains at low pH values to less dense extended coils at higher pH values due to the electrostatic repulsion of the charged side groups allows for an increased total number of counter ions in the chain network and thus the increase in the surface conductivity.

Zhu and Chen[33] used a mix of methods to investigated the blood compatibility of poly(ethylene terephthalate) (PET) films grafted with acrylic acid (AAc) and coupled with chitosan (CS) and *o*-carboxymethylchitosan (OCMCS), respectively. The results demonstrate that surface coupling of OCMCS shows much less platelet adhesive and fibrinogen adsorption compared to the other surface-modified PET films. This effect is ascribed to the suitable balance of hydrophobicity/hydrophilicity, surface zeta potential and the low adsorption of protein.

With the aim of an in depth understanding of the relationship between electric charge at a material surface and protein adsorption and the mechanism of biological integration of materials with tissues, Cai *et al.*[34] studied the influence of the surface chemistry of titanium thin films realized *via* different functional end groups, such as CH, CH_2, NH_2 and COOH, and surface electric charge (zeta potential) properties on protein adsorption and cell proliferation. Ti-COOH samples showed the lowest water contact angles and zeta potential compared to all other samples investigated in this study.[34] Additionally, a tendency for lower protein adsorption was observed on samples that possessed lower zeta potentials of the samples. This was in agreement with cell proliferation tests showing a lower cell proliferation on COOH-terminated titanium films compared with NH_2-terminated titanium films, a finding which was attributed to the difference in the protein adsorption of these samples.

Purely inorganic surfaces have been investigated by Roessler *et al.*[27], focusing on air-formed and anodic oxide layers on the materials Ti_6Al_4V and commercially pure (cp) Ti. For all the oxide layers investigated, IEPs of ~ 4.4 were found. The IEP of the air-formed passive layer on Ti_6Al_4V did not depend on the KCl concentration, indicating that the IEP is identical to the point of zero charge (pzc). Controversially, the charge formation process seems to depend on the chloride ion concentration in the neutral and basic pH region.

Electrokinetic investigations of HA surfaces have been performed by Kawasaki *et al.*[35] in combination with studies on protein adsorption from saliva in comparison with the adsorption of several typical proteins with different electric charges, *i.e.* lysozyme, human serum albumin, β-lactoglobulin and ovalbumin. Adsorption kinetics were investigated by streaming potential measurements of a HA surface in contact with a protein solution, allowing monitoring of changes in the zeta potential of the protein-covered HA surface in real time. The adsorbed amounts show that, as compared to most of the other proteins, the saliva proteins have remarkably low adsorption affinity. The measured values for the electrophoretic mobilities indicate that positively charged proteins in the saliva mixture preferentially adsorb on to the negatively charged HA surface. Preferential uptake of the positively charged saliva proteins during the initial stages of the adsorption process is also concluded from the results of the kinetics experiments.

In summary, electrokinetic investigations using the streaming potential and the current approach form an extremely valuable tool for studying charge-related surface (interaction) processes, especially if applied with the full spectrum of experimental and evaluation techniques. However, for an in depth understanding of surface phenomena, this technique should be combined with other complementary methods such as measurements of the surface energy, protein adsorption investigations, ellipsometric measurements and other surface analytical methods, for example, photoelectron spectroscopy *etc.*

3.6 Principles and Applications of the Kelvin Probe Method to Biomaterals' Surfaces

3.6.1 Principles, Benefits and Disadvantages of the Kelvin Probe Technique

The history of this technique originates in the 19th century with Lord Kelvin's discovery of the contact electricity of metals. Although the instrumentation of this method has been constantly refined, the basic principle relies on the existence of a potential difference outside the surface of two different materials – the Kelvin probe and the sample (both being metals in classical applications) – that have been electrically connected. These are placed in close proximity to form a capacitor. The presence of the potential difference originates in the different Fermi levels that equalize under electrical connection. The electrons from the higher Fermi level start flowing to the other conduction band until the same electrochemical potential is reached in both metals. The movement of charges induces the appearance of a potential difference (Volta potential difference or also named contact potential difference) which is equal to the initial work function difference (Figures 3.7 and 3.8).[36]

Due to small harmonic vibrations of the probe in the direction perpendicular to the tested surface the charge on the plates is changed and induces a current changing proportionally to the amplitude and frequency of that vibration. A backing potential (V_b) is applied to the capacitor to null the output current. In this special case the backing potential will be equal to the voltage between the probe and the sample's surface. Finally, the crucial factor is the proper detection of the current, so that the backing potential can be appropriately adjusted. A broad variety of designs for current detection have been developed in order to improve the quality of surface charge and potential readings (Baikie *et al.*[37] and the references therein). The value of the required backing potential can be measured with a high impedance voltmeter and refers to the work function difference. More detailed background and information about instrumentation is available in several publications.[36–38]

The benefits of this method are based on its relative simplicity and quick and easy operation due to the non-contact and non-destructive operation mode to obtain information regarding relative changes in surface potentials/surface charges. However, when stored at ambient conditions charged surfaces can be modulated due to discharging effects caused by the air. Even under isolating conditions such charge decay was demonstrated in the case of dry pieces of

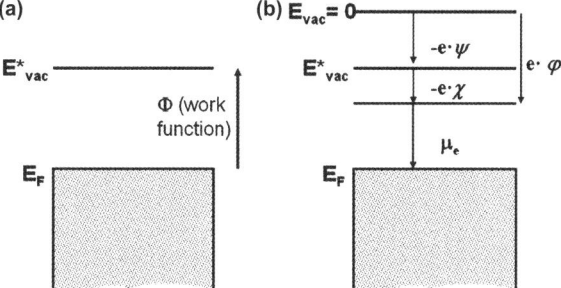

Figure 3.7 Description of energy levels in surface physics. (a) Work function is the minimum work required for the removal of an electron within the samples to a point just outside the sample. (b) For charged samples the position just outside the sample (E*vac) is still ey (with the external Volta potential $y^1 0$) away from the absolute vacuum level Evac. The work function comprises the chemical energy (μ_e, material specific) and the surface potential (c, depending on the specific surface characteristics. Reprinted from ref. 36 with permission from Elsevier, copyright 2007.

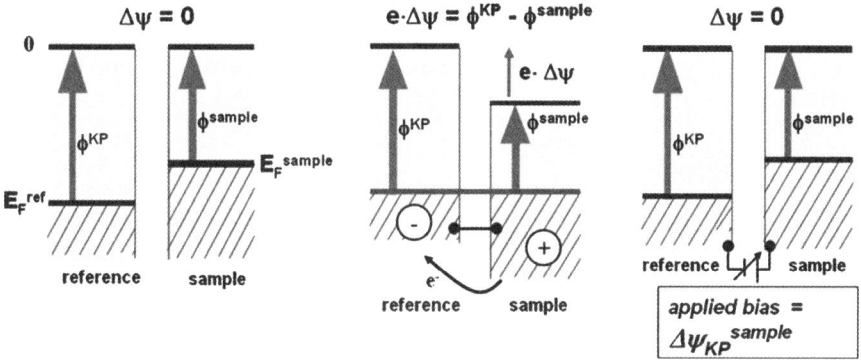

Figure 3.8 Basic principle of classical Kelvin Probe technique. The reference (KP – Kelvin Probe) and the sample are brought into contact. The equilibration of the Fermi levels result in a Volta potential difference due to the correlated charge transfer. An external backing potential, V_b, is applied between samples and Kelvin probe until it compensates this potential difference. Finally, both surfaces are uncharged and the value of the applied backing potential is determined. Reprinted from ref. 36 with permission from Elsevier, copyright 2007.

human skin.[39] Usually, air contains a few charged particles or ions due to ionization, electrical or radioactive radiation, which are attracted to the surfaces and compensate formerly introduced charges. Adsorbed water molecules enhance the charge compensation by ions stemming from dissolved impurities or the autodissociation of water. Due to the strong dependence of the results on the actual conditions (various amount of impurities and relative humidity),

KPM measurements should be carried out in defined conditions (ideally 0% humidity[40] or, in the case of corrosion studies, at 100% humidity).

The traditional KPM mode offers some benefits over the above-mentioned KPFM mode. In KPFM the signal is expected to strongly depend on the tip–sample distance. Usually the measured potential differences on a sample do not represent the real potential differences between tip and sample as the cantilever contribution is difficult to exclude.[36,40] Another advantage of traditional KPM lies in the comparably lower electrical fields that might act on the physical conditions of the samples, in particular, in semiconducting samples.[40]

The KPM technique has been used in corrosion studies and is frequently applied in the automobile industry to monitor the corrosion behaviour of steel under paints, but was expanded to other applications such as adsorption studies, bulk contaminations, surface roughness or surface charge imaging.[37] The technique has been further refined to answer biologically related questions. Baikie *et al.*[37] studied the displacement and biopotentials along and around a plant shoot in response to environmental stimuli. Another promising application is the entirely label free and sensitive analysis of gene arrays as demonstrated by Cheran *et al.*[38]

KPM studies have been performed to obtain additional information about intrinsic changes within atomic doped diamond, such as carbons with nitrogen[41] or silicon,[42] causing significant differences in endothelial cell adhesion on such surfaces. For both treatments, a decrease in work function was accompanied by slight increases of water contact angles and resulted in enhanced cell adhesion.

3.7 Conclusions

The most commonly used charge measurement methods in dry and wet condition have been discussed in this Chapter. In dry condition, a number of methods (TSDC, induction probe, field mill, EFM and KPFM) for electrostatic charge measurement have been reviewed. The TSDC method is capable of measuring the integral charge at the surface as well as in the volume of the sample, but the charge is destroyed by the measurement. On the other hand, induction techniques are sensitive only to the surface charge, but they are usually non-contact and non-destructive. Particularly, SPM techniques allow measurement of surface charge with high lateral (tens of nanometres) and potential ($\sim$mV) resolution. In wet conditions, the streaming potential technique is most commonly used for surface potential measurements, although KPM can also be employed in certain cases.

References

1. R. Gerhard-Multhaupt, *Electrical Insulation*, 1987, **EI-22(5)**, 531.
2. V. N. Kestelman, L. S. Pinchuk, V. A. Goldade, in *Electrets in Engineering: Fundamentals and Applications*, Kluwer Academic Publishers, 2000.
3. K. Okazaki, *Jpn. J. Appl. Phys.*, 1993, **32**, 4241.
4. I. A. Matveeva and L. A. Shekhtman, *Meas. Tech.*, 1985, **27**, 1056.

5. P. E. Secker, *J. Electrostatics*, 1975, **1**(1), 1975.
6. P. Plovera, P. Molinie, A. Soria and A. Quijano, *J. Electrostatics*, 2009, **67(2–3)**, 457.
7. Lord Kelvin, *Philos. Mag.*, 1898, **46**, 82.
8. W. A. Zisman, *Rev. Sci. Instrum.*, 1932, **3**, 367.
9. D. Bonnell, *Scanning Probe Microscopy and Spectroscopy; Theory, Techniques and Applications*, Wiley-VCH, USA, 2nd edn, 2001, p. 493.
10. E. Mayer, H. Josef Hug, R. Bennewitz, *Scanning Probe Microscopy: The Lab on a Tip*, Springer-Verlag, Germany, 2004, p. 210.
11. C. Sommerhalter, T. W. Matthes, T. Glatzel, A. Jäger-Waldau and M. C. Lux-Steiner, *Appl. Phys. Lett.*, 1999, **75**, 286.
12. M. Nonnenmacher, M. P. O'Boyle and H. K. Wickramasinghe, *Appl. Phys. Lett.*, 1991, **58**, 2921.
13. J. M. R. Weaver and D. W. Abraham, *J. Vac. Sci. Technol., B*, 1991, **9**, 1559.
14. J. Belana, M. Mudarra, J. Calaf, J. C. Canadas and E. Menendez, *Electrical Insulation*, 2002, **28**, 287–293.
15. D. K. Burghate, V. S. Deogaonkar, S. B. Sawarkar, S. P. Yawale and S. V. Pakade, *Bull. Mater. Sci.*, 2003, **26**(2), 267–271.
16. J. P. Gittings, C. R. Bowen, A. C. E. Dent, I. G. Turner, F. R. Baxter, S. Cartmell and J. Chaudhuri, *Ferroelectrics*, 2009, **390**, 168–176.
17. N. Horio, M. Hiramatsu, M. Nawata, K. Imaeda and T. Torii, *Vacuum*, 1998, **51**(4), 719–722.
18. A. Kyritsisa, P. Pissisb, J. L. G. Ribellesc and M. M. Pradasc, *J. Non-Cryst. Solids*, 1994, **172–174**, 1041–1046.
19. A. Lamure, M. Martin, M. F. Harmand and C. Lacabanne, *Thermochim. Acta*, 1991, **192**, 313–320.
20. V. M. Gun'ko, V. I. Zarko, E. V. Goncharuk, L. S. Andriyko, V. V. Turov, Y. M. Nychiporuk, R. Leboda, J. Skubiszewska-Zięba, A. L. Gabchak, V. D. Osovskii, Y. G. Ptushinskii, G. R. Yurchenko, O. A. Mishchuk, P. P. Gorbik, P. Pissis and J. P. Blitz, *Adv. Colloid Interface Sci.*, 2007, **131**, 1–89.
21. M. Fois, A. Lamure, M. J. Fauran and C. Lacabanne, *10th International Symposium on Electrets*, 1999, 217–220.
22. Y. Tanaka, T. Iwasaki, M. Nakamura, A. Nagai, K. Katayama and K. Yamashita, *J. Appl. Phys.*, 2010, 107.
23. R. Kato, S. Nakamura, K. Katayama and K. Yamashita, *J. Biomed. Mater. Res., A*, 2005, **74A**, 652–658.
24. S. Nakamura, H. Takeda and K. Yamashita, *J. Appl. Phys.*, 2001, **89**, 5386–5392.
25. M. Gregor, A. A. Ghandi, S. A. M. Tofail, *unpublished data*.
26. J. J. Bikermann, *Kolloid-Z.*, 1935, **72**, 100–108.
27. S. Roessler, R. Zimmermann, D. Scharnweber, C. Werner and H. Worch, *Colloid Surf., B*, 2002, **26**, 387–395; R. J. Hunter, 1988.
28. C. Werner, R. Zimmermann and T. Kratzmuller, *Colloids Surf., A*, 2001, **192**, 205–213.

29. R. J. Hunter, *Zeta Potential in Colloid Science: Principles and Applications*, Academic Press, London, 1981.
30. J. Lyklema, *Fundamentals of Colloid and Interface Science, Vol. 2, Solid–Liquid Interfaces*, Academic Press, London, 1995.
31. R. Zimmermann, T. Osaki, R. Schweiss and C. Werner, *Microfluid. Nanofluid.*, 2006, **2**, 367–379.
32. C. Werner, U. Konig, A. Augsburg, C. Arnhold, H. Korber, R. Zimmermann and H. Jacobasch, *Colloids Surf., A*, 1999, **159**, 519–529.
33. A. Zhu and T. Chen, *Colloids Surf., B*, 2006, **50**, 120–125.
34. K. Cai, M. Frant, J. Bossert, G. Hildebrand, K. Liefeith and K. D. Jandt, *Colloids Surf., B.*, 2006, **50**, 1–8.
35. K. Kawasaki, M. Kambara, H. Matsumura and W. Norde, *Colloids Surf., B.*, 2003, **32**, 321–334.
36. M. Rohwerder and F. Turcu, *Electrochim. Acta*, 2007, **53**, 290–299.
37. I. D. Baikie, P. J. S. Smith, D. M. Porterfield and P. J. Estrup, *Rev. Sci. Instrum.*, 1999, **70**, 1842–1850.
38. L-E. Cheran, S. Sadeghi and T. Michael, *Analyst*, 2005, **130**, 1569–1576.
39. W. Tang, B. Bushan and S. Ge, *J. Vac. Sci. Technol., A*, 2010, **28**, 1018–1028.
40. M. A. Stevens-Kalceff, *Microsc. Microanal.*, 2004, **10**, 797–803.
41. T. I. T. Okpalugo, A. A. Ogwu, A. C. Okpalugo, R. W. McCullough and W. Ahmed, *J. Biomed. Mater. Res. B, Appl. Biomater.*, 2008, **85B**, 188–195.
42. A. A. Ogwu, T. I. Okpalugo, N. Ali, P. D. Maguire and J. A. McLaughlin, *J. Biomed. Mater., B, Appl. Biomater.*, 2008, **85**, 105–113.

CHAPTER 4

Non-linear Characterizations of Surface Charge and Interfacial Morphology

S. B. LANG,*[a] G. A. STANCIU*[b] AND S. G. STANCIU[b]

[a] Department of Chemical Engineering, Ben-Gurion University of the Negev, 84105 Beer Sheva, Israel; [b] University Politehnica Bucharest, Center for Microscopy – Microanalysis and Information Processing, Splaiul Independentei 313, 060042, Bucharest, Romania

4.1 Introduction

This chapter will discuss a special technique of measurement of surface and subsurface charge in electrically modified biomaterials. In addition, it will also discuss the laser-scanning microscopy (LSM) technique used in surface and interfacial characterization of biomaterials and biological materials. Analysis of the space charge or polarization profiles produced by electrical modification can be made using a number of thermal or acoustic techniques.[1,2] Thermal techniques reveal greater detail of the profiles near the surface of materials, whereas acoustic techniques give higher resolution at larger depths. The thermal technique, called the laser intensity modulation method (LIMM), is one of the best-known thermal techniques and will be described in relation to corona-charged biomaterials. Corona charging was selected for two major reasons: (1) it is experimentally relatively simple, requiring minimal facilities, and (2) there are many parameters that can be varied such as polarity, voltage, charging time and temperature. The application of corona charging of biomaterials and the analysis of the polarization and space charge profiles are

RSC Nanoscience & Nanotechnology No. 21
Biological Interactions with Surface Charge in Biomaterials
Edited by Syed A. M. Tofail
© Royal Society of Chemistry 2012
Published by the Royal Society of Chemistry, www.rsc.org

described in the following subsections. The chapter will then describe how LSM is used for morphological analysis of the surface and interface of biomaterials.

4.2   Corona Charging of Biomaterials

4.2.1   Nature of Coronas

A corona is a self-sustaining electrical discharge that occurs when a sufficiently high electrical potential is applied between asymmetric electrodes such as a sharp needle point and a plate.[3] A high electric field is formed around the point, and the surrounding gas (usually air), which is normally insulating, becomes ionized. The resulting ions are forced toward the plate. The voltage required for the onset of the corona depends upon the diameter of the point, and the temperature, pressure and composition of the gas.[4] The nature of the discharge depends upon the polarity of the point. For negative coronas (a negative point), the current consists of a train of very short pulses with a well-defined repetition frequency. These are called Trichel pulses and, in air at atmospheric pressure, they have rapid rise times of the order 1.5 ns.[5] At higher voltages, the negative corona becomes a continuous glow discharge. The positive corona is initially a continuous glow discharge. At sufficiently high voltages, both positive and negative coronas form streamers and eventually complete breakdown of the air occurs. In positive coronas in dry air, O_2^+ and N_2^+ are the major charge carriers, with various hydrated species forming at higher humidities. In addition to the electrons formed in negative coronas, CO_3^- ions are created.[3]

4.2.2   Experimental Apparatus

The corona charging apparatus was constructed in the Department of Electrical Communications Engineering of the Technical University of Darmstadt (Germany). It consisted of a steel cylinder approximately 28 cm in height and 21 cm in diameter which was mounted between two brass plates that were electrically grounded. The cylinder was perforated with 5 mm holes spaced approximately 8 mm center-to-center in order to provide ventilation and to reduce the weight. A hinged door in the cylinder permitted access to the interior. Test materials were placed on a grounded 5 cm brass plate mounted on a pedestal that could be changed in height. A sharp needle formed the high voltage electrode and was connected directly to a high voltage power supply (Spellman Bertan 230). It was capable of delivering up to ± 30 kV, the polarity of which was changed by opening the instrument case and rotating the polarity reversal module by 180°. A wire mesh grid connected to a high voltage supply (Keithley 246) (± 3.1 kV) could be mounted above the sample plate. It was used for charging very thin samples in order to increase the lateral uniformity of charge. The entire apparatus was placed in a laboratory oven for temperature control. As a safety measure, an interlock was constructed in the oven door so that the high voltage power supplies could not be turned on unless

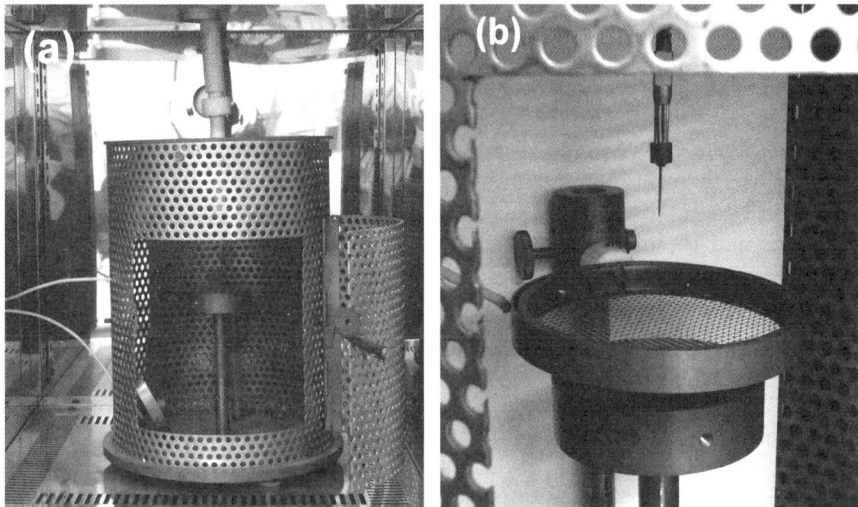

Figure 4.1   (a) Corona charging apparatus in a laboratory oven. (b) Detail showing needle, grid and base plate.

the oven door was closed. Figure 4.1(a) shows a photograph of the apparatus in the oven and Figure 4.1(b) shows the details of the needle, grid and base plate.

4.2.3   Technique of Corona Charging

The materials that were used in the experiments were 30 μm thick polyurethane (PU), 500 nm thick hydroxyapatite (HA) films on silicon wafers, and poly-(vinylidene fluoride) (PVDF) polymer and P(VDF-TrFE) copolymer of thicknesses ranging 3 to 32 μm. Samples either were unmounted or were attached to small aluminum plates by means of double-sided electrically conductive tape (SPI Supplies). The corona needle was adjusted at various distances between 13 and 20 mm above the samples and the oven temperature was set at various values between room temperature and 100 °C. The voltage on the needle was set at a value such that arcing did not occur. Depending upon the type and thickness of the sample, this was as high as 20 kV for negative coronas and 15 kV for positive ones. The charging times were varied between 5 min and 60 min.

If the samples were to be examined with LIMM (see Section 4.3), it was necessary to sputter Au/Pd electrodes on the upper surface. This was affected using a mini sputter coater (Quorom Technologies SC7620). Aluminum masks were used to restrict the electrode diameters to 8 mm. The coater was evacuated and then back-filled with argon. Sputtering was carried out for 9 min using a plasma current of 25 mA. The deposited electrodes had a thickness of approximately 80 nm. Their optical properties were determined using Au/Pd deposits sputtered on glass slides. It was found that 72% of the laser energy was absorbed in the electrodes, 22% was transmitted through the electrodes and 6% was reflected.

4.3 LIMM

4.3.1 Principle and Experimental Implementation of LIMM

LIMM is a technique for the determination of the spatial distribution in the thickness dimension of polarization in polar materials and space charge in non-polar ones. The technique was first suggested by Lang and Das-Gupta in 1981[6] and described in detail in 1986.[7] Various modifications of the experimental technique and the mathematical analysis of the data have been presented.[8–10] The experimental technique is described as follows. LIMM samples were prepared by corona charging as discussed in Section 4.2. A thin opaque electrode was applied to the upper surface by sputtering. The sample was placed on a grounded metal plate and electrical contact was made to the upper surface by means of a microprobe (Karl Suss Dresden GmbH). The surface of the sample was exposed to a laser beam that was modulated sinusoidally in intensity. The energy of the laser beam was absorbed in the electrode and heat diffused into the sample in the form of temperature waves. The waves were attenuated in magnitude and retarded in phase as they traversed the sample. The depth of penetration of the waves was greater for low frequencies of laser beam modulation and lesser for higher frequencies. The result was a non-uniform time-varying temperature distribution that interacted with the local polarization or space charge to produce a pyroelectric current. The real and imaginary components of the generated current were measured at 50 to 100 different logarithmically spaced frequencies and a lock-in amplifier was used to determine the amplitude and phase of the current relative to the phase of the modulated laser beam.

A diode laser (Lisa Laser Products) with an internal modulator was used. It emitted radiation at 685 nm and had a nominal power of 30 mW. The pyroelectric current was amplified with a variable-gain low-noise current amplifier (Femto). Gains of 10^6 to 10^9 V/A were normally used. The phase and amplitude of the voltage output of the current amplifier were measured by means of a lock-in amplifier that also provided the driving voltage for the laser modulator. A Stanford Research Model SR850 lock-in amplifier was used for modulation frequencies between 10 Hz and 100 kHz, and a Stanford Research Model SR844 RF lock-in amplifier for frequencies between 25 kHz and 2 MHz. The SR850 amplifier was used for all samples except for 500 nm films of HA for which the SR844 amplifier was utilized. The test sample and the current amplifier were placed in a tabletop Faraday cage to eliminate external electrical interference. All of the equipment was under computer control using a TestPoint program (Measurement Computing Corp.).

4.3.2 Analysis of LIMM Data

The polarization or electric field/space charge distribution is found from a solution of the LIMM equation, as shown is eqn (1):[7,11,12]

$$I(\omega) = \frac{A}{L} \int_{z_1}^{z_2} \beta(z) \frac{\partial T(z,\omega)}{\partial t} \, dz \tag{1}$$

where

$$\beta(z) = p(z) - (\alpha_x - \alpha_\varepsilon)\, \varepsilon\, \varepsilon_o\, E(z)$$

Here I is the pyroelectric current; ω is the radial frequency; A is the cross-sectional area of the laser beam; L is the sample thickness; T is temperature; t is time; z, z_1 and z_2 are the spatial coordinate and the coordinates of the sample, respectively; p is the pyroelectric coefficient; α_x and α_ε are the temperature dependence of thickness and permittivity, respectively; ε and ε_0 is the sample permittivity and vacuum permittivity, respectively; and E is the electric field. In a polar material, E is zero and the pyroelectric coefficient distribution is determined. This is equivalent to the relative polarization distribution. Absolute values of the polarization distribution require additional information. (The terms pyroelectric coefficient distribution and polarization distribution are used interchangeably herein and in the literature.) In a non-polar material, P is zero and the space charge distribution is determined from eqn (2):

$$\rho = \varepsilon\,\varepsilon_0\, \frac{dE}{dz} \qquad (2)$$

where ρ is space charge.

Eqn (1) is a Fredholm integral equation of the first kind. It is an ill-posed problem with multiple solutions, all of which satisfy the equation within experimental error.[13] In particular, a linear regression least-squares solution will always satisfy eqn (1), but the results will be completely incorrect. It is not mathematically possible to determine the true polarization or space charge distribution. However, a number of solution techniques have been developed that have a high probability of giving the correct distribution. Three methods have been used in the current studies:

1. Scale transformation (SC).[12] This technique is based on a simple set of coordinate transformations using the raw experimental data. Because of its simplicity, it is frequently used and offers a good guide for selection of more sophisticated techniques.
2. Monte Carlo (MC) method.[14] In this technique, a least-squares solution is obtained for a randomly selected set of spatial location values. This process is repeated a large number of times for different sets of randomly selected locations and the results are averaged. The MC method gives very good results for distributions that are very non-uniform.
3. Polynomial Regularization-L Curve (PRM-LC) method.[9,10] This method is based on the fact that the polarization/space charge distributions must be relatively smooth. They can have hills and valleys but sharp peaks or rapid sign changes are unrealistic physically and must be suppressed. It is based on Provencher's 'Principle of Parsimony'[15] that states that the simplest solution may not have full details, but it will not include meaningless artifacts. This technique is useful for distributions that are relatively uniform.

In order to visualize the differences among the techniques, two mathematical functions were chosen to represent typical pyroelectric coefficient distributions. Random noise was added to the functions in order that they would resemble 'true' experimental data. Eqn (1) was used to simulate LIMM data. Then the LIMM equation was solved using each of the three methods. A non-uniform pyroelectric coefficient distribution is illustrated in Figure 4.2(a). Both the MC and the PRM-LC results were very close to the 'true' data, with the MC method being slightly better. The SC results were less satisfactory. A completely uniform distribution is illustrated in Figure 4.2(b). In this case, the PRM-LC and SC methods gave good results and the MC method was completely incorrect. As a general rule in the analysis of experimental data, the MC method was used for non-uniform distributions and the PRM-LC method for more uniform

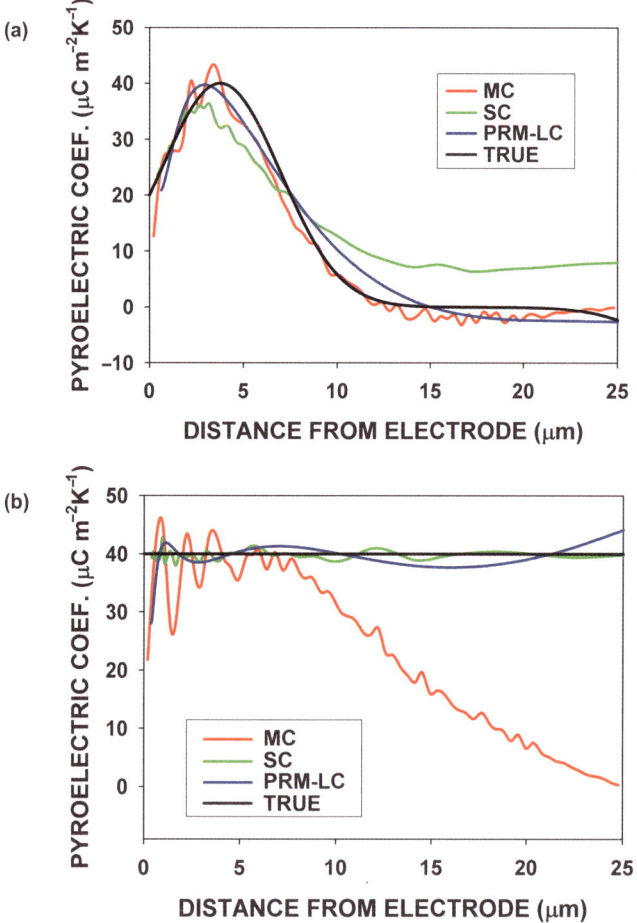

Figure 4.2   Solutions of the LIMM equation using three different methods. (a) Non-uniform pyroelectric coefficient distribution. (b) Uniform pyroelectric coefficient distribution.

distributions. The SC method was usually a good approximation to the correct distribution and thus served as a guide to the selection of either the MC or PRM-LC method. The three methods were implemented using the program Mathematica.[16] Execution of the programs required less than 1 min on a laptop computer with a 2.6 GHz CPU.

In addition to the laser intensity, electrode properties and dimensions of the sample, it was necessary to determine the sample thermal diffusivity. This was done using a modification of LIMM in which the phase retardation of the temperature waves through the sample was determined.[17-19]

4.4 Applications of Corona Charging and LIMM

This section is divided into three parts. In the first part, the space charge distribution is determined in a non-polar material, PU, and the pyroelectric coefficient distribution in a polar material, P(VDF-TrFE). The second part describes studies on corona polarity and charging time of PU. A study of stability of space charge in PU is presented in the third part.

4.4.1 Space Charge in PU and Pyroelectric Coefficient in P(VDF-TrFE) Copolymer

Polarization/space charge distributions could only be determined to a depth in the samples at which the amplitudes of the temperature waves at the lowest frequency used were attenuated to between 5 and 10% of their initial value. This corresponded to a depth of 20 μm for PU and 12 μm for PVDF and the P(VDF-TrFE) copolymer. The 500 nm HA films were sufficiently thin that the attenuation was very small at the lowest frequency used.

A large number of samples were corona charged and analyzed using LIMM. PU is non-polar and electric field and space charge distributions were determined for this material. A 30 μm thick sample of PU was corona charged with a needle voltage of –10 kV for 15 min at room temperature. Figure 4.3(a) shows the LIMM measurements of the real and imaginary components of the pyroelectric current at frequencies between 10 Hz and 50 kHz. A preliminary analysis using the SC method showed that the electric field distribution was very non-uniform. Consequently, the MC method was used to analyze the data. The results in Figure 4.3(b) show the electric field distribution and the space charge distribution. Both the electric field and the space charge distributions changed signs in the region closest to the surface. The space charge did not penetrate to a depth greater than 10 μm with these poling conditions.

A 25 μm thick sample of the P(VDF-TrFE) copolymer was poled at +10 kV for 10 min at room temperature. LIMM measurements were made in the frequency range from 10 Hz to 100 kHz. The real component of the LIMM data was much larger than the imaginary component and was relatively uniform with frequency (Figure 4.4a). This indicated that the polarization distribution was relatively uniform, as confirmed by a SC analysis. The PRM-LC

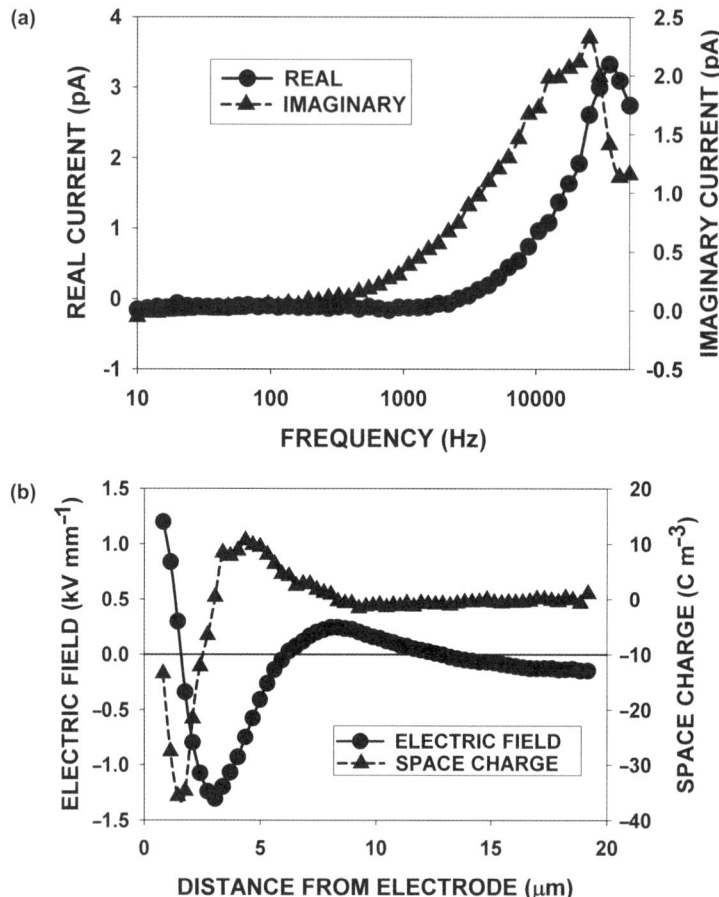

Figure 4.3 (a) LIMM data measured on a 30 μm thick PU sample that was corona charged at –10 kV for 15 min at room temperature. (b) Calculated electric field and space charge distributions.

method was used to analyze the data. The polarization distribution, as shown in Figure 4.4(b), was very uniform except at the very surface of the sample.

4.4.2 Effects of Corona Polarity and Charging Time on Space Charge in PU

Some of the initial experiments were designed to study the effects of the polarity of the corona needle and the influence of charging time. Experiments were made on 30 μm thick PU mounted with double-sided conductive tape on aluminum plates. Two charging times were used, 15 and 60 min, and the temperature was between 20 and 25 °C. In order to avoid comparing large numbers of graphs of space charge, methods were sought for characterizing the

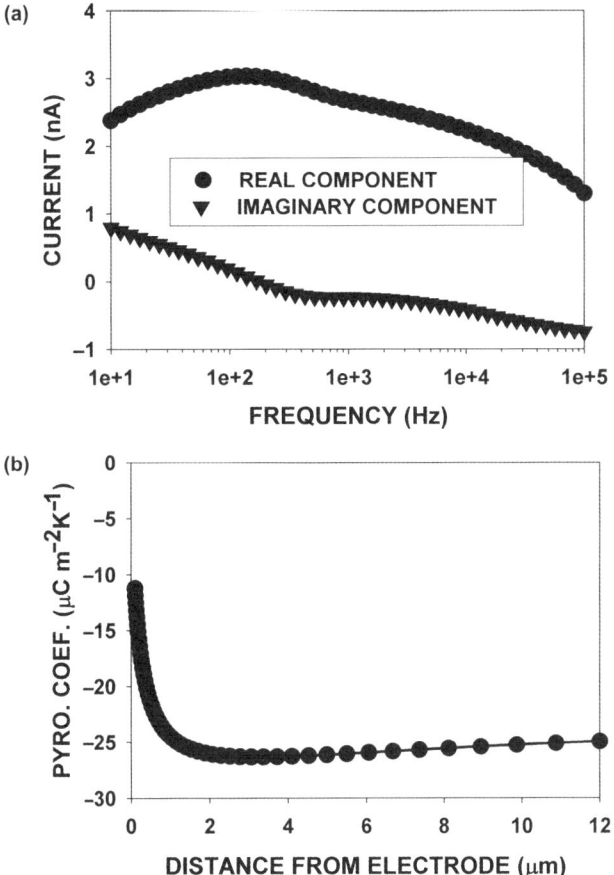

Figure 4.4 (a) LIMM data measured on a 25 μm thick P(VDF-TrFe) copolymer
sample that was corona charged at + 10 kV for 10 min at room temperature. (b) Calculated pyroelectric coefficient distribution.

space charge distributions with a small number of parameters. The parameters
were then presented by means of a bubble chart.

The parameter forming the abscissa of the bubble chart is the total charge
(Q) which was found by integrating the charge $\rho(z)$ over the sample thickness
(L), as shown in eqn (3):

$$Q = \int_0^L \rho(z)\, dz \tag{3}$$

The ordinate of the bubble chart is the calculated surface voltage (V) on the
unelectroded surface that was exposed to the corona, as in eqn (4):

$$V = \frac{L}{\varepsilon\, \varepsilon_0} \int_0^L \rho(z) \left(1 - \frac{z}{L}\right) dz \tag{4}$$

The diameter of the bubble was proportional to the absolute value of the maximum in the space charge versus position graph.

The results are shown in Figure 4.5. Negative corona poling is much more effective than positive poling. The mechanisms are completely different because of the size of the charge carriers, electrons in the case of negative poling and positively charged oxygen and nitrogen ions in the case of positive poling. Poling for 60 min is more effective than for 15 min. However, it was decided that 15 min of poling was deemed adequate for many of the tests that followed.

4.4.3   Stability of Space Charge in PU

It was important to determine if the space charge in PU was stable with time, immersion in water and immersion in PBS solution [buffer solution containing sodium chloride and sodium phosphate (Sigma-Aldrich)]. A preliminary study is described here. A total of 32 samples of PU were prepared and corona charged. Both positive and negative coronas were used. Half of the samples were charged for 1 h at room temperature, the other half at 70 °C. Eight samples were electroded with sputtered Au/Pd and LIMM measurements were made immediately. The other sets of eight each were stored dry, immersed in doubly-distilled water or in PBS solution for periods of up to 4 weeks. Then they were electroded followed by LIMM measurements. The space charge distributions for these samples were characterized by a calculation of the net charge as in eqn (5):

$$Q_{net} = \int_{0}^{L} |\rho(z)| \, dz \qquad (5)$$

where $|\rho(z)|$ is the absolute value of the space charge.

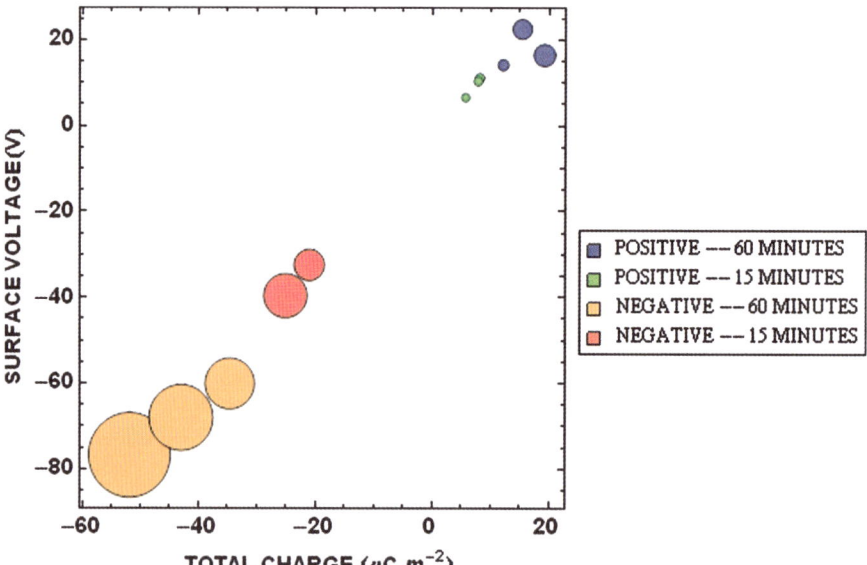

Figure 4.5   Bubble chart showing the effects of corona polarity and charging time.

The results are plotted in the form of a histogram and are shown in Figure 4.6. The samples that were charged at room temperature lost approximately 80% of their charge after storage. It could not be determined from these experiments if the charge loss in water or PBS solution was due to the storage media or simple charge decay. Samples that were charged at 70 °C retained their charge in all of the storage media.

4.5 Interfacial Effects Investigated by LSM

4.5.1 Introduction

The surface properties have an enormous effect on the success or failure of a biomaterial device, and because of this the need for complex biomaterial surface characterization methods is very high. The microscopy techniques used for the analysis of biomaterial surfaces include scanning electron microscopy (SEM), transmission electron microscopy (TEM), scanning probe microscopy (SPM) techniques and LSM techniques.

Probably the most important achievement in regard to the investigation and characterization of interfacial effects was made by discovering confocal laser microscopy (CSLM). It is widely accepted that the confocal microscope was invented by Marvin Minsky, who filed a patent in 1957.[20] The first confocal scanning laser microscope was reported by Sheppard and Choudhury.[21] The main advantages of CSLM compared with conventional optical microscopy are: the ability for non-invasive optical sectioning; and the point spread function (PSF) of a CSLM system which, in the case of a confocal microscope, is approximately the square of the PSF of a conventional microscope, leading to an improvement in resolution by square root of 2. A major improvement for studying both living cells and thick samples was achieved with the development of the multi-photon microscopy based on the idea of Marie Goppert Mayer.[22] The first laser scanning microscope based on two-photon excitation (TPE) was reported by Denk *et al.*[23] In this non-linear method, each molecule is usually excited by two or three photons such that the total energy is equal to the excited molecule energy. TPE has become a powerful method for imaging thick samples, since the excitation (absorption) takes place only in objective focus, and it thus causes minimal harm to the surrounding cells or tissues through the unwanted absorption of short wavelength excitation energy. This important advantage is achieved while still maintaining a good spatial resolution owing to the long wavelength used in the excitation (which is less damaging and penetrates deeper into tissues) and to the efficient fluorescence detection that is spectrally far from the excitation. Other non-linear methods include second harmonic generation (SHG),[24] third harmonic generation (THG)[25] and coherent anti-Stokes Raman scattering (CARS).[26] Stimulated emission depletion (STED) microscopy was conceptually introduced by Hell *et al.*[27] and was demonstrated in practice only a few years ago.[28] The principle of the method is to ensure that the volume that emits fluorescence in the sample is extremely small. This is accomplished by using two pulsed lasers. The first laser has a

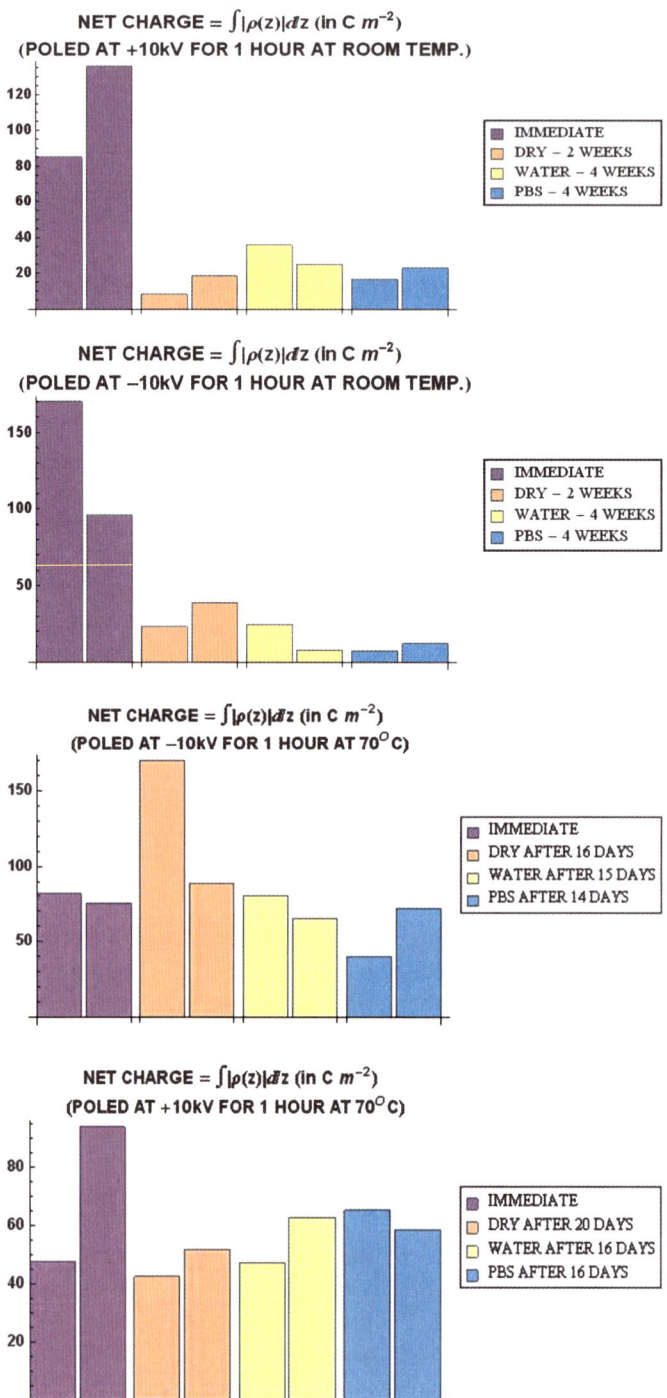

Figure 4.6   Histograms showing the charge stability of PU samples stored in various
media.

wavelength that excites the fluorescent molecules and the second illuminates the sample with a doughnut-like pattern in a wavelength that drives (depletes) the excited states of the fluorescent molecules back to the ground state. The only volume that is left with fluorescent molecules in the excited state is the one corresponding to the hole of the doughnut, from where the fluorescence is actually detected. The main advantages offered by the techniques detailed above are represented by the ability of non-invasive optical sectioning.

In the case of transparent samples all laser-scanning techniques described above are able to investigate samples bulk and interfacial effects. On the basis of the optical sections of the sample, a three-dimensional image may be obtained.

4.5.2 Surface and Interfacial Effects Investigated by Higher Harmonic Generation Imaging in LSM

4.5.2.1 Principle of Imaging Techniques Based on High Harmonics Generation

The interaction of light with the matter may be modeled as an interaction of an oscillating electric field with a dielectric material. As a light wave passes through a dielectric, the electric field component will separate the positive and negative charges in the material. This charge separation establishes internal electric fields that oppose the driving field and generate forces that restore the separated charges back towards equilibrium. The electric fields that result from the polarization of the dielectric are the source of scattering by a dielectric material. At optical frequencies, the oscillations of atomic electron clouds relative to their nuclei are the dominant source of dielectric polarization; more massive particles have higher moments of inertia, too large to permit much physical displacement in the short period of the oscillation of the light. In general, the non-linear polarization for a material can be expressed in eqn (6):

$$P = \chi^{(1)}E + \chi^{(2)}E^2 + \chi^{(3)}E^3 + \dots \tag{6}$$

where P is the induced polarization, $\chi^{(n)}$ is the nth order non-linear susceptibility, and E is the electric field vector of the incident light. Assuming that the electric field varies such that
$E = E_0\sin\omega t$, substituting this definition into the above equation, the electric polarization (Section 4.4.1) can be written as eqn (7):

$$P = \chi^{(1)}E_0\sin\omega t + \chi^{(2)}E_0^2(1-\cos2\omega t) + \chi^{(3)}E_0^3(3\sin\omega t - \sin3\omega t/4) + \dots \tag{7}$$

The first term describes normal absorption and reflection of light, the second, SHG, sum, and difference frequency generation, and the third, both two and three photon absorption, as well as THG. The second term predicts that a dielectric material could scatter light that is the second harmonic of the incident light. The intensity of this scattered light should depend quadratically upon the intensity of the primary light. This effect, and the other effects described by non-linear terms of $E(t)$, are generally insignificant at the intensities of light usually encountered; however, laser techniques can generate sufficiently intense fields to make these effects observable. The first demonstration of SHG was reported in 1961, only one year after the first report of an optical laser.[29]

The second-order (SHG) term of the non-linear hyperpolarizability is proportional to the square of electromagnetic field, with $\chi^{(2)}$ as a proportionality constant. This second-order non-linear susceptibility $\chi^{(2)}$ is a bulk material property that relates to the molecular hyperpolarizability, $\langle\beta\rangle = \chi^{(2)}/N_s$, where N_s is the density of dye molecules and the brackets denote an orientational average. Thus $\langle\beta\rangle$ is maximal for parallel aligned molecular dipoles (*e.g.*, membrane resident dyes) and zero for anti-parallel oriented molecules, the contributions of which interfere destructively due to their phase shift. In bulk solution, one can always find a pair of molecules that are anti-parallel and cancel their respective contributions. This makes it clear why SHG requires a polarizable material with non-centrosymmetric symmetry. These two conditions lead to the following symmetry requirements:

- The molecules are non-centrosymmetric (do not possess an inversion symmetry).
- The molecules in the bulk form are arranged in a non-centrosymmetric structure.

The second condition, requiring a non-centrosymmetric bulk form, permits a second-order non-linear process such as SHG only for the following bulk forms:

- A non-centrosymmetric crystal.
- An interface between two media (an interface is always asymmetric).
- An artificially aligned molecular orientation under the influence of an electric field (such structures are called *electrically poled*).

Thus a solution or a glass, in its natural form being random, does not exhibit any second-order effect. For biological systems, important second-order effects are associated with the interface and with electrical field poling.

The third-order term of the polarization equation also leads to new optical frequency generation by mixing of three waves of the ω_1, ω_2 and ω_3 frequencies. Since all materials have non-vanishing third-order susceptibilities, THG microscopy can be utilized as a general-purpose microscopy technique, with no need for fluorescence labeling or staining.

Figure 4.7 shows a Jablonski diagram of SHG and THG. Higher harmonic generation does not require actual absorption but is based on the non-linear scattering of two low-energy photons that produce one frequency-doubled photon. They also differ from multi-photon absorption in that they are phase preserving, *i.e.* coherent, which causes a non-linear dependence on the molecule density. The properties of higher harmonic generation offer some advantages

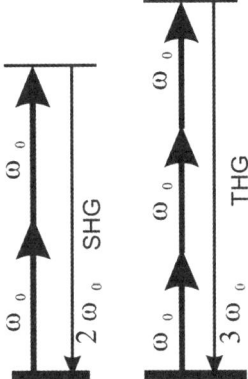

Figure 4.7 Jablonski diagram of second (left) and third (right) harmonic generation.

for live cells or tissues for multi-photon fluorescence microscopy. Harmonic generation does not involve the excitation of molecules, therefore it does not suffer from phototoxicity or photobleaching effects.

Unlike the second-order non-linear optical processes, which can take place only in a non-centrosymmetrically oriented medium, the third-order processes can take place in any medium, random or ordered. Hence, they can be observed in liquids, amorphous solid media or crystalline media. However, because it is a higher-order process, THG is a less efficient process than SHG in a medium in which both can occur. In other words, it will take a more intense optical pulse (higher electric field) to generate THG than that required for SHG.

SHG and THG microscopy both rely on femtosecond laser technology and are easily combined with two-photon microscopy. As an imaging platform we used a confocal microscope upgraded for IR excitation (TPE). The experimental setup is shown schematically in Figure 4.8. A synchronous Ti:sapphire laser with femtosecond pulses at a repetition rate of 80 MHz may be used as a laser source. With an IR port, the laser beam is coupled with the optical system of a confocal microscope. The raster of the laser beam is focused by the microscope objective on the sample surface. The emission wavelength of the laser may be tuned to the range of 700 nm to 1050 nm. The second or the third harmonics are collected by a condenser and measured by a photomultiplier (PMT) appropriate for UV and visible radiation. A filter placed before the PMT separates the second or the third harmonic photons from fundamental light and from the background signal. The PMT signal is amplified, digitized and fed into a computer.

4.5.2.2   Surface and Interfacial Charge Effects and Morphology Investigated by SHG Microscopy

Some of the techniques based on LSM (CSLM, TPE imaging and STED) are very useful for investigating the effects which appear at biological–non-biological

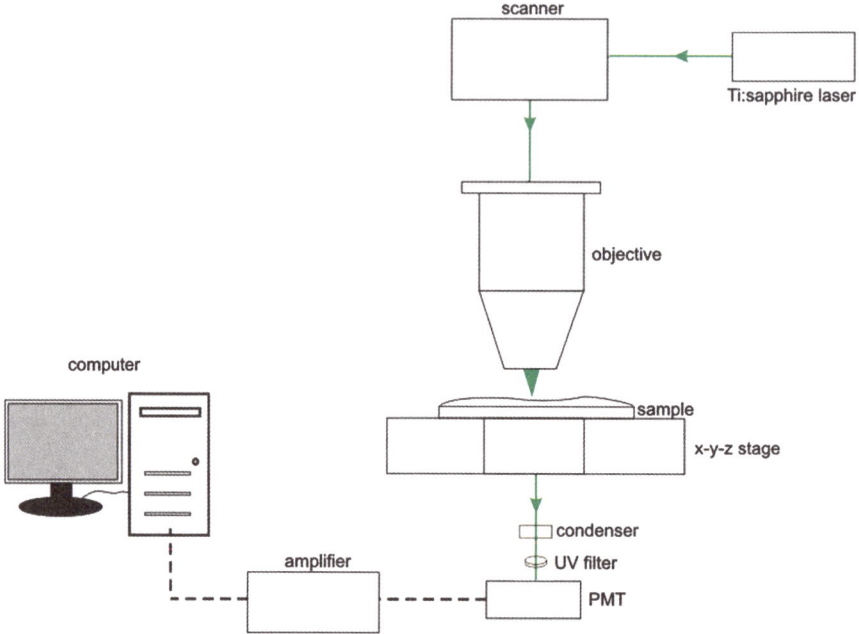

Figure 4.8   Optical setup of a laser-scanning microscope based on second harmonic and third harmonic imaging.

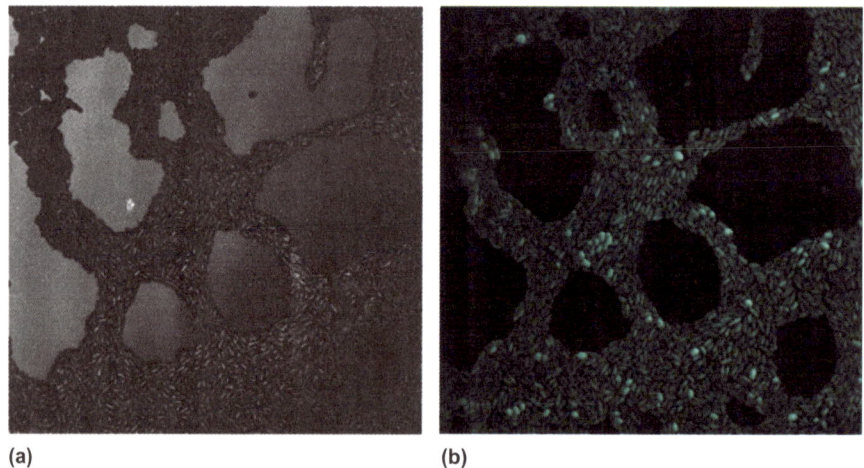

(a)   **(b)**

Figure 4.9   (a) CSLM and (b) SHG images of the interface of *Saccharomyces cere-visiae* and glass surface.

interfaces by using fluorescent molecules (fluorophores). The main disadvantage of these techniques is associated with fluorophore photobleaching and photo-toxicity under laser light exposure.

As an all-optical contactless and non-destructive technique, SHG has been proven to be highly sensitive to surfaces and interfaces. The generation of the second harmonic radiation is possible at surfaces and interfaces where the symmetry of the solid is broken. SHG is sensitive to dangling bonds, interface states and the presence of internally or externally applied electric fields. Second harmonic imaging microscopy has been used for visualizing biomolecular arrays in cells, tissues and organisms.[30]

In Figure 4.9 confocal and SHG images collected at the interface between glass and *Saccharomyces cerevisiae* are shown. It can be seen that the two images are complementary.

As TPE imaging and second harmonic imaging are obtained by using the same microscope and excitation laser, there is a possibility to have complementary information on the same sample.

4.6 Conclusions

Corona poling/charging is a very useful method for introducing polarization into polar materials and space charges into non-polar ones. The apparatus required is relatively inexpensive and a large number of parameters can be varied. LIMM is an excellent technique for the analysis of the polarization/space charge deposited in regions near the surface of the materials by corona charging. Preliminary experiments have been conducted to show the effects of corona polarity, charging time and temperature, and charge loss by storage in different media.

LSM based on second harmonic imaging represents a very useful tool for investigating the surface structure and the interfacial properties of the biological samples. The combined strong scattering and absorption of UV and visible light in the case of confocal microscopy lead to a rapid loss of both intensity and contrast when the penetration depth in the tissues is higher than a few tens of microns into tissue. Unlike confocal microscopy, in the case of the second harmonic imaging the scanning light is able to penetrate much deeper inside the volume of the sample.

References

1. S. Hole and T. Ditchi, *IEEE Trans. Dielectr. Electr. Insul.*, 2003, **10**, 670.
2. J. Lewiner, S. Hole and T. Ditchi, *IEEE Trans. Dielectr. Electr. Insul.*, 2005, **12**, 114.
3. J. A. Giacometti and O. N. Oliveira, Jr, *IEEE Trans. Electr. Insul.*, 1992, **27**, 924.
4. J. J. Lowke and R. Morrow, *Pure Appl. Chem.*, 1994, **66**, 1287.
5. L. B. Loeb, *Phys. Rev.*, 1952, **86**, 256.
6. S. B. Lang and D. K. Das-Gupta, *Ferroelectrics*, 1981, **39**, 1249.
7. S. B. Lang and D. K. Das-Gupta, *J. Appl. Phys.*, 1986, **59**, 2151.
8. S. B. Lang, *Ferroelectrics*, 1991, **118**, 343.
9. S. B. Lang, *IEEE Trans. Dielectr. Electr. Insul.*, 2004, **11**, 3.
10. S. B. Lang, *J. Mater. Sci.*, 2006, **41**, 147.

11. S. Bauer and S. Bauer-Gogonea, *IEEE Trans. Dielec. Elec. Insul.*, 2003, **10**, 883.
12. B. Ploss, R. Emmerich and S. Bauer, *J. Appl. Phys.*, 1992, **72**, 5363.
13. B. L. Phillips, *J. Assoc. Comput. Mach.*, 1962, **9**, 84.
14. E. Tuncer and S. B. Lang, *Appl. Phys. Lett.*, 2005, **86**, 71107.
15. S. W. Provencher, *Comput. Phys. Commun.*, 1982, **27**, 213.
16. *Mathematica*, 2010, Wolfram Research, Inc., Champaign, IL, USA.
17. S. B. Lang, *Ferroelectrics*, 1989, **93**, 87.
18. S. Muensit and S. B. Lang, *Ferroelectrics*, 2003, **293**, 341.
19. S. B. Lang, E. Ringgaard, S. Muensit, X. Q. Wu, J. C. Lashley and Y. W. Wong, *IEEE Trans. Ultrason. Ferroelectr. Freq. Control*, 2007, **54**, 2608.
20. M. Minsky, *US Patent.*, 3013467, 1961.
21. C. J. R. Sheppard and A. Choudhury, *Opt. Acta*, 1977, **24**, 1051–1073.
22. M. Goeppert-Mayer, *Ann. Phys.*, 1931, **9**, 273–295.
23. W. Denk, J. H. Strickler and W. W. Webb, *Science*, 1990, **248**, 73–76.
24. J. N. Gannaway and C. J. R. Sheppard, *Opt. Quant. Electron.*, 1978, **10**, 435–439.
25. Y. Barad, H. Eizenberg, M. Horowitz and Y. Silberberg, *Appl. Phys. Lett.*, 1997, **70**, 922–924.
26. M. D. Duncan, J. Reintjes and T. J. Manuccia, *Opt. Lett.*, 1982, **7**(8), 350–352.
27. S. W. Hell and J. Wichmann, *Opt. Lett.*, 1994, **19**, 780–782.
28. V. Westphal, L. Kastrup and S. W. Hell, *Appl. Phys. B: Lasers Opt.*, 2003, **77**, 377–380.
29. P. A. Franken, A. E. Hill, C. W. Peters and G. Weinrich, *Phys. Rev. Lett.*, 1961, **7**, 118–119.
30. P. Campagnola and L. M. Loew, *Nature Biotechnology*, 2003, **21**, 1336–1360.

Part II
Protein Interactions at the Surface

CHAPTER 5

Immobilization of Enzymes on Porous Surfaces

P. PERIYAT* AND E. MAGNER*

Materials and Surface Science Institute, University of Limerick, Limerick, Ireland

5.1 Introduction

Enzymes can catalyse reactions with high specificity under very mild conditions at normal temperature and pressure.[1] However, they are not necessarily suitable for some applications due to issues such as instability/denaturation and reduced activity under harsh conditions such as those experienced with organic solvents or at high temperatures.[2,3] In industrial processes when enzymes are removed from the reaction mixture, denaturation of the enzyme can occur and may not be cost effective. The optimal solution to these problems is to immobilize the enzyme.[4,5] The main advantages of enzyme immobilization on solid substrates are enhanced enzyme stability, and ease of separation and recovery for reuse while maintaining the activity of the enzyme. This Chapter focuses on recent advances in the development of porous substrates for the immobilization of enzymes. Such substrates can provide sheltered environments for enzymes and have potential applications in biocatalysis, drug delivery and biosensors.

5.2 Factors Influencing Enzyme Immobilization

Enzymes can be immobilized on a variety of porous supports *via* different methods.[3–6] However, successful immobilization depends on a range of factors which are important in maintaining the catalytic activity of the enzyme and

RSC Nanoscience & Nanotechnology No. 21
Biological Interactions with Surface Charge in Biomaterials
Edited by Syed A. M. Tofail
© Royal Society of Chemistry 2012
Published by the Royal Society of Chemistry, www.rsc.org

improving its stability, ease of recovery and reuse of the immobilized enzymes. These factors are discussed below.

5.2.1 Size of Enzymes and Pores

One of the important factors is matching the size of the pores and the enzyme. The size of the enzyme needs to be smaller than the pore diameter of the support to enable the enzyme to enter the pore where the enzyme can be protected from the external environment. In addition, the pore diameter needs to be of sufficient size to enable facile diffusion of the substrate into the pores.

5.2.2 Isoelectric Points of Enzyme, Support and pH

The isoelectric point (pI) refers to the pH at which the material bears a neutral charge. Above or below this pH value, the material will bear an overall negative or positive charge; in the case of an enzyme the charge is mainly located on the surface of the molecule. When electrostatic forces are an important component in the immobilization process, matching of charges can be an important factor in ensuring successful immobilization of the enzyme. An additional factor to consider is the stability of both the enzyme and the support at a given pH. With some supports, the solution pH is not important, for example, the polymeric support, Eupergit®C [a copolymer of *N,N'*-methylene-bi-(methacrylamide), glycidyl methacrylate, allyl glycidyl ether and methacrylamide] is stable over a pH range of 0 to 14 and does not swell or shrink, even upon drastic pH changes in this range.[3,7] Penicillin amidase immobilized on Eupergit®C exhibited high operational stability and maintained 60% of its initial activity, even after 800 cycles.[7] Sol-gel procedures for the encapsulation of biomolecules depend on the pH, as the rate of hydrolysis of the sol-gel precursor for the preparation of the sol mainly depends on the pH.[8] Sol-gel precursors such as tetramethoxysilane (TMOS) and tetraethoxysilane (TEOS) will hydrolyse to form the sol at acid pH values of ~ 2. At this pH, enzymes can be denatured, hence before encapsulation the pH of the system has to be increased to a suitable value.[8]

5.2.3 Surface Functional Groups of Porous Supports and Proteins

By examining the functional groups on the surface of the enzyme, suitable complementary binding groups on the surface of the support can be identified to provide strong interactions for immobilization. Ordered mesoporous silicates and porous silicon substrates offer the opportunity of tuning the chemical composition of the surfaces generating functionalized materials to possess the desired properties. Various organic functional groups such as thiol, amine, chloro, carboxylic acid groups, cyano, alkane and aromatic functional groups can be incorporated into these materials.[9] This functionalization plays an important role in improving the immobilization process.[10] Functionalization after immobilization of the protein can enhance the hydrophobicity and/or amphiphilicity of the support for effective substrate diffusion through the pores. However, the solvent used for this reaction can reduce the activity of the immobilized enzyme while high levels of functionalization can cause pore blocking or reduced pore diameters.

5.3 Porous Materials Used for Enzyme Immobilization

A range of porous supports have been used for enzyme immobilization and include ordered mesoporous silicates (MPS) such as MCM-41 and SBA-15, mesoporous carbon,[11] sol-gels,[12] porous silicon,[13] polymeric hydrogels,[14] organic microparticles[15] and controlled pore glass (CPG),[16] whereas, recently, materials such as TiO$_2$[17] and carbon nanotubes[18] have been utilized as supports. Each of these materials has unique advantages and disadvantages.

Ordered mesoporous supports have the advantage of large surface areas, ordered porous structures with narrow pore size distributions, regular repeating mesoporous structures, mechanical stability and ease of preparation. These properties make them attractive materials for the facile immobilization of the enzyme by adsorbing or entrapping the biomolecules within the pores. Mesoporous silicates and carbons can immobilize smaller enzyme effectively; however, until recently it was not feasible to immobilize larger (10 nm diameter) enzymes and biomolecules. Moreover the structural stability of mesoporous silica adsorbents under aqueous condition can be poor. On the other hand, sol-gel matrices are thermally stable, highly porous and can accommodate large enzymes. This method can prevent leaching of the enzyme due to effective entrapment. However, these materials have brittle structures and enzymes encapsulated during the sol-gel synthesis can be denatured and the diffusion of bulky substrates can be restricted.

Porous silicon possesses large surface areas with small volumes, with controlled pore sizes. Modification of the surface of porous silicon is facile, making it an attractive material for enzyme immobilization, particularly for biosensing applications. However, the material is formed *via* electrochemical etching[13,19] of crystalline silicon in aqueous HF solution, a method of preparation that is not easy to perform compared to other porous supports such as mesoporus silicates and carbon.

Hydrogels and organic microparticles can immobilize enzymes/whole cells effectively, however, free enzymes, owing to their smaller size, can diffuse out of the gel matrix. CPG has also been successfully used for the enzyme immobilization, possessing the advantages of high mechanical strength, thermal stability, resistance to organic solvents or acidic conditions. CPG can be prepared with a wide range of porosities and pore sizes[16,20] and can withstand high pressures without compression or collapse. However, the preparation of CPG is expensive and high surface area materials can only be prepared at the expense of reduced pore volumes.

The above support materials are synthesized using a range of methods such as surfactant template synthesis, electrochemical, the sol-gel method, hydrothermal and polymeric methods *etc.* Porous supports such as ordered mesoporus silicates and carbon materials are mainly synthesized by surfactant template synthesis. These porous support materials are characterized using methods such as small angle X-ray diffraction (XRD), Brunauer-Emmett-Teller (BET) measurements, transmission electron microscopy, scanning electron microscopy; Fourier-transform infrared (FTIR), resonance Raman, solid-state NMR spectroscopy and carbon, hydrogen and nitrogen (CHN) elemental analysis. A number of reviews describing enzyme immobilization using porous supports and the application of these immobilized enzymes in biocatalysis have been published recently.[3,4,11,21–23]

5.4 Immobilization on Porous Support

Invertase was the first enzyme to be immobilized in 1916.[24] The immobilized enzyme retained the same activity as the native enzyme. There have been numerous reports on the immobilization of enzymes on a wide range of porous supports.[2–23] However, a general strategy for enzyme immobilization has not been described, with the development of a successful method of immobilization relying on trial and error approaches. Hudson *et al.*[21] recently established a methodology for the immobilization of an enzyme on a porous support, in particular mesoporous silicates. This approach can be applied to any porous support. Successful immobilization of an enzyme on a support requires a detailed understanding of the properties of the support and the enzyme used, which involves matching the physicochemical properties of the support and the protein surface. The correct selection of these properties and the successful interaction between the support and enzyme can provide an immobilized enzyme with high operational and storage stability. The amount of enzyme adsorbed on a porous support depends on factors such as the pore diameter, pore volume, surface characteristics, isoelectric point, morphology and particle size of the support material used. It will also depend on the characteristics of the enzyme such as its isoelectric point, the presence of surface functional groups and of hydrophilic/hydrophobic regions, and the stability of the enzyme under the immobilization conditions used.[21] This general protocol was successfully used for the immobilization of a range of enzymes (chloroperoxidase, lipase and alanine racemase) on mesoporous silicates (SBA-15, MCM-41 or functionalized silicates *etc.*).[10,21,25] This method of immobilization maintained the catalytic activity of the enzyme, improving its stability, ease of recovery and reuse.

5.5 Current Methods for Enzyme Immobilization

There are three main different methods in which one can immobilize an enzyme: (i) by adsorption of the enzyme on a solid support, (ii) entrapment/encapsulation of enzyme on insoluble beads or microspheres and (iii) immobilization by covalent binding/cross-linking of an enzyme. In (i) immobilization of the enzyme occurs by physisorption *via* electrostatic, hydrophobic/

hydrophilic and van der Waals interactions. The immobilized enzyme is situated inside the pores or on the outer surface of the support. A range of different enzymes have been successfully immobilized on mesoporous silicates[3,4,11,21,22] and carbon[26–29] using this method.

Entrapment/encapsulation has been mainly utilized with sol-gel materials and polymers.[30,31] The polymeric network formed during the sol-gel reaction traps the bio-molecule in the porous network. Proteins such as myoglobin, cytochrome *c* and glucose oxidase have been immobilized using this method.[32–34]

Covalent binding/cross-linking entails the formation of covalent bonds between the enzyme and the support. Cross-linked enzyme aggregates (CLEAs) are covalently bonded to a matrix through a chemical reaction. CLEA formation is a well-established technique where the addition of salts, water-miscible organic solvents or non-ionic polymers to an aqueous solution of protein leads to their precipitation as physical aggregates of protein molecules without any denaturation.[3,22,35] In this method of immobilization the CLEAs are prepared by aggregating free enzyme and then followed by cross-linking using a di-functional agent such as glutaraldehyde. For example, lipase from *Burkholderia cepacia* (BCL) was cross-linked with bovine serum albumin in the presence of dextrin using sol-gel materials.[36]

Modifications of these methods have also been reported in literature. For example, Gaffney *et al.*[37] reported a modified method in which an inorganic porous support SBA-15 is functionalized with Ni^{2+} for the successful immobilization of a model protein, His_6-tagged Spi, and Lee *et al.*[38] reported immobilization of α-chymotrypsin cross-linked with glutaraldehyde on hierarchically ordered mesocellullar mesoporous silica materials (HMS). Each of these methods has its own advantages and disadvantages and these are summarized in Table 5.1.

Table 5.1 List of immobilization methods and materials used with associated advantages and disadvantages.

Method	Materials	Advantages	Disadvantages	Ref.
Adsorption on solid support	Mesoporus silicates, carbon	1. No need for enzyme pre-treatment or chemical modification. 2. Easy to perform.	1. Immobilization is highly dependent on pH, solvent and temperature. 2. Large enzymes are difficult to immobilize inside the pores.	3–6, 21–23, 26–29
Covalent binding/ cross-linking	Functionalized silicates	1. Possibility of leaching and denaturing of enzyme in aqueous media is minimal. 2. No protein contamination occurs.	1. Enzyme is chemically modified and can be denatured. 2. Immobilization is not uniform due to non-uniform functionalization on the support.	21–23, 37,38

5.6   Characterization of Immobilized Enzymes

The success of the immobilization of an enzyme depends on the enzyme loading, the activity of the immobilized enzyme and the degree to which it can be reused. The amount of protein/enzyme adsorbed on the porous support can usually be monitored by an indirect approach by quantifying the concentration of protein in the supernatant before and after immobilization on the support using a UV-visible spectrophotometer.[39] Immobilized enzymes are characterized by attenuated total reflection-FTIR (ATR-FTIR), X-ray photon spectroscopy (XPS) and pore analysis of the support materials used for immobilization before and after enzyme immobilization. Temperature-programmed desorption coupled with a mass spectrometer (TPD-MS) for quantifying protein adsorption in a porous material has been described recently.[40]

5.7   Applications

These immobilized enzymes have applications in different fields such as biocatalysts, drug delivery, biosensors and others.

5.7.1   Biocatalysis

Biocatalysis is the most important field for the application of immobilized enzymes. A large number of biocatalytic reactions described in the literature that utilize immobilized enzymes use standardized activity assays rather than describing applications of biocatalytic reactors in the synthesis of compounds. However, there are a number of industrial applications of immobilized enzymes. For industrial applications, immobilized enzymes are mainly used in esterification reactions (*e.g.* proteases, amylases, amidases and esterases), resolution of recemic mixtures and hydrolysis reactions (esterases and proteases).[22,23] α-Amylase immobilized on SBA-15, mesocellullar foams (MCFs) and MCM-41 have been used for the hydrolysis of starch.[41] Another example is the immobilization of penicillin acylase for the production of 6-aminopenicillanic acid (a precursor for many semi-synthetic penicillins).[42] A 'nanoporous reactor' based on trypsin immobilized on cyano-functionalized mesoporous silicates for efficient proteolysis has been described.[43] A range of other applications examples are described in recent reviews.[3,21–23]

5.7.2   Drug Delivery

Inorganic porous supports can act as successful carriers for the controlled release of drugs and also for the protection of protein-based drugs from degradation/denaturation. Mesoporous silicates, *e.g.* SBA-15, have sufficiently wide pore diameters that can enhance the release of the hydrophobic drugs itraconazole[44] and vancomycin.[45] MCM-41 materials have been used for the controlled release of the anti-inflammatory drugs ibuprofen[46] and aspirin.[47] Further examples are described in a recently published review which examines drug storage and release using different mesoporous silicates such as M41S, MSU and HMS.[48] However, the biological response of these materials has not explored. Recent reports show that these materials exhibit low toxicity at low concentrations,[49] which appears to increase at higher concentrations.[50]

5.7.3 Biosensors

Enzymes immobilized on porous supports are suitable for applications as biosensors. Sol-gel-based immobilized enzymes have been successfully used in the detection of oxygen[51] and glucose.[34] Recently, glucose oxidase immobilized in a mesocellullar carbon form was used as glucose biosensor with high sensitivity.[52] Porous silicon materials have been used for the conductivity-based detection of catechol[53] using immobilized tyrosinase. Other potential applications of the immobilized enzyme include the separation of proteins and digestion, tissue engineering and diseases diagnosis.[54,55]

5.8 Conclusions

In the immobilization of enzymes on porous supports, three different immobilization methods *viz* (i) adsorption on a support (ii) entrapment/encapsulation and (iii) covalent binding/cross-linking have been discussed. In addition, the factors affecting immobilization and application of immobilized enzymes in biocatalysis, drug delivery and as biosensors were discussed. Novel concepts continue to appear, a recent example being BSA[56] and fibrinogen[57] immobilization on functionalized TiO_2. These type of novel biocatalysts would be also be useful in the field of pharmaceuticals, disease diagnosis and fine chemical synthesis. However, many literature reports utilize standardized activity assays rather than biocatalytic reactors for the synthesis of compounds. Further attention should be given to the performance of immobilized biocatalysts in real industrial practice.

References

1. A. L. Lehninger, in *Biochemistry*, Worth Publishers, New York, p. 217.
2. A. Schmid, J. S. Dordick, B. Hauer, A. Kiener, M. Wubbolts and B. Witholt, *Nature*, 2001, **409**, 258.
3. R. A. Sheldon, *Adv. Synth. Catal.*, 2007, **349**, 1289.
4. M. Hartmann, *Chem. Mater.*, 2005, **17**, 4578.
5. D. Moelans, P. Cool, J. Baeyens and E. F. Vansant, *Catal. Comm.*, 2005, **6**, 307.
6. J. F. Diaz and K. J. Balkus, *J. Mol. Catal. B: Enzymatic*, 1996, **2**, 115.
7. E. Katchalski-Katzir and D. M. Kraemer, *J. Mol. Catal. B: Enzym.*, 2000, **10**, 157.

8. B. C. Dave, B. Dunn, J. S. Valentine and J. I. Zink, in *Immobilised Biomolecules in Analysis*, ed. T. Cass and F. S. Ligler, Oxford University Press, New York, 1998, p. 113.
9. F. Hoffmann, M. Cornelius, J. Morell and M. Froba, *Angew. Chem.*, 2006, **118**, 3290; F. Hoffmann, M. Cornelius, J. Morell and M. Froba, *Angew. Chem. Int. Ed.*,2006, **45**, 3216.
10. S. Hudson, J. Cooney, B. K. Hodnett and E. Magner, *Chem. Mater.*, 2007, **19**, 2049.
11. H. H. P. Yiu and P. A. Wright, *J. Mater. Chem.*, 2005, **15**, 3690.
12. (a) I. Gill and A. Ballesteros, *J. Am. Chem. Soc.*, 1998, **120**, 8587; (b) J. D. Brennan, *Acc. Chem. Res.*, 2007, **40**, 827.
13. M. J. Sailor and E. C. Wu, *Adv. Fun. Mater.*, 2009, **19**, 3195.
14. (a) Y. Yi, S. Kermasha, L. L'Hocine and R. Neufeld, *J. Mol. Catal. B: Enzym.*, 2002, **19**, 319; (b) K. Y. Lee and S. H. Yuk, *Prog. Polym. Sci.*, 2007, **32**, 669.
15. E. S. Lee, M. J. Kwon, H. Lee and J. J. Kim, *Int. J. Pharm.*, 2007, **331**, 27.
16. H. H. Weetall, *Science*, 1969, **66**, 615.
17. (a) S. J. Bao, C. M. Li, J. F. Zang, X. Q. Cui, Y. Qiao and J. Guo, *Adv. Mater.*, 2008, **18**, 591; (b) Y. Astuti, E. Topoglidis, A. G. Cass and J. R. Durrant, *Anal. Chim. Acta*, 2009, **648**, 2.
18. K. Jiang, L. S. Schadler, R. W. Seigel, X. Zhang, H. Zhang and M. Terrones, *J. Mater. Chem.*, 2004, **14**, 37.
19. A. Salis, F. Cugia, S. Setzu, G. Mula and M. Monduzzi, *J. Coll. Inteface. Sci.*, 2010, **345**, 448.
20. Ö. Alptekin, S. S. Tükel, D. Yildirim and D. Alagöz, *J. Mol. Catal. B: Enzym.*, 2009, **58**, 124.
21. S. Hudson, J. Cooney and E. Magner, *Angew. Chem. Int. Ed.*, 2008, **47**, 8582.
22. U. Hanefeld, L. Gardossib and E. Magner, *Chem. Soc. Rev.*, 2009, **38**, 453.
23. M. Hartmann and D. Jung, *J. Mater. Chem.*, 2010, **20**, 844.
24. J. M. Nelson and E. G. Griffin, *J. Am. Chem. Soc.*, 1916, **38**, 1109.
25. (a) X. Hu, S. Spada, S. White, S. Hudson, E. Magner and J. G. Wall, *J. Phys. Chem. B*, 2006, **110**, 18703; (b) D. Goradia, J. Cooney, B. K. Hodnett and E. Magner, *Biotechnol. Prog.*, 2006, **22**, 1125; (c) H. Essa, E. Magner, J. Cooney and B. K. Hodnett, *J. Mol. Catal. B: Enzym.*, 2007, **49**, 61.
26. A. Vinu, C. Streb, V. Murugesan and M. Hartmann, *J. Phys. Chem. B*, 2003, **107**, 8297.
27. A. Vinu, M. Miyahara and K. Ariga, *J. Phys. Chem. B*, 2005, **109**, 6436.
28. A. Vinu, K. Z. Hossain, G. Satishkumar, V. Sivamurugan and K. Ariga, *Stud. Surf. Sci. Catal.*, 2005, **156**, 631.
29. M. Hartmann, A. Vinu and G. Chandrasekar, *Chem. Mater.*, 2005, **17**, 829.
30. N. Nassif, O. Bouvet, M. N. Rager, C. Roux, T. Coradin and J. Livage, *Nat. Mater.*, 2002, **1**, 42.
31. D. Avnir, S. Braun, O. Lev and Ottolnghi, *Chem. Mater.*, 1994, **6**, 1605.

32. L. M. Ellerby, C. R. Nishida, F. Nishida, S. A. Yamanaka, B. S. Dunn and J. S. Valentine, *Science*, 1992, **255**, 1113.
33. B. C. Dave, H. Soyez, J. M. Miller, B. Dunn, J. S. Valentine and J. I. Zink, *Chem. Mater.*, 1995, **7**, 1431.
34. S. A. Yamanaka, F. Nishida, L. M. Ellerby, C. R. Nishida, B. Dunn and J. S. Valentine, *Chem. Mater.*, 1992, **4**, 495.
35. D. L. Brown and C. E. Glatz, *Chem. Eng. Sci.*, 1966, **47**, 1831.
36. P. Hara, U. Hanefeld and L. T. Kanerva, *J. Mol. Catal. B: Enzym.*, 2008, **50**, 80.
37. D. A. Gaffney, S. O'Neill, M. C. O'Loughlin, U. Hanefeld, J. C. Cooney and E. Magner, *Chem. Commun.*, 2010, **46**, 1124.
38. J. Lee, J. Kim, J. Kim, H. Jia, M. I. Kim, J. H. Kwak, S. Jin, A. Dohnalkova, H. G. Park, H. N. Chang, P. Wang, J. W. Grate and T. Hyeon, *Small*, 2005, **1**, 744.
39. S. Hudson, E. Magner, J. Cooney and B. K. Hodnett, *J. Phys. Chem. B*, 2005, **109**, 19496.
40. R. Gadiou, E. A. Dos Santos, M. Vijayaraj, K. Anselme, J. Dentzer, G. A. Soares and C. Vix-Guterl, *Colloids Surf. B*, 2009, **73**, 168.
41. P. H. Pandya, R. V. Jasra, B. L. Newwalkar and P. N. Bhatt, *Microporous Mesoporous. Mater.*, 2005, **77**, 67.
42. L. W. Powell, in *Industrial Enzymology*, ed. T. Godfrey and S. West, MacMillan, London, 2nd edn, 1996, p. 267.
43. L. Qiao, Y. Liu, S. P. Hudson, P. Yang, E. Magner and B. Liu, *Chem.-Eur. J.*, 2008, **14**, 151.
44. R. Mellaerts, C. A. Aerts, J. V. Humbeeck, P. Augustijns, G. V. Mooter and J. A. Martens, *Chem. Commun.*, 2007, 1375.
45. Q. Yang, S. Wang, P. Fan, L. Wang, Y. Di, K. Lin and F. Xiao, *Chem. Mater.*, 2005, **17**, 5999.
46. (a) M. Vallet-Regi, A. Ramila, R. P. del Real and J. Perez-Pariente, *Chem. Mater.*, 2001, **13**, 308; (b) P. Horcajada, A. Ramila, J. Perez-Pariente and M. Vallet-Regi, *Microporous Mesoporous Mater.*, 2004, **68**, 105.
47. A. Ramila, B. Munoz, J. Perez-Pariente and M. Vallet-Regi, *J. Sol-Gel Sci. Technol.*, 2003, **26**, 1199.
48. S. Wang, *Micorporous Mesoporous Mater.*, 2009, **117**, 1.
49. (a) J. Lu, M. Liong, J. I. Zink and F. Tamanoi, *Small*, 2007, **3**, 1341; (b) Y. H. Son, M. Park, Y. B. Choy, H. R. Choi, D. S. Kim and K. C. Park, *Chem. Commun.*, 2007, **27**, 2799.
50. (a) S. P. Hudson, R. F. Padera, R. Langer and D. S. Kohane, *Biomaterials*, 2008, **29**, 4045; (b) S. R. Blumen, K. Cheng, M. E. Ramos-Nino, D. J. Taatjes, D. J. Weiss and C. C. Landry, *Am. J. Respir. Cell Mol. Biol.*, 2007, **36**, 333.
51. K. E. Chung, E. H. Lan, M. H. Davison, B. S. Dunn, J. S. Valentine and J. I. Zink, *Anal. Chem.*, 1995, **67**, 1505.
52. D. Lee, J. Lee, J. Kim, J. Kim, H. B. Na, B. Kim, C. Shin, J. H. Kwak, A. Dohnalkova, J. W. Grate, T. Hyeon and H. S. Kim, *Adv. Mater.*, 2005, **17**, 2828.

53. P. S. Chaudhari, A. Gokarna, M. Kulkarni, M. S. Karve and S. V. Bhoraskar, *Sens. Actuators B*, 2005, **107**, 258.
54. Y. J. Han, G. D. Stucky and A. Butler, *J. Am. Chem. Soc.*, 1999, **121**, 9897.
55. H. Fan, Y. Hu, C. Zhang, X. Li, R. Lv, L. Qin and R. H. Zhu, *Biomaterials*, 2006, **27**, 4573.
56. C. K. Simi and E. Abraham, *Colloids Surf. Sci. B*, 2009, **7**, 319.
57. K. Cai, M. Frant, J. Bossert, G. Hildebrand, K. Liefeith and K. D. Jandt, *Colloids Surf. Sci. B*, 2006, **50**, 1.

CHAPTER 6

Fibrous Proteins Interactions with Modified Surfaces of Biomaterials

C. WOLF-BRANDSTETTER* AND D. SCHARNWEBER

Max Bergmann Center of Biomaterials, Institute of Materials Science, TU Dresden, Budapester Str. 27, 01069 Dresden, Germany

6.1 Introduction

A key process in the integration of implants and scaffolds into the host tissue is the interaction of the host tissue cells with the implant surface. *In vivo* cells exist in a highly complex environment. The behaviour of cells is determined by information received from soluble factors, from other cells and from the physical network they are embedded within. This network that provides structure and support is called the extracellular matrix (ECM), a visco-elastic milieu containing a large quantity of biological information. The ECM plays an important role in guiding development, maintaining homeostasis and directing regeneration.

The term ECM comprises a large and very heterogeneous set of components that are assembled locally into an ordered, highly site-specific network. By these differences in composition the ECM takes an active part in regulating the cellular processes and responses in specific situations and tissues. ECM structure varies widely depending on the tissue and developmental stage with the collagen fibril as the central building block that is modified by other collagen types or non-collagenous components. Collagens are the most abundant

RSC Nanoscience & Nanotechnology No. 21
Biological Interactions with Surface Charge in Biomaterials
Edited by Syed A. M. Tofail
© Royal Society of Chemistry 2012
Published by the Royal Society of Chemistry, www.rsc.org

proteins in the body. They are modified in their structure and function by glycoproteins and proteoglycans (PGs); these usually consist of a protein core to which one or more glycosaminoglycan (GAG) chains are attached. Members of these different groups not only associate with each other in a defined manner, they also specifically interact with cells and soluble proteins.

One very basic characteristic of the structural protein components of the ECM is that they confer cell adhesion, which is of vital importance for almost all cells, and the first use made of ECM components – mainly of collagen and fibronectin – was to increase cell adhesion to biomaterial surfaces not optimally suited to it. Although this was a successful approach, it does not take into account the true functions and possibilities of the ECM. Cell adhesion to the ECM is a complex process and is conveyed by specific adhesion receptors, enabling the cells to recognize their surroundings and respond to them. Consequently, adhesion is intimately coupled to signal transduction, with most adhesion receptors functioning as signalling molecules. Engagement of these different receptors gives rise to a wide variety of intracellular signals that, in turn, influence proliferation, differentiation and apoptosis.

The direct interaction of the ECM with cellular receptors is not the only mechanism for influencing cells, as many of the structural components associate with growth factors and cytokines. These soluble factors are potent regulators of cell function, and the matrix components in turn regulate their function: storing, activating or inactivating them, protecting them from degradation and retaining them at one place.

Last, but not least, the ECM provides the necessary structural integrity and load-bearing functions, such as in case of bone. Composites of synthetic and natural polymers, alone or with bioactive ceramics like hydroxyapatite or certain glasses, further modified by other constituents of the ECM, have been developed as scaffold materials with a broad range of mechanical strengths and porosities for possible use in tissue engineering or direct application as implants in cardiovascular, cartilage or bone surgery.

Other fibrous proteins which naturally do not occur within the ECM (fibroin and keratin) are included in the types of materials used for implant coatings or scaffold preparation due to their excellent biomechanical properties, but also because of their reported presence of some cell adhesion amino acid sequences such as RGD or LDV.[1]

6.2 Collagen Structure and Properties

Collagens are the most abundant among the structural proteins. In humans 28 collagens and collagen-like proteins have been identified to date.[2,3] Generally, the term collagen encompasses all proteins containing regions characterized by the repeating amino acid motif (Gly-X-Y) that form a right-handed triple-helical structure and have a role in tissue assembly or maintenance. The collagen family can be subdivided into the following: fibrillar collagens, fibril-associated collagens (FACITs), beaded filament forming collagens, basement membrane collagens, transmembrane collagens, collagens forming hexagonal networks and anchoring fibrils, and multiplexins.[4] The most abundant collagens are those of the fibril-forming subfamily, being approximately 90% of the total collagen,[5] and these are the ones of the most interest for application in artificial ECMs (aECMs).

The fibril-forming collagens are collagen types I, II, III, V, XI, XXIV and XXVII, and they are characterized by their ability to assemble into highly organized aggregates with a typical suprastructure, the quarter-staggered fibril array with diameters between 25 and 400 nm. This results in a banding pattern with the so-called D-periodicity between 64–67 nm based on the staggered arrangement of the collagen monomers (Figure 6.1).

Collagen type I is the most abundant, as well being the best studied. It is the major collagen of bone, tendons, skin, ligaments, cornea, vasculature and many interstitial connective tissues.

All fibril-forming collagens closely resemble each other. They are composed of three peptide chains called α-chains, forming triple-helical molecules of 300 nm in length flanked by the telopeptides, non-helical sequences approximately 20 amino acids in length. The larger N- and C-terminal propeptides are

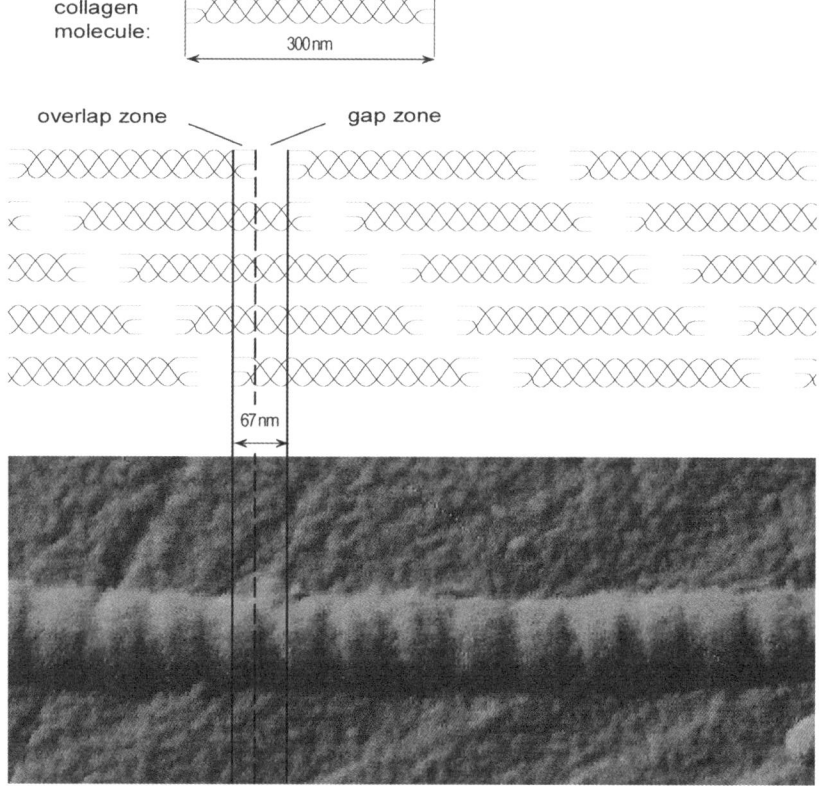

Figure 6.1

removed immediately before fibril formation. Type II and type III each has three identical $\alpha1(II)$ and $\alpha1(III)$ chains, respectively, type I of two identical $\alpha1(I)$ and one $\alpha2(I)$ chain, whereas for collagen V and XI all three chains differ.

In context with surface modification collagen type I is reported to promote cell adhesion of osteoblasts,[6] affect signal transduction by binding to the integrins $\alpha1\beta1$ and $\alpha2\beta1$[7] and modulate the expression of the osteoblastic phenotype.[8,9] The interaction of osteoblasts with type I collagen leads to enhanced alkaline phosphatase activity,[10] osteoblast-associated gene expression[8,11,12] and the stimulation of mineralization.

Collagen is widely used for coating biomaterials surfaces either alone or in combination with other constituents of the ECM. Furthermore, collagen is frequently applied as a component in three-dimensional scaffolds for tissue engineering or bone substitutes, mostly combined with inorganic minerals to provide sufficient mechanical stability. The further sections of this chapter will focus solely on the applications, including coatings.

6.3   Interaction of Monomeric Collagen at Surfaces

Coatings of biomaterials using tropocollagen as the basic unit of the collagen fibrils offer the benefit of simple and quick coating techniques.

The immobilization of proteins in general can be conducted *via* pure adsorption or utilizing covalent cross-linking either within the protein layer,[13] as well as via active groups originally present or additionally generated on top of the materials surfaces.[14,15] While covalent immobilization gives rise to a stably bound protein at the surface, it requires the additional application of coupling chemistry, resulting in non-physiological linkages, causing in a random orientation of bound molecules, as well as random incorporation of amino acid residues for the forming of covalent links within the protein to the biomaterials surface. Most of the proposed methods for covalent linkage cannot be guided in a way to prevent unphysiological cross-links within the protein layer. This can be disadvantageous in case of labile proteins which might lead to diminished protein stability or to the partial or even complete loss of activity due to the chemical change of active sites necessary for proper binding of cell receptors or binding of other components such as PGs or growth factors. To overcome these limitations, biologically mediated methods became the focus of recent investigations.[16]

Adsorptive immobilization is a comparably simple process that can be performed at physiological conditions but may also lead to conformational changes of less stable proteins. The driving forces that lead to protein adsorption at surfaces are based on energy gains that result mainly from (a) hydrophobic interactions, in which case parts of the protein and the surface dehydrate; (b) electrostatic interactions, which depend on pH and ionic strength, and (c) structural changes in the molecule, which may elicit entropy or enthalpy contributions. Protein adsorption will only take place if the basic criterion of the reduction of the free energy of the complete system is fulfilled.

Due to the number of effects which partially result in oppositional contributions, no general rules can be established and each system has to be tested.[17]

To understand the phenomena occurring during tropocollagen adsorption, it is helpful to imagine the geometrical situation at the surface. On the basis of the dimensions of the tropocollagen molecule (300 nm in length with a thickness of 1.5 nm) the collagen mass which would correspond to a monolayer assuming a flat tightly packed arrangement would be 0.1 µg cm^{-2}. The maximal adsorbed amount obtained by pure adsorption was reported to be 0.8–0.9 µg cm^{-2}.[18,19] This means that at high concentrations of collagen in the adsorption solution the molecules are only able to adsorb with a part/some parts of its entire length on the substrate surface. The contact area of a single molecule which interacts with one-tenth of its length amounts to approximately 45 nm^2. In contrast, the minimum areas of interface (ΔA) which are required to stabilize adsorbing molecules against desorption range between 1 to 2 nm^2 for various proteins and are nearly independent of the size or molecular weight of the proteins.[20] Once 'connected' with the surfaces the molecules can optimize their interfacial interactions by time-dependent orientation and potential conformational changes. Hence collagen possesses outstanding potential for a sufficient contact area due to its high surface area to volume ratio. Furthermore, large parts of the molecules are free and should be available for interactions with solution components. Applying low collagen concentrations, the molecules have the opportunity to form a lot of interactions with the surface which should result in an increased contact area and a diminished length of accessible parts of collagen molecules, as verified with quartz microbalance adsorption studies.[18] The adsorption process of collagen at titanium oxide surfaces have been shown to be nearly independent of pH and ionic strength, revealing a binding mechanism based mainly on hydrophobic interactions. These can be overlaid by effects that diminish the repulsion forces between individual protein molecules (for instance a pH value near the isoelectric point or a low ionic strength[18]).

Protein adsorption has been shown to be irreversible in a number of investigations[21–23] by simple incubation in water or buffer solutions. Adsorbed proteins exhibit a great number of interaction sites, all of which have to disengage for the protein to desorb. This is statistically improbable. However, exchange is a different matter. Therefore collagen desorption during incubation in PBS has to be regarded as an exchange of loosely bound collagen with adjacent molecules. Remaining molecules gain more interaction sites by displacing other molecules. Starting from high initial adsorbed amounts, the remaining collagen amount tended towards the theoretical amount of the monolayer.[18]

Higher amounts of immobilized 'non-assembled' collagen coatings than the above-mentioned 0.8–0.9 µg cm^{-2} as reported in Kim *et al.*[13] (40–80 µg cm^{-2}) have to be regarded as undefined layers of collagen molecules where only a minor part of the molecules can gain contact with the material's surface while protein–protein interactions have to be assumed that can partly resemble the interactions within fibrillar structures. Minimum stability of such loosely connected protein layers requires covalent bindings, which was shown to increase resistance against desorption as well as enzymatic degradation.[13]

6.4 Immobilization Strategies for Fibrillar Collagen

Coating of biomaterials surfaces with fibrillar collagen requires the generation of fibrils that is accomplished by two main ways. One utilizes the natural self-assembly potential of collagen that allows for an *in vitro* formation of fibrils, and the other is based on the application of insoluble collagen suspensions. The collagen used in the majority of both cases is extracted from tissues. In principle the application of recombinant-derived collagen would be possible,[24] but the fibril formation with such collagen is often impaired due to insufficient post-translational processing. Although collagen monomers are soluble in dilute acid, most collagen in tissues is enzymatically cross-linked via the telopeptides. Acid extraction yields the non-cross-linked portion with intact telopeptides. Another extraction method uses the protease pepsin, digesting the non-helical telopeptides and thus breaking the link and giving higher yields. The collagen fraction that is non-enzymatically cross-linked via the helical part is not accessible by either method, but can be used in the mentioned insoluble slurries.

As the utilization of insoluble fibrils by direct recovery from native tissue can include undefined amounts of other constituents (different collagen types, PGs *etc.*), preparing fibrils *in vitro* using the natural self-assembly potential of purified collagen allows for the generation of a more defined collagen matrix with regards to the components as well as the structure.

The self-assembly of monomers into fibrils is initiated by raising the pH and temperature of the acidic solution into the physiological range. The assembly pattern is determined by the charge pattern on the monomer surface, but the fibril formation itself is an entropy-driven process, in which surface-exposed hydrophobic residues are buried within the fibril, thereby increasing the entropy of the solvent. Fibril assembly proceeds via nucleation and growth, and classically shows three phases: an initial lag phase with no change in turbidity, a rapid growth phase and a plateau region. During the lag phase, small numbers of molecules associate to form metastable nuclei, upon which further molecules accrete during the growth phase. The initial interaction seems to be a pair of 4D-staggered molecules with a short N- and C-terminal overlap, mediated by the telopeptides. Growth in fibril length then occurs via longitudinal and lateral interactions.[25]

In vitro fibrillogenesis has been studied by a number of authors, most notably by Williams *et al.*[26] who determined conditions under which the resultant fibril morphology was optimized, with a clear D-periodic banding pattern (mix together on ice 1 mg mL^{-1} collagen and 5 mM acetic acid with an equal volume of 60 mM TES, 60 mM sodium phosphate and 270 mM NaCl, pH 7.3, then raise the temperature to 30 °C). Most preparations of collagen fibrils are based on modifications of this method. The resulting collagen gel can be used as such, freeze-dried or homogenized, and treated like insoluble collagen slurry.

The process of fibril formation and the structure of the resultant fibrils are influenced by a large number of parameters: *in vivo*, these include the collagen type or types, the extent of procollagen processing and the presence of non-collagenous components; *in vitro*, there are some additional parameters to be considered such as temperature, pH and buffer composition.[27,28] In the buffer, pH, phosphate concentration and ionic strength are deciding parameters. Depending on the pH, early subfibrils are formed over a comparatively wide range; low pH possibly stabilizes intermediate states.[29] The presence of phosphate is important for producing well-banded fibrils, but increasing phosphate over the optimum concentration results in less and smaller fibrils of a different morphology. This effect is probably based on phosphate binding to collagen; a fibril associated with phosphate may be hindered in the alignment of monomers, as the charge distribution of the monomer is changed, altering the nature of collagen molecular interactions.[30]

Ionic strength also influences the electrostatic interactions of the monomers, but in a way differing from phosphate, as an increase can – in the case of collagen type I – increase the size of the resultant fibrils. This is probably based in part on a facilitation of superfibrillar bundling, again due to changes in electrostatic interactions. Another part may be due to the slower growth rate, as this too gives rise to fibrils of increased diameter,[31] which can also be achieved by lowering the formation temperature.[25]

The collagen preparation used also has a marked effect on fibrillogenesis. Intact telopeptides as in acid-extracted collagen help to initiate fibrillogenesis, producing long cylindrical fibrils, while for collagen with degraded telopeptides self-assembly is slower, and the fibrils are slightly thicker and less stable.

Two main methods for immobilization of fibrillar collagen onto biomaterials surfaces were proposed and successfully realized. The first approach is based on the *in situ* fibrillogenesis directly at the surface of the material. This can be accomplished with starting from densely adsorbed tropocollagen molecules providing parts of the molecules available for further fibrillar growth.[18] This method required repeated addition of fresh monomeric molecules while simple exposure of the surface to *in situ* reconstituting fibrils resulted in fibrillar growth adjacent but with poor attachment to the surface. The output of a similar approach could be significantly improved by covalent binding of the first monomeric molecules via maleic anhydride,[15] resulting in a stably bound *in situ*-assembled fibrillar network without further coupling chemistry.

The utilization of separately formed fibrils or, as already mentioned, insoluble collagen slurries is by far the mostly applied method to coat surfaces with fibrillar collagen.

Suspensions of fibrillar collagen show a high tendency to bind to a variety of hydrophobic or moderate hydrophobic surfaces including metals and polymers. This process is independent of pH or ionic strength, which is an indication for a binding *via* hydrophobic interactions.

Adsorbed coatings will not be desorbed by salt-containing buffers, such as PBS, or protein-containing solution,[32] but exhibit a low resistance against proteolytic degradation (using bacterial collagenases).[13,14] The degradation resistance could be increased by covalent cross-linking. It was shown that the total remaining amount of immobilized collagen was only influenced by the presence of amino groups on the material surface, to which the collagen could be cross-linked. The density of such amino groups was of less importance, possibly due to the high number of potential immobilization sites in fibrillar collagen.[14]

6.5 Applications of Other Fibrous Proteins

Elastin is another important ECM protein, which is found in connective tissue where it provides elasticity. As a result, elastin is most abundant in organs where elasticity is of major importance, such as blood vessels, elastic ligaments, lung, skin and elastive cartilage.[33] It is a highly insoluble, cross-linked polymer synthesized *in vivo* from soluble tropoelastin precursors ($\sim$72 kDa) by a variety of cells, including smooth muscle cells, endothelial cells, fibroblasts and chondrocytes.[34] Elastin exhibits a quite uncommon amino acid composition with approximately 75% hydrophobic residues (glycine, alanine and valine residues). Similar to collagen, glycine accounts for one third of the amino acid residues and proline is present in high amounts. The tropoelastin molecule typically consists of two types of domains: (i) hydrophobic domains with many glycine, valine, alanine and proline residues which often occur in repeats of several amino acids, and (ii) hydrophilic domains with many lysine and alanine residues playing an important role in formation of tetrafunctional cross-links. Three hydrophobic domains flanking two cross-linking domains are sufficient to support this self-assembly process.[35] Four lysine residues from tropoelastin molecules react with each other to form desmosin and isodesmosine, the most abundant among some other lysine-based cross-links.[36]

A number of biophysical properties are important for the above-mentioned role of elastin in the body to provide elasticity. The reader is referred to reviews by other authors, as this topic is of more interest for applications in scaffold designing.[37,38]

Elastin has been applied in biomaterials in various forms, such as insoluble elastin, soluble (hydrolyzed) preparations, repeated elastin-like sequences produced by synthetic or recombinant applications, with the majority of applications in creating three-dimensional scaffolds.[34]

Elastin and elastin-degradation products, as well as synthetic elastin-like polymers, are reported to bind to the elastin–laminin receptor and hence to elicit elastin synthesis, chemotaxis, inhibition of vascular calcification, pro-liferation of fibroblasts among others.[34] In particular, the induction of elastin synthesis is assumed to accelerate tissue healing in skin. Hence the application as a coating material has also been explored. Solubilized elastin has been used to improve the biocompatibility of polyethylene terephtalate (PET).[39] The potential of some tropoelastin fragments to inhibit thrombosis was successfully utilized for the coating of synthetic materials for vascular grafts.[40]

Silk fibroin occurs naturally as the core protein of silk produced by *Bombyx mori* silkworms, as well as by spiders. It is regarded as one of the strongest natural fibres, which is attributed to the chemical structure of the protein itself. The main structural feature consists of a repetitive Gly-Ala-Gly-Ala-Gly-Ser sequence which gives rise to its self-assembly into an anti-parallel β-sheet structure. Such β-sheets are highly crystalline and stabilize the protein structure by strong intra- and intermolecular hydrogen bonds. The extent of β-sheet structure can be controlled via physical or chemical methods, thus leading to materials with controlled crystallinity and degradation rate.

Silk fibroin is of special interest due to its extraordinary mechanical properties, rather than due to its unmodified biochemical features, and has been extensively used in applications of sutures, drug delivery matrices and a variety of three-dimensional scaffolds for ligament, bone and cartilage engineering.

Prior to use in biomedical application the main other constituent of silk, a glue-like protein called sericin, has to be removed to avoid inflammatory responses *in vivo*. This is mainly achieved by extraction (degumming). Further treatments are necessary as there can be dissolution at elevated temperature including concentrated salts or dissolution in organic solvents that leads to regenerated forms.[41]

The presence of several reactive amino acid residues allows for further chemical modifications to tune the interaction with the living system and have been reported in more detail by Murphy and Kaplan.[41] Such reported modifications comprise coupling reactions *via* cyanuric chloride and glutaraldehyde, and, mainly by carbodi-imide chemistry, amino acid modifications (arginine masking, sulfation of tyrosine and azo-modifications of tyrosine) or grafting reactions.

Coatings consisting of chemically modified fibroin have been frequently investigated with the aim to fine tune hydrophilicity for optimization of growth of mammalian cells.[41,42] A number of groups have utilized fibroin as basic material to immobilize specific bioactive groups, such as RGD sequences,[43] or growth factors [bone morphogenetic protein 2 (BMP-2)].[44]

Incorporation of sulfate groups[45,46] has been demonstrated as a promising approach to increase biocompatibility of biomaterials in blood contact.

Keratins are a large family of structural proteins found in the cytoskeleton and mainly in the protective tissues of vertebrates. Wound healing, drug delivery and tissue engineering have been the subject of keratin-related research, leading to a number of patents addressing, in particular, its wound healing potential.[47] Human hair keratins are classified into three broad groups: α-, β- and γ-keratins. The α-keratins serve as the main structural component of human hair and possess an α-helical tertiary structure, comparably low in sulfur content. The β-keratins are primarily protective and form the majority of the cuticle. γ-Keratins are globular, higher in sulfur content but lower in molecular weight than the other types, and serve as disulfide cross-linker that enables the formation of the cortical superstructure. Soluble keratins are removed from this structure by breaking down the disulfide cross-links. An oxidation step followed by treatment with denaturing agents allows the extraction of soluble α- and β-keratins, which can further purified by filtrations and dialysis. In addition to the mechanical properties causing the potential for the formation of hydrogels or foams, keratins are also reported to contain intrinsic sites of cellular recognition that mimic the ECM.[47] Certain keratin-based biomaterials have been shown to be mitogenic and chemotactic for a variety of cell types[47,48] and to mediate changes in gene expression that are consistent with the promotion of wound healing.

6.6 Towards aECMs

An aECM is a matrix consisting of isolated ECM components that have been reconstituted *in vitro* to construct a microenvironment that mimics the ECM in its ability to guide morphogenesis in tissue repair and engineering. It does not mean perfectly reconstructing the host tissue, but using those parts of the whole complex that elicit a specific response. This biomimicry may include, depending on the purpose, biochemical composition, fibrillar structure and mechanical properties, as well as bioadhesive characteristics, proteolytic susceptibility and growth factor-binding capacity.[49]

Due to its significant role in the different ECM types of the human body, collagen is widely used as a basis for further modifications by other natural constituents of the ECM. There are a few examples of materials which do not occur naturally that have been adapted into aECMs. Some of the most important and frequently applied compounds will be described below. For a more comprehensive overview, the reader is conferred to Bierbaum and Scharnweber.[50]

Glycoproteins are a group of proteins that is quite heterogeneous. They exist in various forms and posses multiple binding domains for interactions with collagen and PGs as well as with cell surfaces. The binding domains contain specific amino acid sequences, such as the well-known RGD sequence, which interact with cell surface receptors and serve as adhesion recognition signals. The most prominent members are fibronectin, vitronectin, laminin and thrompospondin.

PGs are molecules composed of a core protein substituted with covalently linked sulfated GAG chains that are linked to the core protein via serine residues. They can be subdivided into the small leucine-rich PGs (SLRP) and those PGs of a large molecular weight. These include the hyaluronic acid (HA)-binding hyalectans and non-HA-binding PGs, among which are the basement membrane and cell surface PGs. SLRPs are a large family of proteins consisting of core proteins with leucine-rich repeats (LRR) and at least one GAG side chain.[51] SLRPs are important regulators of various biological processes through their ability to interact with different cell surface receptors, cytokines and growth factors, while their interaction with the fibril-forming collagens regulates fibril assembly and diameter, which is important for the proper assembly of tissues. The interaction is mediated in most cases by the protein core, whereas the GAG chains are found in the interfibrillar space[52] and contribute to matrix integrity by binding other ECM components such as tenascin-X.[53] Collagen-binding SLRPs include decorin, fibromodulin, lumican, biglycan, keratocan and osteoglycan, with decorin, biglycan and lumican being the best characterized. The mechanism of action appears to be an interaction with collagen that leads to a coating of the fibril with the PG, as has been shown for decorin. Low decorin concentrations at the fibril tip allow for an increase in length while maintaining a constant fibril diameter. Biglycan interacts with collagen without influencing fibril diameter, and may be involved in organizing assembly of type IV collagen into networks.

GAGs are linear polysaccharides able to interact with numerous proteins and modulate their activities. They fall into two classes: (i) sulfated GAGs comprising chondroitin sulfate, dermatan sulfate, keratin sulfate, heparin and heparan sulfate, and (ii) the non-sulfated GAG hyaluronan, which is the only one to occur without a core protein.

GAG chains are made up of repeating disaccharide units comprising mainly of acetylated amino sugars (*N*-acetyl-galactosamine or *N*-acetyl-glucosamine) and uronic acids (D-glucuronic acid or L-iduronic acid). For further details, the reader is referred to Bierbaum and Scharnweber.[50]

The specificity of GAG-protein binding is defined by a number of parameters that are determined by the structural fit between the GAG and the protein on the basis of the three-dimensional structure of the GAG. Most important are the ionic interactions of the sulfate and carboxylate groups of the GAG with the basic amino acids of the protein, although van der Waals contacts also play a role. They depend on the number, position and distribution (O- or N-) of the sulfate groups, the length of the sugar chain,[54–56] as well as the three-dimensional structure of the polysaccharide backbone and resulting orientation of the sulfate groups. Although many of the functions of PGs are determined by the core protein, the GAG chains are responsible for many more important interactions, *i.e.* with a number of growth factors.

GAGs present the largest diversity among biological macromolecules as they vary in basic saccharide compositions, acetylation, N- and O-sulfatation. They play a critical role in assembling protein–protein complexes, such as growth factor–receptor complexes, and thus are directly involved in initiating cell signalling events. Furthermore, GAGs can potentially immobilize and sequester proteins or present them to the appropriate sites for activation. The positioning of the protein-binding oligosaccharide motif along the GAG chain determines if an active signalling complex is assembled at the cell surface or an inactive complex is sequestered into the matrix.[57]

Non-collagenous components have been introduced into collageneous matrices by adsorption to preformed fibrils or by co-precipitation in an acidic milieu,[58] resulting in amorphous precipitates. Preparing fibril *in vitro* using its natural self-assembly, as described in Section 6.3, allows the generation of a matrix that more resembles the process *in vivo*. Fibrillogenesis can take place in presence of collagen-binding components, resulting in integration of the non-collagenous matrix as well as modification of collagen organization.

Depending on the nature of the compounds (other collagen types, glycoproteins or PGs) and on the amount used, as well as the solution conditions, the amount incorporated and the structure of the obtained fibrils may vary broadly.[31,59]

Interestingly, one of the most widely used combinations is collagen in combination with GAGs, although GAGs, with the exception of HA, are normally bound to a core protein. GAGs interact with collagen mainly due to their overall negative charge. Hence the ionic strength of the buffer has significant effects on incorporated amount as well as on fibril morphology. The utilization of pure GAGs circumvents some problems with immunogenicity of the whole PGs and opens an interesting field to tailor the properties of the incorporated polysaccharide, as, for example, by oversulfatation of HA, leading to increased growth factor binding.[60]

The proposed immobilization methods of aECMs on biomaterials' surfaces are identical to those reported for pure fibrillar collagen in Section 6.4. In general, as long as collagen is the main component of such a ECMs and collagen fibrils were obtained during the preparation procedure, the immobilization behaviour seems to be mainly determined by the collagen part.[50] Most publications do not characterize the interactions between the aECM and the surface, but concentrate on the specific impact on growth factor binding and release or cell responses.

6.7   Conclusions

Although the ECM has been recognized as an important factor determining cell response, clinical applications utilizing an aECM for implant coatings do not exist to date. Other surface techniques modifying physicochemical parameters, such as wettability and (closely related) the surface energy or surface roughness, are still the method of choice in clinical practice. Recent research in the field of aECMs is focused on the tailoring of the properties of some crucial parts of the ECM such as engineered proteins or chemically modified GAGs, offering the opportunity to generate comparably simple systems to construct defined microenvironments for surface–cell interactions.

References

1. A. Tachibana, Y. Furuta, H. Takeshima, T. Tanabe and K. Yamauchi, *J. Biotechnol.*, 2002, **93**, 165–170.
2. K. E. Kadler, C. Baldock, J. Bella and R. P. Boot-Handford, *J. Cell Sci.*, 2007, **120**, 1955–1958.
3. S. Ricard-Blum and F. Ruggiero, *Pathol. Biol. (Paris)*, 2005, **53**, 430–442.
4. J. Heino, *Bioessays*, 2007, **29**, 1001–1010.
5. K. Gelse, E. Poschl and T. Aigner, *Adv. Drug Deliv. Rev.*, 2003, **55**, 1531–1546.

6. L. F. Cooper, B. Handelman, S. M. McCormack and A. D. Guckes, *Int. J. Oral Maxillofac. Implants*, 1993, **8**, 264–272.
7. Y. Takeuchi, K. Nakayama and T. Matsumoto, *J. Biol. Chem.*, 1996, **271**, 3938–3944.
8. S. Celic, Y. Katayama, P. J. Chilco, T. J. Martin and D. M. Findlay, *J. Endocrinol.*, 1998, **158**, 377–388.
9. A. G. Andrianarivo, J. A. Robinson, K. G. Mann and R. P. Tracy, *J. Cell. Physiol.*, 1992, **153**, 256–265.
10. L. Masi, A. Franchi, M. Santucci, D. Danielli, L. Arganini, V. Giannone, L. Formigli, S. Benvenuti, A. Tanini, F. Beghe, M. Mian and M. L. Brandi, *Calcif. Tissue Int.*, 1992, **51**, 202–212.
11. M. P. Lynch, J. L. Stein, G. S. Stein and J. B. Lian, *Exp.Cell Res.*, 1995, **216**, 35–45.
12. S. Shi, M. Kirk and A. J. Kahn, *J. Bone Miner. Res.*, 1996, **11**, 1139–1145.
13. H. W. Kim, L. H. Li, E. J. Lee, S. H. Lee and H. E. Kim, *J. Biomed. Mater. Res. A*, 2005, **75**, 629–638.
14. R. Muller, J. Abke, E. Schnell, D. Scharnweber, R. Kujat, C. Englert, D. Taheri, M. Nerlich and P. Angele, *Biomaterials*, 2006, **27**, 4059–4068.
15. K. Salchert, U. Streller, T. Pompe, N. Herold, M. Grimmer and C. Werner, *Biomacromolecules*, 2004, **5**, 1340–1350.
16. L. S. Wong, F. Khan and J. Micklefield, *Chem. Rev.*, 2009, **109**, 4025–4053.
17. W. Norde and J. Lyklema, *J. Biomater. Sci. Polym. Ed.*, 1991, **2**, 183–202.
18. C. Wolf-Brandstetter, *BIOmaterialien*, 2007, **8**, 32–39.
19. A. Silberberg, *Modelling of Protein Adsorption*, Plenum Press, New York, 1985.
20. F. Macritchie, *Adv. Protein Chem.*, 1978, **32**, 283–326.
21. P. van Dulm and W. Norde, *J. Colloid Interface Sci.*, 1983, **91**, 248–255.
22. J. D. Andrade and V. Hlady, in *Biopolymers/Non-Exclusion HPLC*, ed. Autorenkollektiv, Akademie Verlag, Berlin, 1987, pp. 1–64.
23. A. Bentaleb, Y. Haikel, J. C. Voegel and P. Schaaf, *J. Biomed. Mater. Res.*, 1998, **40**, 449–457.
24. C. Yang, P. J. Hillas, J. A. Baez, M. Nokelainen, J. Balan, J. Tang, R. Spiro and J. W. Polarek, *BioDrugs*, 2004, **18**, 103–119.
25. D. J. S. Hulmes, *Collagen Diversity, Synthesis and Assembly*, Springer Science and Business Media, New York, 2008.
26. B. R. Williams, R. A. Gelman, D. C. Poppke and K. A. Piez, *J. Biol. Chem.*, 1978, **253**, 6578–6585.
27. A. O. Brightman, B. P. Rajwa, J. E. Sturgis, M. E. McCallister, J. P. Robinson and S. L. Voytik-Harbin, *Biopolymers*, 2000, **54**, 222–234.
28. U. Freudenberg, S. H. Behrens, P. B. Welzel, M. Muller, M. Grimmer, K. Salchert, T. Taeger, K. Schmidt, W. Pompe and C. Werner, *Biophys. J.*, 2007, **92**, 2108–2119.
29. J. R. Harris and A. Reiber, *Micron*, 2007, **38**, 513–521.
30. E. L. Mertz and S. Leikin, *Biochemistry*, 2004, **43**, 14901–14912.
31. S. Bierbaum, R. Beutner, T. Hanke, D. Scharnweber, U. Hempel and H. Worch, *J. Biomed. Mater. Res. A*, 2003, **67**, 421–430.

32. U. Geissler, U. Hempel, C. Wolf, D. Scharnweber, H. Worch and K. Wenzel, *J. Biomed. Mater. Res.*, 2000, **51**, 752–760.
33. F. W. Keeley, C. M. Bellingham and K. A. Woodhouse, *Philos. Trans. R. Soc. Lond. B. Biol. Sci.*, 2002, **357**, 185–189.
34. W. F. Daamen, J. H. Veerkamp, J. C. van Hest and T. H. van Kuppevelt, *Biomaterials*, 2007, **28**, 4378–4398.
35. C. M. Bellingham, M. A. Lillie, J. M. Gosline, G. M. Wright, B. C. Starcher, A. J. Bailey, K. A. Woodhouse and F. W. Keeley, *Biopolymers*, 2003, **70**, 445–455.
36. H. Umeda, M. Takeuchi and K. Suyama, *J. Biol. Chem.*, 2001, **276**, 12579–12587.
37. D. W. Urry and T. M. Parker, *J. Muscle Res. Cell Motil.*, 2002, **23**, 543–559.
38. B. Li and V. Daggett, *J. Muscle Res. Cell Motil.*, 2002, **23**, 561–573.
39. S. Dutoya, A. Verna, F. Lefebvre and M. Rabaud, *Biomaterials*, 2000, **21**, 1521–1529.
40. K. A. Woodhouse, P. Klement, V. Chen, M. B. Gorbet, F. W. Keeley, R. Stahl, J. D. Fromstein and C. M. Bellingham, *Biomaterials*, 2004, **25**, 4543–4553.
41. A. R. Murphy and D. L. Kaplan, *J. Mater. Chem.*, 2009, **19**, 6443–6450.
42. A. R. Murphy, P. St John and D. L. Kaplan, *Biomaterials*, 2008, **29**, 2829–2838.
43. U. Hersel, C. Dahmen and H. Kessler, *Biomaterials*, 2003, **24**, 4385–4415.
44. V. Karageorgiou, L. Meinel, S. Hofmann, A. Malhotra, V. Volloch and D. Kaplan, *J. Biomed. Mater. Res. A*, 2004, **71**, 528–537.
45. Y. Tamada, *Biomaterials*, 2004, **25**, 377–383.
46. X. Ma, C. Cao and H. Zhu, *J. Biomed. Mater. Res. B Appl. Biomater.*, 2006, **78**, 89–96.
47. M. E. Furth, A. Atala and M. E. Van Dyke, *Biomaterials*, 2007, **28**, 5068–5073.
48. P. Sierpinski, J. Garrett, J. Ma, P. Apel, D. Klorig, T. Smith, L. A. Koman, A. Atala and M. Van Dyke, *Biomaterials*, 2008, **29**, 118–128.
49. J. A. Hubbell, *Curr. Opin. Biotechnol.*, 2003, **14**, 551–558.
50. S. Bierbaum and D. Scharnweber, *Artificial Extracellular Matrices to Functionalize Biomaterial Surfaces*, Elsevier, Amsterdam, 2010.
51. L. Schaefer and R. M. Schaefer, *Cell Tissue Res.*, 2010, **339**, 237–246.
52. R. V. Iozzo, *J. Biol. Chem.*, 1999, **274**, 18843–18846.
53. R. Merline, R. M. Schaefer and L. Schaefer, *J. Cell Commun. Signal*, 2009, **3**, 323–335.
54. S. Ashikari-Hada, H. Habuchi, Y. Kariya, N. Itoh, A. H. Reddi and K. Kimata, *J. Biol. Chem.*, 2004, **279**, 12346–12354.
55. F. H. Blackhall, C. L. Merry, M. Lyon, G. C. Jayson, J. Folkman, K. Javaherian and J. T. Gallagher, *Biochem. J.*, 2003, **375**, 131–139.
56. J. Tapon-Bretaudiere, D. Chabut, M. Zierer, S. Matou, D. Helley, A. Bros, P. A. Mourao and A. M. Fischer, *Mol. Cancer Res.*, 2002, **1**, 96–102.

57. R. Sasisekharan, R. Raman and V. Prabhakar, *Annu. Rev. Biomed. Eng.*, 2006, **8**, 181–231.
58. J. S. Pieper, A. Oosterhof, P. J. Dijkstra, J. H. Veerkamp and T. H. van Kuppevelt, *Biomaterials*, 1999, **20**, 847–858.
59. C. Wolf-Brandstetter, A. Lode, T. Hanke, D. Scharnweber and H. Worch, *J. Biomed. Mater. Res. A*, 2006, **79**, 882–894.
60. V. Hintze, S. Moeller, M. Schnabelrauch, S. Bierbaum, M. Viola, H. Worch and D. Scharnweber, *Biomacromolecules*, 2009, **10**, 3290–3297.

CHAPTER 7

Antibody Immobilization on Solid Surfaces: Methods and Applications

X. HU,[a,b] I. B. O'CONNOR[a] AND J. G. WALL[*a]

[a] National University of Ireland, Galway, Microbiology and Network of Excellence in Functional Biomaterials, University Road, Galway, Ireland; [b] Dalian University, Medical School, Dalian Development Zone, Dalian, China

7.1 Introduction

The immunoassay is one of the most important analytical methods for clinical diagnosis and biochemical studies because of its extremely high specificity and sensitivity. Numerous studies have demonstrated that the proper immobilization of the antibody component is a critical step in the development of sensitive, robust antibody-based immunoassay devices. Ideally, the immobilization method should result in antibody molecules that are correctly oriented on the substrate, stably (preferably covalently) attached, form a monolayer and are unmodified by the reaction chemistry used to generate the linkage.[1] Early immunosensors utilized polyclonal mixtures of antibodies purified from the serum or ascites of immunized animals, but monoclonal antibody (mAb) preparations with a single binding specificity, and subsequently a variety of engineered variants, are now preferred. Meanwhile immobilization of the antibody component has matured from simple adsorption or covalent cross-linking, which are still commonly used in many assays, to incorporate highly

RSC Nanoscience & Nanotechnology No. 21
Biological Interactions with Surface Charge in Biomaterials
Edited by Syed A. M. Tofail
© Royal Society of Chemistry 2012
Published by the Royal Society of Chemistry, www.rsc.org

specialized techniques to optimize the orientation, density and stability of the antibody moiety for enhanced analyte binding.

In this Chapter, we will focus on two major variables of the immunoassay – the nature of the antibody moiety and its attachment to the sensor layer – and discuss how they can be manipulated to improve detection of a target analyte.

7.2 Antibodies and Antibody-Derived Fragments Used in Immunoassays

Immunological techniques that require antibodies – immunoassays, immuno-sensors and immunopurification – originally typically used polyclonal antibodies (pAbs) that are purified from immunized animals. While such assays have been successfully used for more than 30 years in both clinical diagnostics and research, pAbs present a number of significant problems in assay development. Critically, different batches can vary extensively while their poly-reactivity – antibodies against antigens to which the animal was previously exposed may also be present – can lead to false signal detection in a resultant assay. The cost of scale-up and ethical issues related to animal experimentation have also contributed to the current preference for monospecific mAb preparations. mAbs are derived from a single B-lymphocyte, *i.e.* containing a single, usually highly specific reactivity. These are almost always IgGs (Figure 7.1) and the development of hybridoma technology, providing continuously replicating, IgG-secreting B-lymphocytes has led to their availability in virtually unlimited quantities from cell culture. This overcomes inter-batch variability and poly-specificity associated with pAbs, as well as obviating the need for further animal immunization once a high-affinity antibody-secreting hybridoma has been successfully identified. While mAbs have been the staple affinity reagents in protein identification [enzyme immunoassay (EIA), immunoblotting], purification (immunoprecipitation, immunoaffinity chromatography) and characterization (EIA) over the past 30 years, a variety of smaller antibody fragments with improved properties such as higher affinities have more recently become available through developments in protein engineering and expression. A number of these are discussed below in the context of their suitability for covalent and/or oriented surface immobilization for improved biological activity or stability.[2]

7.2.1 Overview of Antibody Structure

The typical antibody molecule or immunoglobulin is roughly Y-shaped and consists of two identical heavy and two identical light chains held together by disulfide bridges. Each heavy chain contains the C_H1, C_H2 and C_H3 constant domains as well as a V_H variable domain, whereas each light chain consists of a single C_L constant domain and a V_L variable domain (Figure 7.1). The stem of the molecule is comprised entirely of heavy-chain constant domains and functions in interaction with immune response effectors, such as the

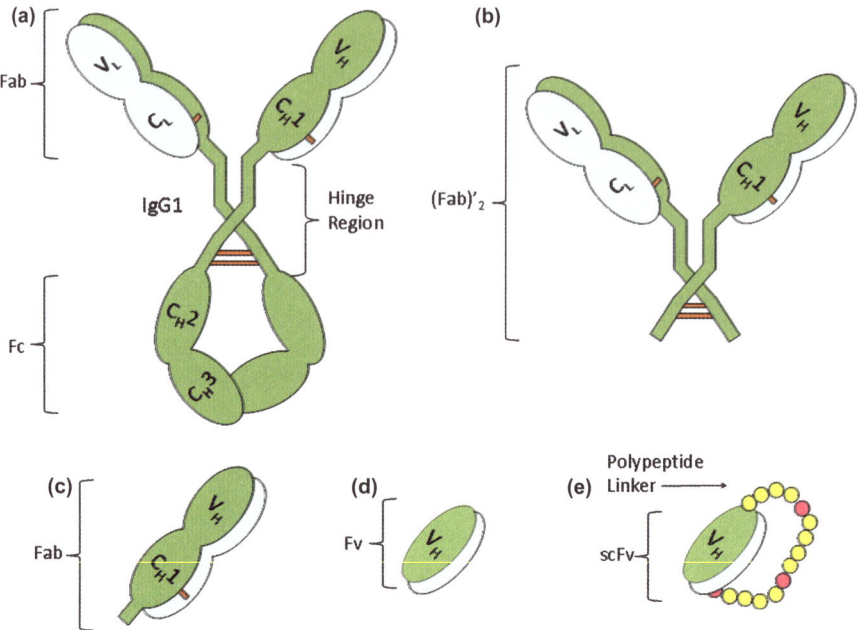

Figure 7.1 Structure of a typical IgG antibody and derived antigen-binding fragments. (a) IgG with heavy chains in dark and light chains in lighter
shading. Disulfide bridges are illustrated as bars connecting the domains.
(b) (Fab)′$_2$, in which the Fc of the IgG stem has been removed by enzymatic digestion. (c) Fab fragment. (d) Fv fragment, consisting of the IgG
V_H and V_L domains. (e) scFv in which the V_H and V_L domains are
covalently attached by a short peptide linker, typically of 15 amino acid
residues.

complement cascade or Fc-receptor-containing cytotoxic cells, whereas the V_H
and V_L domains of the two arms are responsible for binding the vast plethora
of potential antigens that the diverse antibody library may come across. Within
the V_H and V_L domains, relatively conserved frameworks regions form a
structural scaffold from which three hypervariable loops per domain [com-
plementarity-determining regions (CDRs)] extend to form a highly specific
binding pocket in which antigen molecules are bound. This separation of
function is an ingenious design which allows antibodies to fulfill their dual
role to identify a multitude of would-be invaders and then target them for
elimination by conserved antigen-specific or non-antigen specific processes.

   This domain architecture also clearly lends itself to engineering of component fragments of antibodies which retain all or part of the original activities of
the molecule. In *in vitro* applications, such as immunosensing or protein purification where cytotoxicity is not a concern, molecules containing only the
binding moieties clearly might be advantageous due to their smaller size and
consequent potential higher coating densities on support scaffolds.

7.2.2 Antigen-Binding Antibody Fragments

Fragmentation of mAbs has been largely driven by researchers' desire for antibody moieties that demonstrate high-affinity and high-specificity target binding but in many applications – both *in vitro* and *in vivo* – without the need for interaction with immune effector functions. Fragmentation was initially carried out through proteolysis of the parent IgG using pepsin to yield a bivalent $F(ab')_2$ and a proteolysed Fc stem or with papain to generate an intact Fc domain (again typically discarded) and two monovalent Fab arms, each containing a single structurally intact antigen-binding pocket from the parent mAb.

Recombinant DNA technology, however, allows the cloning of antibody-encoding genes and thus the production of individual immunoglobulin domains or, *e.g.*, functional, target-binding Fv fragments containing the antibody V_H and V_L domains (Figure 7.1).[3] Unfortunately, V_Hs and V_Ls generally exhibit a dramatically reduced interaction energy in stand-alone Fvs due to the lack of a covalent linkage between them (in the whole antibody the domains are held together by a $C_H1–C_L$ disulfide bridge in the Fab arm), whereas Fvs are also often produced in very poor yields. This led to a frenzy of fragment engineering over the past 20 years, largely to overcome expression or stability problems or to bestow additional reactivities for improved cytotoxicity *in vivo* or immobilization *in vitro*.[4]

In terms of stability and expression, virtually all protein variants have now ceded position to the single-chain Fv (scFv) fragment in which the V_H and V_L are covalently linked by a short, flexible peptide linker of typically 15–20 amino acids (Figure 7.1). scFvs usually retain the binding properties of the parent antibody molecule and may exhibit advantages *in vivo* such as significantly improved tissue penetration due to their reduced size. Fvs or scFvs with attached cytotoxins or radiolabels have also emerged as viable options for drug delivery or imaging, respectively, while controlled immobilization has been greatly improved by the addition of flexible peptide tags which allow oriented, sometimes covalent, attachment of fragments (see Section 7.3). Meanwhile, any remaining issues of poor expression or folding in the recombinant expression system can often be addressed by manipulation of the protein folding machinery of the host cell[5] or by targeted engineering of the protein itself.[6,7]

Other engineered fragments to have emerged over the past 15 years include multivalent molecules such as diabodies and triabodies,[8] although these will likely find greater applications *in vivo* than in *in vitro* analytical applications, and bi-specific molecules that recognize different antigens (anathema to traditional immunologists!) suitable for, *e.g.*, recruitment of cytotoxic T-cells to a tumour-specific antigen but again likely to be of limited relevance to *in vitro* applications. Reducing still further the size of ligand-binding fragments has been a consistent theme throughout the literature,[9,10] but the scFv currently remains the smallest antibody fragment broadly accepted due to its relative ease of production in recombinant expression systems such as *Escherichia coli* and the normally comparable affinity and specificity of its binding pocket with the parent mAb from which it is derived.

7.2.3   Recombinant Protein Expression Systems

A critical feature of any recombinant protein is the ability to produce it in relatively large amounts in biological systems. The most commonly used expression hosts for recombinant proteins are bacterial, yeast and mammalian cells. Antibodies present a number of particular expression challenges, notably their multiple disulfide bonds, the requirement to assemble the whole immunoglobulin molecule from four component chains (Figure 7.1) and their Fc glycosylation that is required for immune system communication.

The bacterium *E. coli* is the staple of most recombinant protein expression laboratories because of its extensive characterization, simple fermentation, short doubling time, and uncomplicated nutritional and sterility requirements.[11] Disulfide bonds are stably formed in polypeptides upon export to its oxidizing periplasmic space,[3] but the 15 disulfide bridges of whole antibody molecules constitute a considerable bottleneck to the host cell folding machinery. Furthermore, the inability of the bacterium to carry out glycosylation precludes its use in the production of therapeutic whole antibodies for *in vivo* use, although, of course, not for most *in vitro* applications. As a result, *E. coli* is typically the preferred organism for antibody fragments, whereas whole molecules must usually be expressed in eukaryotes. scFvs, in particular, can be produced relatively easily in *E. coli* cells, although high-level expression of antibody fragments, similar to any recombinant protein, can lead to the accumulation of large, insoluble, non-functional aggregates of protein, whereupon the arsenal of techniques available to the recombinant protein technologist to improve expression comes into play.[12]

The major advantage of eukaryotic cells such as immortalized B lymphocytes (myelomas) or Chinese hamster ovary (CHO) cells is their ability to produce more 'native-like' human proteins for *in vivo* applications.[13] This advantage comes at the cost of slower growth rates, considerably higher costs (growth medium, sterility requirements, *etc.*) and reduced ease of protein engineering, but is offset by their ability to glycosylate Fc domains in a manner compatible with recognition by immune effector functions *in vivo*. As other eukaryotic expression hosts that glycosylate proteins, such as yeasts, insect cells or even plants, do so with a pattern of sugar monomers often very different from that of mammalian cells, therapeutic antibodies must be expressed in CHO cells and myelomas, although the yeasts *Saccharomyces cerevisia* or *Pichia pastoris*, in particular, find some use in immunoglobulin production for *in vitro* applications.[13]

7.2.4   Advantages of Recombinant Protein Expression

Recombinant protein expression typically offers three major advantages over harvesting of naturally occurring proteins: increased availability, improved safety and the potential to engineer the protein of interest. The increased availability of recombinant antibody fragments over animal-derived whole antibodies is obvious, while improved safety and purity is achieved by the use of typically non-pathogenic, non-toxin-producing bacterial producers and protein-tagging methods to ensure purification to near homogeneity of the recombinant antibody fragments. The point at which recombinant expression comes into its own, however, is in engineering of the antibody fragment: scFvs alone can be produced in a huge variety of different constructs (with different domain order, various linker lengths, differing recombinant tags, various location of tags, *etc.*; Figure 7.2) for improved expression or physicochemical properties such as binding kinetics and stability. The incorporation of short peptide tags at the N- or C-terminus was initially designed as a tool for immunodetection and ease of purification of the translated product[14] and this now practically ubiquitous technique also has significant potential in protein immobilization, as described below.

7.2.5 Use of Antibody Fragments in Immobilization

The over-riding advantage of recombinant antibodies in immobilization is the ability to modify their properties by protein engineering. scFv dimensions of the order of 5 nm $\times$ 4 nm $\times$ 4 nm compare favourably with whole antibodies of approximately 15 nm $\times$ 7 nm $\times$ 3.5 nm in terms of achieving higher packing densities on surfaces with concomitant potential for increased detection sensitivities or purification efficiencies. Importantly, also, the smaller fragment size may allow the use of novel porous supports with very large surface areas but

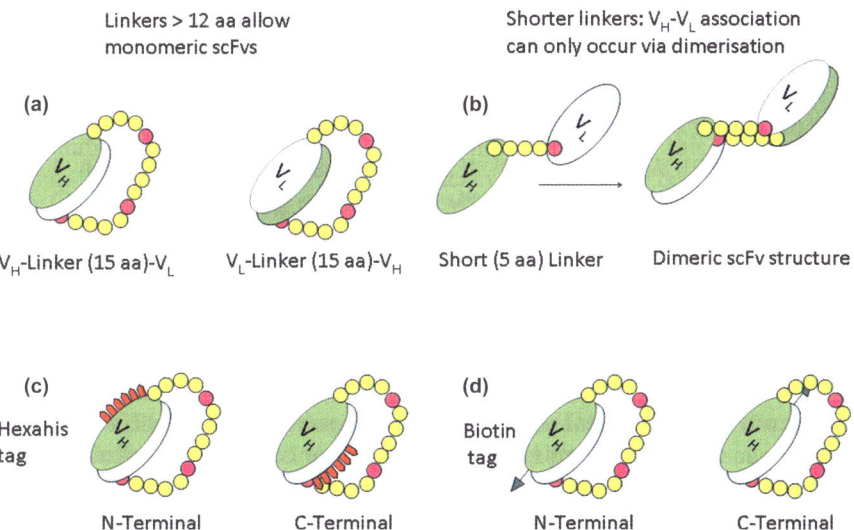

Figure 7.2 Common scFv antibody formats. (a) V_H-linker-V_L and V_L-linker-V_H formats containing the typical 15-amino-acid (aa) [(Gly$_4$Ser)$_3$] polypeptide linker joining the domains. (b) V_H-linker-V_L scFv with a 5-amino-acid linker. (c) N- and C-terminally hexahistidine-tagged scFvs. (d) N- and C-terminally biotin-tagged scFvs.

pore sizes too small for entry of immunoglobulins[15] in detection or purification of small molecule analytes.

In many *in vitro* applications, high-affinity binding is critical for sensitive detection or efficient purification. A powerful arsenal array of targeted[16] or random[17] molecular techniques now allows enhancement of the affinity of binding pockets, allowing for thermodynamic limitations, of course, by several orders of magnitude.[18] In practice, this allows the use of naïve antibody libraries as fragments can be matured to the nanomolar affinity range typically available from immunized libraries. Similarly fragments can be engineered to reduce cross-reactivities and so facilitate measurement of single molecular species or to increase stability for increased duration of performance in their immobilized state. The latter has particular application in immunoaffinity chromatography,[18] but is increasingly important also in the development of low-cost re-useable immunosensors.

Other improvements in antibody fragments for immobilization include the introduction of synthetic linker groups by genetic engineering for improved interaction with the support matrix. These can be relatively simple charged motifs[19] or biotin or multihistidine tags, nominally for improved isolation but which can also be used to mediate extremely strong and potentially targeted attachment to (strept)avidin- or Ni^{2+}-coated surfaces, respectively (Figure 7.2).[20] The addition of a C-terminal cysteine residue to scFvs can also achieve the same result by allowing correctly oriented coupling to gold-coated (piezo)sensor interfaces.[21]

7.3 Methods of Antibody Immobilization

The most commonly used methods of attachment of antibodies to solid supports have been simple adsorption or covalent cross-linking, although a variety of modified antibodies with novel properties for improved attachment are increasingly available (Figure 7.3). Ideally, immobilization should yield antibody molecules that are correctly oriented on the substrate, stably (preferably covalently) attached, arranged in a monolayer and unmodified by the immobilization procedure.[1]

7.3.1 Adsorption

Simple physical adsorption of antibodies on to surfaces such as polystyrene, poly(vinylidene fluoride) (PVDF) or nitrocellulose is the most commonly used method of immobilization in standard techniques such as enzyme immunoassays and immunoblotting. The main advantages of this approach are its ease and low cost, relatively high antibody-binding capacity and the fact that it involves no manipulation of the antibody.[22] The disadvantages, however, are numerous: many proteins denature when adsorbed on to a solid surface, resulting in a disordered or inaccessible binding site in as many as 90% of adsorbed molecules;[23] as attachment occurs via non-site-specific interactions,

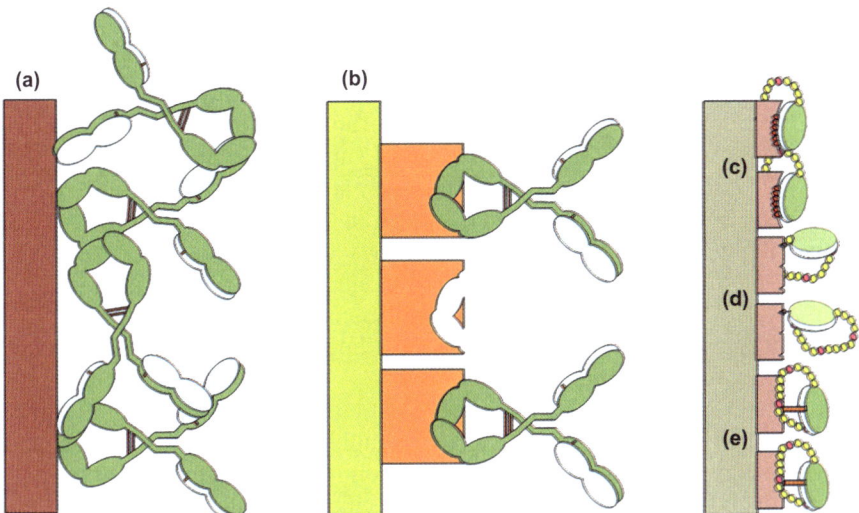

Figure 7.3 Immobilization of whole antibody molecules or scFv fragments via: (a) adsorption; (b) Protein A- or Protein G-mediated attachment; (c) hexahistidine tag–Ni^{2+} interaction; (d) biotin tag–streptavidin interaction; (e) covalent attachment using, *e.g.*, a disulfide bridge formed with an engineered tag in the scFv.

binding pockets are randomly oriented on the substrate, may be inaccessible to analyte and adsorption is not reproducible between analyses (Figure 7.3); the reversible nature of the non-covalent attachment leads to desorption of the antibody and reduced stability over extended usage; the detection sensitivity may be reduced due to non-specific adsorption of other proteins; and deposition of multiple layers of protein is common, leading again to reduced sensitivity and higher antibody production costs.[1] Nevertheless, adsorption remains the method of choice for many single-use immobilization techniques, particularly for qualitative evaluation of antibody–antigen interactions.

7.3.2 Antibody-Binding Proteins

One approach to specifically and directionally immobilize antibodies is via antibody-binding proteins. Protein A and Protein G are naturally occurring bacterial proteins that bind antibody Fc regions. Antibody immobilization via pre-immobilized Protein A or G thus achieves proper orientation of the antibody layer (Figure 7.3), leading, in many cases, to higher detection sensitivities than in sensors produced using random covalent attachment methods.[24] It also avoids the need for engineering of the antibody, although it necessitates an additional immobilization step and is unlikely to achieve high antibody-loading capacities due to the increased dimensions of the immobilized layer. Furthermore, the interaction between the capture protein and antibody is reversible, whereas covalent attachment of the antibody remains preferable for long-term stability.

7.3.3   Covalent Attachment

Covalent attachment methods overcome the problem of protein leaching from the support matrix, but not, in many cases, that of antibody orientation. At its most basic level, the terminal amino or carboxyl group can be coupled via relatively simple chemistry to a variety of supports, such as aldehyde- and epoxy-activated substrates, resulting in highly stable displayed proteins. Cross-linking of solvent-exposed amino acids can also occur, however, as can cross-linking within the antibody molecules themselves. Other approaches include modification of antibodies with affinity tags such as biotin for near-covalent interactions with (strept)avidin-coated surfaces ($K_d = 10^{-15}$ M). However, the random orientation of the antibody molecules on the transducer again limits the availability of binding sites.

A number of strategies have emerged that either exploit naturally occurring moieties such as carbohydrate chains or disulfide bridges, usually on whole or enzymatically digested antibodies, or utilize motifs engineered into antibody fragments to facilitate their directed, site-specific immobilization. In the former case, chemical treatment is required prior to immobilization, such as oxidation of carbohydrate groups in the Fc stem to generate aldehydes that react with amine or hydrazide groups on the transducer surface.[25] Alternatively, whole antibodies can be digested with pepsin and chemically reduced to yield (Fab')s – Fab fragments with an exposed sulfhydryl group from the immunoglobulin hinge region (Figure 7.1). These can be cross-linked simply and rapidly to thiol groups on the matrix to provide highly oriented antigen-binding sites with high accessibility for antigens and improved analyte recognition over randomly adsorbed molecules (Figure 7.3).[26,27]

7.3.4   Recombinant Tags

Recombinant tags can now be added to antibody fragments by genetic engineering in such a routine manner that the approach is usually preferred to chemical modification. The addition of cysteine residues at the C-terminal end of a fragment allows for simple, oriented immobilization via exposed sulfhydryl groups (Figure 7.3) or scFvs can be produced with a tag such as a leucine-zipper motif that interacts with a partner motif on the support matrix to form a tightly bound and correctly oriented heterodimeric pair.[28] Another relatively common approach is the expression of antibody fragments with an attached protein such as glutathione-S-transferase or maltose-binding protein, followed by immobilization via a binding partner or specific antibody. This approach also often leads to improved solubility and increased yields of difficult-to-express proteins in the recombinant expression system, but the bulky nature of the binding partners somewhat undermines the advantages of using small-sized antibody fragments to achieve higher coating densities.

One of the most popular immobilization approaches is the engineering of short tags, often only 4–8 amino acid residues, into the antibody fragment of interest. These motifs may be recognized by an antibody which can itself be immobilized,[29] although supports generated in this manner can be prone to leaching or denaturation under harsh regeneration conditions. The multi- (typically hexa-)histidine tag is now the most commonly used tag for immo- bilization of antibody fragments as the N- or C-terminal motif can be easily immobilized on a divalent cation (usually Ni^{2+} but sometimes Co^{2+} or Cu^{2+})- coated or -chelated surface (Figure 7.3). The technique is used extensively for affinity purification of recombinant proteins, particularly with a relatively robust and reuseable Ni^{2+}-nitrilotriacetic acid (Ni^{2+}-NTA) support matrix, while numerous sensors have also been developed on glass plates deposited with gold,[30] activated polystyrene[31] or PEG-coated Si(III)[32] using the approach. The technology seems set to remain the dominant approach to antibody fragment purification and to increase in popularity in immobilization applications in research laboratories over the next decade due, in no small part, also to the commercial availability of numerous expression systems for tagging proteins with the hexahistidine motif, reporter antibodies for its detection and pre- prepared matrices for ease of immobilization of tagged proteins.

7.4 Applications of Immobilized Antibodies

The many and varied applications of surface-immobilized antibodies range from 'traditional' techniques such as EIA or immunoaffinity chromatography in medical diagnostics, environmental monitoring, and food and public safety to novel and emerging applications in microarrays for disease profiling and the generation of cell-targeted nanomedicines.

7.4.1 Immunoassay Techniques

Immunoassays are the dominant method of analyte detection using antibodies or derived fragments. Solid-phase assays such as EIA, radioimmunoassay and fluorimmunoassay have been widely used to detect target analytes as diverse as peptides, proteins, carbohydrates and lipids, secondary metabolites such as vitamins, hormones and antibiotics, whole bacteria, yeasts and viruses, and inorganic compounds such as metals and a variety of drug types. Sandwich assays, in particular, are commonly used as they typically exhibit detection sensitivities from 0.1 to 50 ng ml^{-1} and can be routinely performed in microtitre plates containing 96, 384 or even 1536 sample wells.[33] Although recognition of the advantages of uniformly oriented antibodies and the ease of engineering of recombinant antibody fragments is, slowly, seeing the emergence of more sophisticated EIAs based on directed attachment of the antibody moiety, adsorption of mAbs on to polystyrene- or nitrocellulose-type supports remains the norm in the field for now.

7.4.2 Immunosensors

Immunosensors are devices in which an immobilized antibody (usually mouse or rabbit) binds a specific analyte and translates a signal *via* a transducing element to provide a simple read-out. They have the potential to provide a rapid, low cost and highly sensitive alternative to traditional detection approaches such as HPLC or mass spectrometry and can be used *in situ* in the absence of specialized equipment or highly trained personnel.

Immunosensors are particularly suited to the emerging field of point-of-care (POC) diagnostics due to their potential to provide simple, rapid (within minutes) and often multi-target detection *in situ* in the absence of state-of-the-art molecular analytical techniques or sophisticated laboratory facilities. They have been used to detect tumour-specific antigens and microbial pathogens in clinical samples as well as food spoilage agents, environmental toxins[5,15] and drugs of abuse,[34] amongst numerous other analytes. Several forms of direct detection of analyte using optical [total internal reflection spectroscopy, ellipsometry and surface plasmon resonance (SPR)], electrochemical (potentiometric, amperometric and conductometric), thermometric or microgravimetric methods have been described for laboratory-based methods,[35] although the use of signal-generating labels allows more versatile and frequently more sensitive analyte detection. While recombinant systems can provide continuous, standardized supplies of Fabs and scFvs for use in diagnosis and monitoring, these have yet to corner a significant market share of the critical clinical diagnosis field (tens of billions of US dollars per annum) that is still dominated by the use of whole antibodies as the recognition element. Many hundreds of devices are now commercially available, most for use in clinical settings but numerous relatively inexpensive immunotests are also widely available for home or workplace use for, *e.g.* screening for drugs of abuse. For POC devices, the read-out is typically colorimetric, with a critical threshold concentration of analyte (often < 10 ng) triggering a positive reaction. Although devices used in medical screening are almost invariably single-use, it is desirable in fields such as food or environmental monitoring to develop re-useable sensors which could be regenerated on-site by simple washing, in which case antibodies must clearly be covalently immobilized (see Section 7.3) to avoid leaching during regeneration.

Lateral flow immunoassays (LFIAs) are small, often dipstick-format, immunochromatographic assays particularly suited to testing out of the laboratory environment. The first commercial example was the home pregnancy test kit and the general format has since been widely applied in medical diagnosis, veterinary, food, agriculture, environment and even biodefence applications.[36,37] LFIAs are effectively immunoassays in which an applied sample moves along a reaction pad by capillary action towards lines of immobilized antibody or antigen. Detection is rarely quantitative and typically occurs by the development of coloured lines in a result window, but LFIAs are extremely versatile, very simple to use and interpret and a hugely successful application of immobilized antibodies.

7.4.3 Affinity Chromatography

Immunoaffinity chromatography is a powerful technique in which biological samples are passed over immobilized antibodies, usually in a column format, to purify a target analyte from complex heterogeneous mixtures. After the column is washed clear of unbound material, elution is typically achieved by changing the pH or salt concentration or adding free analyte.[38] Antibodies are immobilized using a variety of different coupling chemistries, including attachment via sulfhydryl and carboxyl groups but also via commercially available Protein G-bound agarose or Ni^{2+}-coated resins (see Section 7.3). Purification can be greater than 10 000-fold in a single step process and covalent immobilization methods are usually preferred to increase stability of the support over the continued cycles of binding and elution necessary in most applications.

7.4.4 Drug Delivery and Nanomedicine

Immobilized antibodies have found exciting new applications in the form of antibody-conjugated nanoparticles for use in magnetic resonance imaging, sensing, cell sorting, bioseparation and cell transfection.[39] The development of improved cell-specific recognition and targeted drug delivery remains the main driving force, although applications such as coating of biomaterials with cell-targeting antibodies for increased biocompatibility or improved wound healing also exhibit immense potential in the development of next generation nano-therapeutics. Nanoparticles are linked almost invariably to nano-sized antibody fragments such as scFvs or single domains rather than whole immunoglobulin molecules and immobilization is frequently *via* disulfide bonds formed with additional cysteine residues added on to the recombinant fragments[40] as the efficacy of cell targeting and nanoparticle uptake is critically dependent on the *in vivo* stability and correct orientation of the attached antibodies.

7.4.5 Antibody Microarrays

DNA microarrays have been in use for over a decade to provide a snapshot of cell-wide gene expression, *e.g.*, in healthy *versus* diseased samples to gain an insight into the molecular basis of numerous diseases. Today, there is great interest in multiplexed, high-throughput antibody microarrays to provide information on protein–protein interactions and post-translational modifications, in particular, which cannot be obtained by analysing messenger RNA. In antibody microarrays, antibodies (typically fragments) are attached to a glass slide or other chip, usually in hundreds, rather than the thousands currently obtainable with DNA chips. With their suitability to high-throughput assays, miniaturization and automation, they provide great potential to carry out rapid, highly sensitive, pre-clinical screening of urine or serum samples for panels of pathogens[41] or tumour-specific proteins.[42] Antibody arraying is greatly facilitated by the availability of Ni^{2+}- and streptavidin-coated slides for effective (and correctly oriented) coupling,[43] whereas fragments are used to achieve high coating densities as the major limitation to the sensitivity or broad applicability of chips is the number of correctly immobilized binding sites that can be achieved.[44] Combinatorial antibody library methods may also find application in identification of disease marker patterns using such array methods. Although a relatively new technology, antibody microarrays have the potential to play an important role in front-line disease diagnostics and drug discovery over the coming decades.

7.4.6 Other Applications

Novel biophysical applications of immobilized antibodies that have emerged over the past number of years include SPR and quartz crystal microbalance (QCM). These techniques offer two major advantages over traditional immunoassays: label-free detection and real-time reporting. SPR has become a relatively mainstream technique for investigation of biomolecular interactions and interaction kinetics[45] with antibodies or fragments immobilized on gold-coated chips via Ni^{2+} or –SH groups and analytes detected by a change in the resonance energy.[46] QCM, similarly, measures changes in mass (down to 0.1 µg) by detecting changes in the frequency of a quartz crystal resonator and, again, labelling of antibody or antigen is not required and results are generated in real-time.

Atomic force microscopy (AFM) is typically used for analysis and imaging of surface topographies, but by attaching antibodies to the cantilever in the correct orientation atomic force microscopes can be converted into single-molecule molecular recognition force microscopes (MRFM) to study interactions at the single molecule level.[47] Attachment of the antibody can be via engineered disulfide bridges and the result is an exquisitely sensitive mechanism of studying how antibody molecules bind their specific antigens.

7.5 Conclusions

Antibodies immobilized on solid matrices have been a mainstay of analyte detection, quantification and purification techniques over the past 30 years. Technological developments have overcome traditional limitations of antibody polyreactivity, poor device stability and random orientation of binding pockets to yield stable, highly sensitive, highly specific and enormously diverse immunoplatforms. Although the field seems certain to continue to be dominated by medical immunodiagnostic devices in the short- to medium-term, at least in terms of commercial importance, emerging applications in nanomedicine, antibody arrays, real-time binding analysis and single molecule interaction investigations suggest hugely exciting opportunities for immobilized antibodies into the future.

References

1. Y. Jung, J. Y. Jeong and B. H. Chung, *Analyst*, 2008, **133**, 697.
2. D. Saerens, L. Huang, K. Bonroy and S. Muyldermans, *Sensors*, 2008, **8**, 4669.
3. A. Skerra and A. Plückthun, *Science*, 1988, **240**, 1038.

4. R. Rapley, *Mol. Biotechnol.*, 1995, **3**, 139.
5. X. Hu, L. O'Hara, S. White, E. Magner, M. Kane and J. G. Wall, *Protein Expression Purif.*, 2007, **52**, 194.
6. A. Knappik and A. Plückthun, *Protein Eng.*, 1995, **8**, 81.
7. X. Hu, R. O'Dwyer and J. G. Wall, *J. Biotechnol.*, 2005, **120**, 38.
8. P. Holliger and P. J. Hudson, *Nat. Biotechnol.*, 2005, **23**, 1126.
9. C. Hamers-Casterman, T. Atarhouch, S. Muyldermans, G. Robinson, C. Hamers, E. B. Songa, N. Bendahman and R. Hamers, *Nature*, 1993, **363**, 446.
10. K. H. Roux, A. S. Greenberg, L. Green, L. Strelets, D. Avila, E. C. McKinney and M. F. Flajnik, *Proc. Natl. Acad. Sci. U.S.A.*, 1998, **95**, 11804.
11. M. Rai and H. Padh, *Curr. Sci.*, 2001, **80**, 1121.
12. O. Kolaj, S. Spada, S. Robin and J. G. Wall, *Microb. Cell Fact.*, 2009, **8**, 9.
13. N. Ferrer-Miralles, J. Domingo-Espín, J. L. Corchero, E. Vázquez and A. Villaverde, *Microb. Cell Fact.*, 2009, **8**, 17.
14. R. Witzgall, E. O'Leary and J. V. Bonventre, *Anal. Biochem.*, 1994, **223**, 291.
15. X. Hu, S. White, S. Spada, S. Hudson, E. Magner and J. G. Wall, *J. Phys. Chem. B*, 2006, **110**, 18703.
16. L. K. Denzin and E. W. Voss Jr, *J. Biol. Chem.*, 1992, **267**, 8925.
17. R. A. Irving, A. A. Kortt and P. J. Hudson, *Immunotechnology.*, 1996, **2**, 127.
18. S. Jung, A. Honegger and A. Plückthun, *J. Mol. Biol.*, 1999, **294**, 163.
19. Z. Shen, H. Yan, Y. Zhang, R. L. Mernaugh and X. Zeng, *Anal. Chem.*, 2008, **80**, 1910.
20. Y. S. Lo, D. H. Nam, H. M. So, H. Chang, J. J. Kim, Y. H. Kim and J. O. Lee, *ACS Nano*, 2009, **3**, 3649.
21. N. Backmann, C. Zahnd, F. Huber, A. Bietsch, A. Plückthun, H. P. Lang, H. J. Güntherodt, M. Hegner and C. Gerber, *Proc. Natl. Acad. Sci. U.S.A.*, 2005, **102**, 14587.
22. W. M. Albers, I. Vikholm, T. Viitala and J. Peltonen, in *Handbook of Surfaces and Interfaces of Materials*, ed. H. S. Nalva, Academic Press, 2001, vol. 5, p 1.
23. I. H. Cho, E. H. Paek, H. Lee, J. Y. Kang, T. S. Kim and S. H. Paek, *Anal. Biochem.*, 2007, **365**, 14.
24. R. Danczyk, B. Krieder, A. North, T. Webster, H. HogenEsch and A. Rundell, *Biotechnol. Bioeng.*, 2003, **84**, 215.
25. J. P. Gering, L. Quaroni and G. Chumanov, *J. Colloid Interface Sci.*, 2002, **252**, 50.
26. I. Vikholm and W. M. Albers, *Langmuir*, 1998, **14**, 3865.
27. M. M. Billah, C. S. Hodges, H. C. W. Hays and P. A. Millner, *Bioelectrochemistry*, 2010, **80**, 49.
28. K. Nakanishi, T. Sakiyama, Y. Kumada, K. Imamura and H. Imanaka, *Curr. Proteomics*, 2008, **5**, 161.
29. J. W. Slootstra, D. Kuperus, A. Plückthun and R. H. Meloen, *Mol. Divers.*, 1997, **2**, 156.

30. R. Vallina-García, M. M. García-Suárez, M. T. Fernández-Abedul, F. J. Méndez and A. Costa-García, *Biosens. Bioelectron.*, 2007, **23**, 210.
31. L. R. Paborsky, K. E. Dunn, C. S. Gibbs and J. P. Dougherty, *Anal. Biochem.*, 1996, **234**, 60.
32. T. Cha, A. Guo, J. Jun, D. Pei and X. Y. Zhu, *Proteomics*, 2004, **4**, 1965.
33. Corning Inc., *Microplate Selection Guide. For assays and drug discovery*, Corning Inc., MA, USA, 2007, p. 1.
34. F. S. Ligler, A. W. Kusterbeck, R. A. Ogert and G. A. Wemhoff, in *Biosensor Design and Application*, ACS Symposium Series, American Chemical Society, 1992, vol. 511, p. 73.
35. P. B. Luppa, L. J. Sokoll and D. W. Chan, *Clin. Chim. Acta*, 2001, **314**, 1.
36. R. Wong and H. Tse, *Lateral Flow Immunoassay*, Humana Press, New York, 2009.
37. R. O'Kennedy, S. Townsend, G. G. Donohoe, P. Leonard, S. Hearty and B. Byrne, *Anal. Letts.*, 2010, **43**, 1630.
38. A. Subramanian, *Mol. Biotechnol.*, 2002, **20**, 41.
39. M. Arruebo, M. Valadares and Á González-Fernández, *J. Nanomaterials*, 2009, Article ID 439389, doi:10.1155/2009/439389.
40. C. J. Ackerson, P. D. Jadzinsky, G. J. Jensen and R. D. Kornberg, *J. Am. Chem. Soc.*, 2006, **128**, 2635.
41. M. Uttamchandani, J. L. Neo, B. N. Ong and S. Moochhala, *Trends Biotechnol.*, 2009, **27**, 53.
42. H. Jin and R. C. Zangar, *Cancer Biomark.*, 2010, **6**, 281.
43. C. Steinhauer, C. Wingren, F. Khan, M. He, M. J. Taussig and C. A. K. Borrebaeck, *Proteomics*, 2006, **6**, 4227.
44. K. Even-Desrumeaux, D. Baty and P. Chames, *Mol. BioSyst.*, 2010, **6**, 2241.
45. W. Huber and F. Mueller, *Curr. Pharm. Des.*, 2006, **12**, 3999.
46. W-C. Tsai and P-J R. Pai, *Microchim. Acta*, 2009, **166**, 115.
47. H. Gao, X. X. Zhang and W. B. Chang, *Front. Biosci.*, 2005, **10**, 1539.

Part III
Cellular Interactions with Abiotic Surfaces

Interactions of Bone-forming Cells with Electrostatic Charge at Biomaterials' Surfaces

U. HEMPEL,*[a] C. WOLF-BRANDSTETTER[b] AND
D. SCHARNWEBER[b]

[a] Technische Universität Dresden, Faculty of Medicine, Institute of
Physiological Chemistry, TU Dresden, Fiedlerstr. 42, 01307 Dresden,
Germany; [b] Max Bergmann Center of Biomaterials, Institute of Materials
Science, TU Dresden, Budapester Str. 27, 01069 Dresden, Germany

8.1 Introduction

In vitro cell culture experiments are useful tools for the study of molecular mechanisms of cell–cell or cell–extracellular matrix (ECM) interactions or for the evaluation of the biocompatibility of new materials. Primary cells from the tissues of interest and appropriate cell lines are abundantly available. Single-cell-type experiments are widely accepted, but represent a very simplified model that does not consider numerous cell-to-cell communications which take place between different tissue cell types. The advantage of single-cell-type experiments is that the cellular reaction can be addressed explicit to particular cell type and that specific molecular mechanisms can be studied. Obviously, the preference should be given to primary cells; however, they have limitations in proliferative capacity and often have significant donor-dependent variations.

RSC Nanoscience & Nanotechnology No. 21
Biological Interactions with Surface Charge in Biomaterials
Edited by Syed A. M. Tofail
© Royal Society of Chemistry 2012
Published by the Royal Society of Chemistry, www.rsc.org

Currently, the trend is towards co-culture approaches where direct cell–cell interactions can be studied and different cell types are used as a type of internal 'drug-delivery system'. Such experiments have to be planned quite carefully; controls are of great importance, and also data analysis and interpretation are always challenging.

This Chapter deals with the mechanisms of cell–ECM interactions, aspects of normalization of the *in vitro* data and osteoblast differentiation. Adhesion to non-physiological substrates will be discussed from a cellular point of view, not including the input of various surface/material properties as topography, chemistry, charge, elasticity *etc.* on the adhesion process. Section 8.3 describes practical methods for direct and indirect cell counting, which are partially suitable also for the determination of proliferation rate. Manifold biomaterials have been developed for application in hard tissue and have been tested *in vitro* with osteoblasts or mesenchymal osteoblast precursor cells. A short overview of osteogenic differentiation and the *in vitro* characterization of these cells is described in Section 8.4. Section 8.5 summarizes the state-of-the-art electrically modified bioceramics and their effects on bone cells and tissues.

8.2   Cell Adhesion

The first contact of anchorage-dependent cells with their physiological environment, ECM or neighbouring cells, as well as with non-natural substrates such as cell culture dishes, biomaterials, *etc.*, decides the further fate of a cell. Once cells have past the adhesion process in an adequate rate and manner, they are encouraged to proliferate, differentiate and fulfill all of their functions. Attachment, adhesion and spreading occur in manifold well-organized hierarchical steps. The earliest stage of cell-surface contact (time ≤ 30 min) is mediated by their pericellular coating, especially by glycosaminoglycans such as chondroitin sulfate or hyaluronan, via electrostatic interaction with the substrate.[1–3] This weak, transient adhesion provides as opportunity for the cells to decide whether or not they will adhere; and when they precede to the commitment to adhere at which particular site this will be. Experiments with chondroitinase ABC- or hyaluronidase-treated cells showed a reduced number of adherent cells after such a treatment.[1,3] Within the initial adhesion/spreading (time ≥ 30 min), transient small dot-like structures, named focal complexes, are formed. Within the formation of the focal complex, the distance of cells to the substrate is reduced from 1 μm at the beginning to approximately 15 nm after some minutes, whereas the contact area increases continuously.[1,4] The focal complexes are formed preferentially at the leading edges of the cells and comprise, amongst others, vinculin and phosphorylated proteins such as focal adhesion kinase (FAK), Src and paxillin (Figure 8.1).

The connection of intracellular adhesion-induced protein complexes to the extracellular environment is realized by transmembranal receptors, called integrins, which sense the presence of extracellular ligands (serum proteins and specific ECM components). In their active form integrins are heterodimers

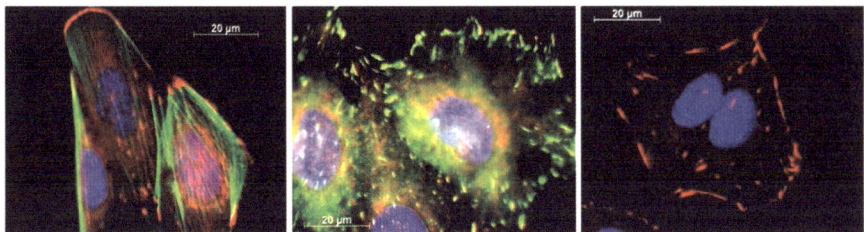

Figure 8.1   Immunofluorecence staining of SaOS-2 (human osteosarcoma cell line) 24 h after plating on glass. F-actin stress fibres are visualized in green, vinculin in red and nuclei in blue (left-hand panel); phospho-FAK-Y397 in red, vinculin in green and nuclei in blue (middle panel); phospho-paxillin-Y31 in red and nuclei in blue (right-hand panel).

of α- and β-subunits, where combinations of 18 α-chains and 8 β-chains offer a huge diversity in ECM recognition. For example, fibronectin can interact with 20 different integrins heterodimers.[5] The ECM itself is a complex network, which permanently adapts to cellular requirements, and is formed and remodeled in a strictly regulated manner. In accordance with the sequential appearance of different ECM proteins, cells have to adapt their integrin expression continuously to the actual ECM situation in order to form strong compact anchors or to achieve migration by detachment/attachment processes. Focal complexes mature into focal adhesions, which are complex structures which are actively involved in cytoskeletal rearrangement.[6,7]

Adhesion-induced, integrin-mediated interaction of cells with the ECM leads to formation of 'cytoskeletal' complexes with cytoplasmatic integrin domains and certain (at least 12) adaptor proteins, *e.g.* vinculin, α-actinin, talin, tensin, zyxin, and paxillin, which support F-actin organization (Figure 8.2).[8]

Rho GTPase, actin-related protein 2/3 (Arp2/3) and vasodilator-stimulated phosphoprotein (VASP) are involved in F-actin polymerization as signalling and adaptor molecules. Kanchanawong *et al.*[9] recently obtained, by interferometric photo-activated localization microscopy (iPALM), a hierarchical architecture model of integrin-based cell adhesions with clear separated functional zones from outside-to-inside integrin signaling layer (FAK, paxillin and talin), a force transduction layer (vinculin and zyxin), an actin regulatory layer (VASP and α-actinin) and mature actin stress fibres. As reviewed by Zaidel-Bar and Geiger,[8] currently approximately 180 players (mostly proteins) are involved in the adhesion process; the number of potential interactions of all of the players is given as about 750. In addition to cytoskeletal rearrangement, adhesion induces a wide variety of signalling networks and translates extracellular information into messages for gene expression, cell cycle regulation and differentiation. As a consequence of adhesion, the auto-phosphorylation of FAK at Tyr-397 creates a docking site for phospho-Src(Tyr-416). By association with phospho-FAK(Tyr-397), phospho-Src(Tyr-416) is able to phosphorylate additional tyrosine residues of FAK (FAK has at least nine putative phosphorylation sites). Phospho-FAK(Tyr-861/Tyr-925) interacts with the

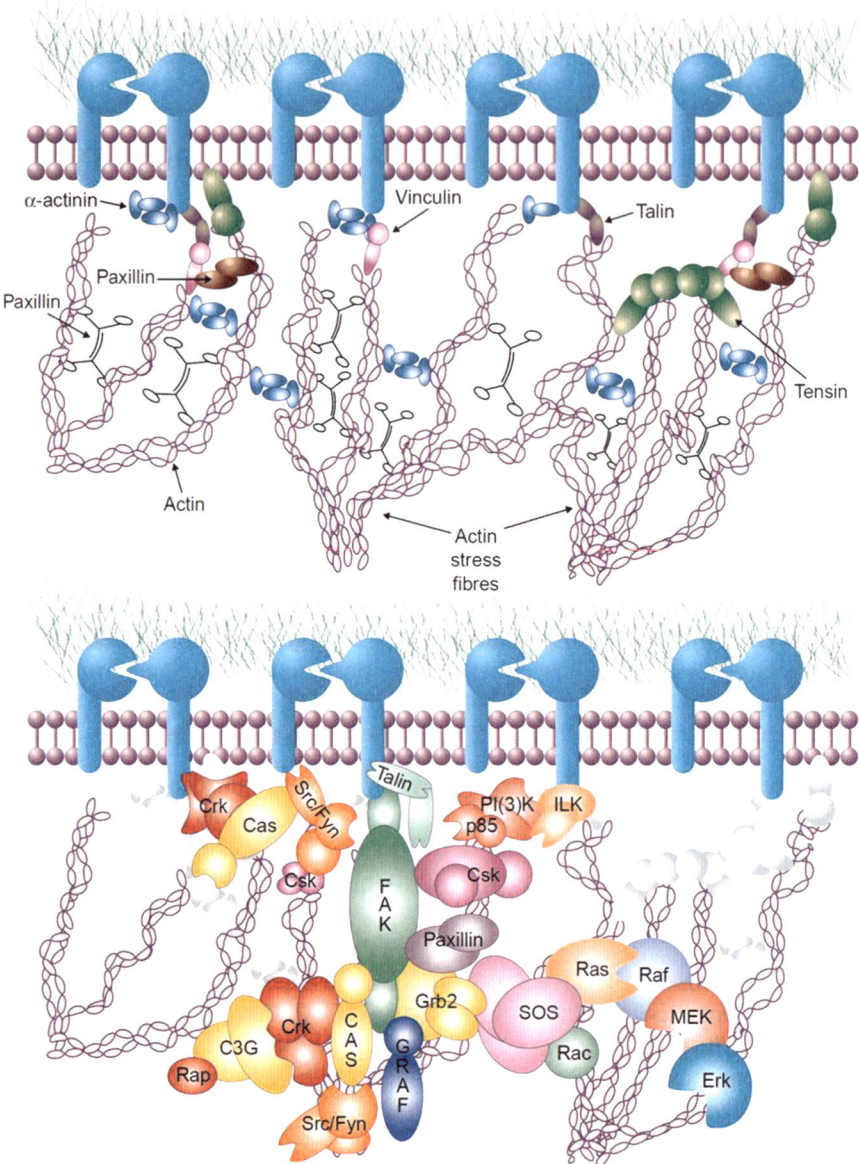

Figure 8.2 Schemes of adhesion induced cytoskeletal rearrangement (upper panel) and adhesion-induced signaling pathways (lower panel). Reprinted from C. K. Miranti and J. S. Brugge, *Nat. Cell Biol.*, 2002, **4**, 83–90, with permission.

adaptor Grb-2 and induces, *via* the SOS pathway, the activation of the Ras/Raf-1/mitogen-activated protein kinase kinase (MEK)/extracellular-regulated kinase (ERK) pathway (Figure 8.2).[10] The ERK pathway is also involved in regulation of proliferation and differentiation. In particular, the differentiation of mesenchymal stromal cells (MSCs) into bone-forming osteoblasts is under the control of RUNX2/Cbfa1 as master transcription regulator. RUNX2/Cbfa1 is a substrate of ERK1/2. The disruption of the signal cascade from FAK over mitogen-activated protein kinase (MAPK) to RUNX2/Cbfa1, *e.g.* by FAK small interfering RNA (siRNA) or in FAK$^{-/-}$ knockout mice, resulted in impaired osteogenic differentiation of MSCs and delayed bone regeneration.[11,12] In contrast, osteogenic differentiation is enhanced when cells are provided with single components of their native ECM or with complete native ECM which they are able to adhere to specifically and form strong focal adhesions.[13,14]

In summary, the first contact of cells with any kind of ECM (non-physiological substrates, single ECM components, native ECM, *etc.*) induces a cascade of protein–protein interactions, protein phosphorylation/dephosphorylation and signal transduction pathways which all occur in regulated manner and determine the further commitment of cells. For the investigation of the adhesion process not only are qualitative methods used [protein co-localization studies and imaging methods such as scanning electron microscopy (SEM), atomic force microscopy (AFM), iPALM and fluorescence correlation spectroscopy (FCS)], but adhesion kinetics can also provide helpful information.

8.3 Cell Proliferation

For *in vitro* experiments, the determination of cell number or any related value is mandatory to evaluate cellular functions such as proliferation or to normalize specific cellular parameters (protein content, enzyme activity, *etc.*). The choice of an adequate method depends on whether cells are intended for further use and mostly should stay intact, or if they are disused following the experiment. Another aspect is whether the information regarding cell number dynamics or actual steady-state values are of interest. Intact whole cells can be counted directly on cell culture dishes when they are forming monolayers and have a clear separated shape. For this approach, randomly picked areas of a complete culture dish are analysed using a microscope. Cell number is mainly determined after enzymatic or mechanical detachment of adherent cells then counting of the cells in the aliquots of the cell suspension. This only works properly when no cell aggregates are formed and all of the cells are completely detachable. The counting is performed by eye or with automated systems (cell counters and FACS can be used). Before counting, the cells can be stained, for example, with fluorescent cell trackers, with calcein acetoxymethyl ester (AM) (stains intact whole cells by esterase activity) or with Trypan Blue in order to distinguish between live and dead cells.

In addition to whole cells, cell organelles, first of all the nuclei, are used for cell number determination. The computer-assisted counting of fluorescently stained nuclei [4',6-diaminine-2-phenylindole (DAPI) and Hoechst 33258/33342 (bisbenzimides) staining, which all bind to the minor groove of AT-rich DNA regions] in fixed cells is a suitable method that results in reliable values.[15] The rule 'one cell, one nucleus' is valid for most cell types, with the exception of red blood cells, thrombocytes, multi-nucleated giant cells or osteoclasts. The total amount of DNA amount in cells can be determined after cell lysis and DNA quantitation using DNA-interacting dyes and defined DNA amounts for calibration. Such fluorescence-based kits, *e.g.* PicoGreen® assay, are commercially available. Another integrating, indirect nuclei-counting method uses Methylene Blue [N,N,N',N'-tetramethylthioninchloride or methyl(en)thioniniumchloride; 3,7-bis(dimethylamino)-phenothiaziniumchloride], a dye which interacts with RNA/DNA, and slightly with the cytoplasm and macromolecular structures.[16] Fixed cells are stained and, after washing and solubilization, the amount of bound dye is measured photometrically at λ_{655} (Figure 8.3).[17] The amount of bound Methylene Blue correlates well with cell number; however, each cell (line) needs an individual calibration curve.

In addition to whole-cell- or cell-organelle-based direct and indirect counting methods, many cellular functions and metabolic reactions are used for normalization. All *in vitro* cell cultures are dependent on glucose (or other monosaccharides) as an energy source. During glycolysis, glucose is metabolized into pyruvate by numerous cytosolic enzymes. The free energy released in this pathway is used to form ATP. Depending on the oxygen supply, pyruvate is either reduced to lactate or converted into acetyl-coenzyme A by pyruvate dehydrogenase in order to provide NADH for the respiratory chain. The constitutively expressed lactate dehydrogenase (LDH, EC1.1.1.27) reversibly catalyses the formation of lactate from pyruvate. The cellular enzyme activity can easily be measured after cell lysis and reflects indirectly the number of cells. Furthermore, the determination of LDH activity in conditioned cell culture medium (extracellular LDH, occurs only when cell membrane integrity is damaged) allows the estimation of cytotoxicity/viability. For this approach, it is necessary to measure both cellular and extracellular LDH activity in the same assay. The ratio of cellular LDH to total (cellular + extracellular) LDH represents the viability of the cells. The test principle is: oxidation of added lactate to pyruvate by LDH in the presence of NAD^+. This reaction is coupled with the reduction of a tetrazolium dye (iodonitrotetrazolium salt is mainly used) into a formazan by NADH and NADH dehydrogenase (diaphorase) and

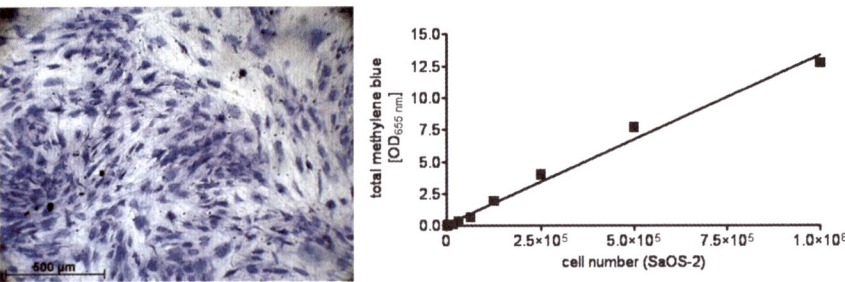

Figure 8.3   SaOS-2 (human osteosarcoma cell line) cell layer, stained with Methylene Blue (left-hand panel) and linear calibration (r^3 = 0.9876) for 10^3–10^6 SaOS-2 cells after Methylene Blue staining (right-hand panel).

the recycling of NAD^+ by that enzyme (Figure 8.4). The absorbance of the formazan is determined photometrically at λ_{490}.

Widely used methods which work with intact cells are the so-called tetrazolium dye assays. Living cells are able to reduce tetrazolium dyes by consumption of reduced coenzymes as NAD(P)H or $FADH_2$. Formazan formation depends on glycolytic, as well as on mitochondrial, activity [mitochondrial succinate dehydrogenase ($FADH_2$) is partially involved in the reaction].[18] A number of cell-permeant dyes are available; the formazan products are either insoluble in aqueous medium {MTT [3-(4,5-dimethyltioazol-2-yl)-2,5-diphenyltetrazolium bromide], λ_{570}, Figure 8.5} or water-soluble {XTT [2,3-bis-(2-methoxy-4-nitro-5-sulfophenyl)-2H-tetrazolium-5-carboxanilide], λ_{450}; MTS [3-(4,5-dimethylthiazol-2-yl)-5-(3-carboxymethoxyphenyl)-2-(4-sulfophenyl)-2H-tetrazolium], λ_{490}; WST-1 [4-(3-(4-iodophenyl)-2-(4-nitrophenyl)-2H-5-tetrazolio)-1,3-benzol-disulfonate] λ_{450}; WST-8 [2-(2-methoxy-4-nitrophenyl)-3-(4-nitrophenyl)-5-(2,4-disulfophenyl)-2H-tetrazolium], λ_{450}}. All water-soluble formazans allow the cell culture to continue after removal of the tetrazolium/formazan solution, or this method can be combined with the measurement of other parameters in the same sample. The amount of formazan formed correlates with the cell number, however, each cell (line) needs an individual calibration curve because of individual metabolic activity/capacity (Figure 8.5).

The determination of proliferation rate needs a dynamic approach, in which the 'first-day' value reflects the actual cell number and says nothing about proliferation rate before not a second time point was analyzed. Well-established methods are [³H]thymidine incorporation and, as a non-radioactive version, the BrdU (5-bromo-2'-deoxyuridine) assay. Alternatively, the described tetrazolium salt-based assays can be used. Figure 8.6 demonstrates an *in silicio*

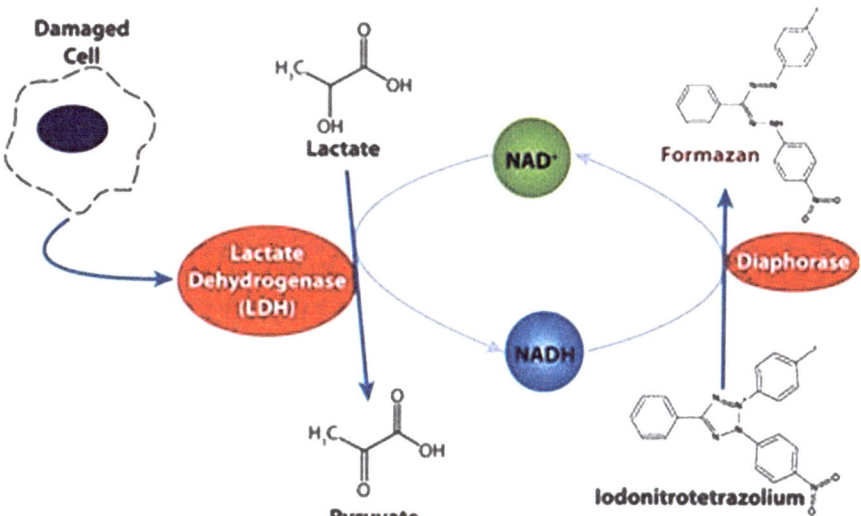

Figure 8.4 Scheme of LDH activity test (from www.gbiosciences.com).

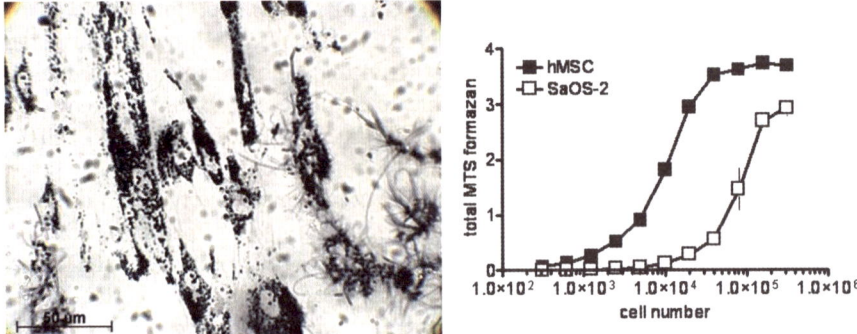

Figure 8.5 Water-insoluble MTT formazan crystals in SaOS-2 (human osteosarcoma cell line) after 2 h incubation with 1.2 mM MTT (left-hand panel). Calibration curve for MTS formazan of either human MSCs (hMSC) or SaOS-2 cells (right-hand panel).

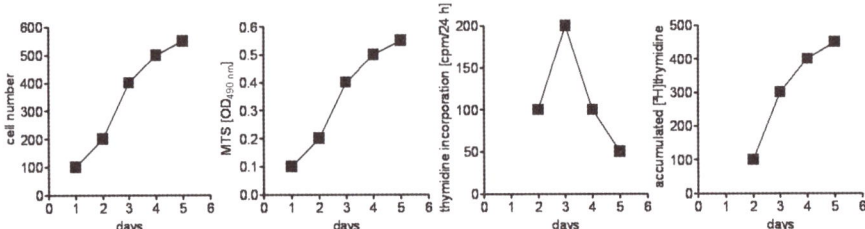

Figure 8.6 *In silico* demonstration of cell number counting, and the corresponding MTS formazan formation, [³H]thymidine incorporation over 24 h and additive [³H]thymidine values (left to right).

approach as how data from direct cell counting, the MTS assay and [³H]thymidine incorporation correspond to each other.

Cell cycle analysis by FACS gives information about how many cells are actually in M, G1/2, and S phases; however, this is only a 'snap-shot' analysis.

Last, but not least, the determination of total cellular protein is a sufficient method for normalization, although sensitivity, susceptibility and possible inaccuracy of the different methods (Bradford, Lowry and Biuret) should be taken into account.

To summarize, all of the described methods are commonly used and tested. They withstand a daily laboratory routine with high throughput and do not require specialized equipment. Figure 8.7 shows that most of the presented parameters correlate quite well with each other and therefore could be replaced by each other. Some of the analyses can be performed in parallel from the same cell culture/lysate; *e.g.* each combination of water-soluble tetrazolium dye assay, LDH activity measurement, protein determination and DNA quantitation is possible.

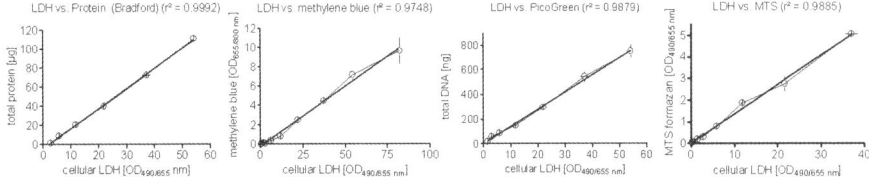

Figure 8.7   Correlation of cellular protein, Methylene Blue staining, DNA content and metabolic activity (MTS assay) *versus* cellular LDH activity (left to right). Analyses were performed with SaOS-2 cells (human osteosarcoma cell line).

8.4   Cell Differentiation

For bone formation *in vivo*, MSCs are recruited from the bone marrow to the site of bone remodelling in which the gaps, produced by bone-resorbing osteoclasts, have to be filled (Figure 8.8). The main function of osteoblasts in the bone remodelling process is to synthesize and organize bone ECM. Bone ECM comprises approximately one-third organic molecules and approximately two-thirds calcium phosphate, mainly hydroxyapatite (HAP). Osteoblasts have to synthesize collagen (it forms approximately 95% of the organic bone ECM) and, to a minor extent, polymeric sugars and non-collagenous proteins such as osteonectin, osteopontin, osteocalcin, osteoadherin and bone sialoprotein. The latter support regulated mineralization of organic ECM. The mineralization process requires, first of all, a high local concentration of calcium ions. This is carried out by the calcium-binding motif (γ-carboxylic chelate structure) of osteocalcin. The mineralization process also needs a local increase of phosphate ions, which are provided by tissue non-specific alkaline phosphatase (TNAP). This enzyme is expressed in the osteogenic differentiation process. It is localized in the osteoblast membrane; with ongoing mineralization TNAP is released with matrix vesicles into bone ECM.[19] When mineralization is completed, the osteoblasts vanish from the bone surface into the bone matrix as osteocytes, undergo apoptosis or stay on the osteoid as resting bone-lining cells. The whole bone remodelling process is under control of certain endocrine, as well as paracrine and autocrine, signals and further requires the direct interaction of osteoblasts with osteoclasts and their precursors, respectively.[20,21]

In vivo bone remodelling process takes place manifold at different sites of bone, so that all of the time osteoblasts at different differentiation stages are present. In *in vitro* experiments the cells can be more or less synchronized by the plating procedure and progress together through an approximate 3/4 week differentiation process. As demonstrated in Figure 8.9, after adhesion (0–1 day) osteoblasts start to proliferate (day 1 until about day 3) and to synthesize ECM components (about day 3 decreasing until day 11).[22] In the differentiation phase, the cells express most of the bone-specific/typical parameters (TNAP and osteoproteins; starting with addition of osteogenic supplements, ongoing until approximately day 22). Afterwards, mineral accumulation increases (starting about 2.5–3 weeks after seeding).[23]

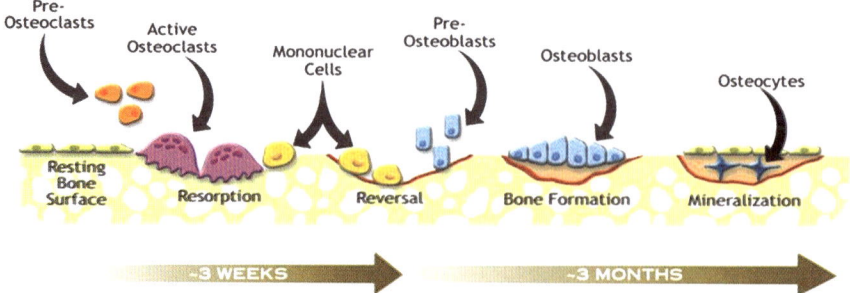

Figure 8.8   Scheme of *in vivo* bone remodelling and estimated time periods for different processes. (Adapted from: www.umich.edu/news/Releases/2005/Feb05/img/bone.jpg).

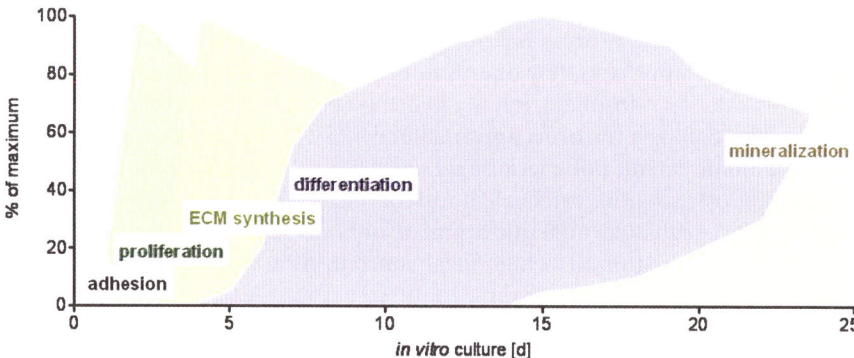

Figure 8.9   Scheme of *in vitro* osteogenic differentiation. With a slight shifting in the time axis, this scheme is valid for primary osteoblasts and osteoblast cell lines.

Depending on the source of primary cells or the preferred cell line, the differentiation process *in vitro* requires additives such as vitamin C to support collagen synthesis, an organic phosphate source as a substrate for TNAP, mainly β-glycerophosphate, and, for some cell lines, a synthetic glucocorticoid such as dexamethasone is mandatory. Glucocorticoids are mainly necessary in pre-osteoblasts/premature osteoblasts (cell lines) (*e.g.* bone marrow-derived osteoblasts, MG-63 cell line) for induction of specific gene expression and to initiate the differentiation process. MSCs, which have features of multi-potent stem cells, are a good source for osteoblasts.[24]

MSCs are available in sufficient number, proliferate *in vitro* and can differentiate into osteoblasts under established protocols.[25] This differentiation procedure, however, results in more or less 'artificial' osteoblasts, since dexamethasone induces other genes as well as osteogenic genes; the glucocorticoid also represses gene expression of, for example, procollagen, osteocalcin and osteoprotegerin, and activates expression of, for example, peroxisome

Table 8.1 TNAP activity of human MSCs (hMSCs) at day 15. Osteogenic differentiation was induced with 300 μM ascorbate, 10 mM β-glycerophosphate and 10 nM dexamethasone. Significance is indicated by [a] ($P < 0.001$); $n = 368$.

	Mean	S.E.M.	Minimum	Maximum
	(mU mg of protein^{-1})			
Uninduced hMSCs	131.5	± 132.9	1	885.5
Induced hMSCs	301.9[a]	± 233.8	2	1773

proliferator-activated receptor γ (PPARγ), an adipocyte-derived transcription factor.[26] MSCs dedifferentiate *in vitro*; after approximately 8–10 subcultures they lose their osteogenic differentiation competence. Another aspect has to be considered: primary cells reveal high donor-dependent differences in differentiation ability and also vary strongly in the expression levels of certain parameters. As an example, the variability in TNAP activity of approximately 350 donors is given in Table 8.1.

Primary adult osteoblasts can be easily obtained from spongiosa-derived bone pieces by outgrowth in culture medium; however, they are strongly limited in proliferative capacity. Although cell lines are much easier to handle and give better reproducible results, preference should be given to tissue-derived, species-specific primary cells and their early sub-cultures.

In summary, for the *in vitro* evaluation of osteogenic differentiation (*e.g.* response to substrates, exposure to substances, and the study of regulation or molecular mechanisms of differentiations) many cell lines and even primary cells are available. In order to obtain reliable and robust *in vitro* results, it is always important to collect, first of all, as much information as possible about the cells and to stay with the parameter of interest in the correct time window.

8.5 Impact of Charged Surfaces on Cell Behaviour *In Vitro* and *In Vivo*

Nearly a century ago, the first experiments were performed that revealed the piezoelectric properties of bone.[27] Electric potentials occur in bone tissue due to mechanical loading and are expected to guide/regulate the mechanical adaptation of bone. The detailed mechanism of this regulation is not yet understood. However, the piezoelectric properties of collagen in bone, as well as movement of ionic fluids, are reported to be involved in this process.

Thus new approaches in biomaterial development included the generation of electrical charges in implant materials to improve bone healing, utilizing the proposed stimulation of bone-forming cells by either polarizing an ionic material with the application of an external electrical DC field or by generation of charges due to mechanical stress in piezoelectric materials.

The reader is referred to a comprehensive review by Baxter *et al.*[28] concerning background and current knowledge derived from *in vitro* and *in vivo* cell behaviour. (See also Chapter 1 for more information on introducing charge in bioceramics.)

Due to the comparably new approach, information about the effects of electrically charged surfaces on proteins, cells or tissues are rare. Tarafdar *et al.*[29] analyzed adsorption and release of bovine serum albumin on different electrically polarized calcium phosphates (β-tricalcium phosphate or HAP were used). They found that the protein was bound in multilayers, with the highest amount on positively polarized surfaces. Bovine serum albumin was released from the positively poled surfaces, where increasing release was seen with decreasing the amount of stored charge.

Polarization of HAP increased not only the surface charge, but also the wettability and surface energy. Negatively polarized surfaces were able to attract calcium ions, which accelerated apatite nucleation, whereas positively polarized surfaces were found to suppress apatite nucleation by accumulation of chloride ions.[30] Thus it should be possible to direct apatite nucleation, surface properties and cellular reactions by the type of surface charge.[31]

In vitro experiments with electrically polarized biphasic calcium phosphates comprising β-tricalcium phosphate and HAP showed that negatively polarized surfaces supported osteoblast adhesion and spreading. Osteoblasts on those negatively polarized HAP materials proliferated faster and showed an enhanced ECM deposition.[32] Although two studies have confirmed the beneficial impact of negatively polarized surfaces,[31,33] Nakamura *et al.*[34] demonstrated that polarized HAP, independently of whether the surface was negatively or positively charged, induced an accelerated spreading and migration of osteoblasts (mouse cell line). However, negatively polarized HAP demonstrated slightly better results for spreading in comparison to the positively polarized surface. In their study of osteoblast attachment and

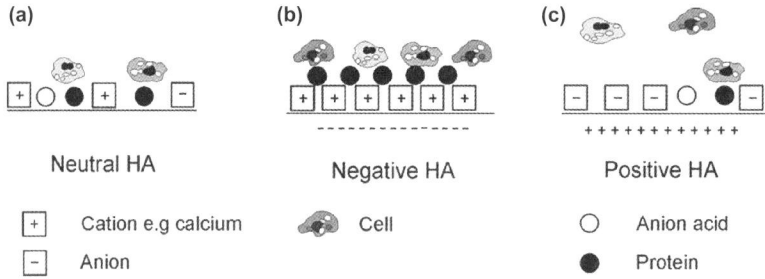

Figure 8.10   Schematic drawing of the interaction between the polarized surfaces and physiological solutions. (a) On the neutral material, inorganic ions, amino acids and proteins float and adsorb on HAP (HA). (b) On the negative material cations, particularly calcium, are selectively adsorbed and form a bone-like apatite layer. This apatite layer encourages the adsorption of proteins and the attachment of cells. (c) On the positive material, anions and negatively charged species are selectively adsorbed (from Baxter *et al.*[28]).

proliferation on polarized HAP, Kumar *et al.*[35] confirmed the finding that polarized HAP had a positive effect compared to unpoled HAP on these cellular processes, but the promotion of adhesion, attachment and proliferation was independent of the kind of charge. However, the opposite effect was reported by Kizuki *et al.*[36] who found increased osteoblast-like (MC3T3) cell numbers on positively polarized HAP surfaces compared to the negatively polarized surfaces after 2 days of cell culture.

In vivo experiments performed with polarized HAP/β-tricalcium phosphate showed an enhanced bone growth in comparison to unpolarized material after implantation in femoral diaphysis of rabbits.[37] Increased amount of bound fibrin were found on positively, as well as negatively, polarized surfaces when compared to the unpolarized surfaces. This increased fibrin layer was assumed to be responsible for improved osteoconduction originating from an increase in platelet adhesion and activation. These findings concur with data from Itoh *et al.*[38] who observed enhanced osteoblast activity on both the positively and negatively polarized HAP surfaces implanted in calvarial bones of rats that was accompanied by a decrease in osteoclast activity.

8.6 Conclusions

Despite some earlier investigations highlighting the preferential impact of negatively polarized surfaces on bone-related cells *in* vitro, more recent studies have revealed improved effects on bone healing-related events for both types of polarized surfaces *in vitro* and *in vivo* compared to untreated surfaces. Some explanations are given which include the different adsorption of ions and proteins, with a special emphasize on calcium attraction at negatively poled surfaces[28,33,39] and direct fibrin adsorption via electrostatic attraction of the – COOH groups[39] at the positively polarized surfaces (see Figure 8.10 for the proposed model[28]).

However, these models rely on assumptions that have not yet proven by surface analytics.

All of the mentioned studies revealed a significant positive impact on bone-related cells. Not a single one of these studies included solely heat-treated controls, although contact angle measurements were reported to result in decreased hydrophobicity due to the polarizing treatment.[30] The increase in wettability due to the overall treatment might account for some of the observed effects. Our own preliminary investigations support the theory of surface energy-related effects obtained by pure heating of β-tricalcium phosphate/HAP samples.

Furthermore, hardly any surface-related methods have been used to date for the characterization of polarized samples. The common thermally stimulated depolarization current (TSDC) measurements generate results only related to bulk properties. Hence more detailed research is necessary to understand the observed effects.

References

1. M. Cohen, D. Joester, B. Geiger and L. Addadi, *ChemBioChem*, 2004, **5**, 1393–1399.
2. S. P. Evanko, M. I. Tammi, R. H. Tammi and T. N. Wight, *Adv. Drug Deliv. Rev.*, 2007, **59**, 1351–1365.
3. E. Zimmermann, B. Geiger and A. L., *Biophys. J.*, 2002, **82**, 1848–1857.
4. O. Medalia and B. Geiger, *Curr. Opin. Cell Biol.*, 2010, **22**, 659–668.
5. L. A. Hidalgo-Bastida and S. H. Cartmell, *Tissue Eng. Part B Rev.*, 2010, **16**, 405–412.
6. E. Cukierman, R. Pankov and K. M. Yamada, *Curr. Opin. Cell Biol.*, 2002, **14**, 633–639.
7. K. M. Yamada, R. Pankov and E. Cukierman, *Braz. J. Med. Biol. Res.*, 2003, **36**, 959–966.
8. R. Zaidel-Bar and B. Geiger, *J. Cell Sci.*, 2010, **123**, 1385–1388.
9. P. Kanchanawong, G. Shtengel, A. M. Pasapera, E. B. Ramko, M. W. Davidson, H. F. Hess and C. M. Waterman, *Nature*, 2010, **468**, 580–584.
10. C. K. Miranti and J. S. Brugge, *Nat. Cell Biol.*, 2002, **4**, E83–E90.
11. J. B. Kim, P. Leucht, C. A. Luppen, Y. J. Park, H. E. Beggs, C. H. Damsky and J. A. Helms, *Bone*, 2007, **41**, 39–51.
12. R. M. Salasznyk, R. F. Klees, W. A. Williams, A. Boskey and G. E. Plopper, *Exp. Cell Res.*, 2007, **313**, 22–37.
13. R. M. Salasznyk, W. A. Williams, A. Boskey, A. Batorsky and G. E. Plopper, *J. Biomed. Biotechnol.*, 2004, **2004**, 24–34.
14. G. Tour, M. Wendel and I. Tcacencu, *Tissue Eng. Part A*, 2011, **17**, 127–137.
15. U. Hempel, T. Hefti, M. Kalbacova, C. Wolf-Brandstetter, P. Dieter and F. Schlottig, *Clin. Oral Implants Res.*, 2010, **21**, 174–181.
16. A. Scutt, A. Berg and H. Mayer, *Anal. Biochem.*, 1992, **203**, 290–294.
17. U. Fischer, U. Hempel, D. Becker, S. Bierbaum, D. Scharnweber, H. Worch and K. W. Wenzel, *Biomaterials*, 2003, **24**, 2631–2641.
18. M. V. Berridge, A. S. Tan, K. D. McCoy and R. Wang, *Biochemica*, 1996, **4**, 14–19.
19. Z. Xiao, C. E. Camalier, K. Nagashima, K. C. Chan, D. A. Lucas, M. J. de la Cruz, M. Gignac, S. Lockett, H. J. Issaq, T. D. Veenstra, T. P. Conrads and G. R. Beck, Jr, *J. Cell Physiol.*, 2007, **210**, 325–335.
20. B. F. Boyce and L. Xing, *Arch. Biochem. Biophys.*, 2008, **473**, 139–146.
21. S. Harada and G. A. Rodan, *Nature*, 2003, **423**, 349–355.
22. D. Becker, U. Geissler, U. Hempel, S. Bierbaum, D. Scharnweber, H. Worch and K. W. Wenzel, *J. Biomed. Mater. Res.*, 2002, **59**, 516–527.
23. M. P. Lynch, C. Capparelli, J. L. Stein, G. S. Stein and J. B. Lian, *J. Cell. Biochem.*, 1998, **68**, 31–49.
24. C. Nombela-Arrieta, J. Ritz and L. E. Silberstein, *Nat. Rev. Mol. Cell Biol.*, 2011, **12**, 126–131.
25. M. F. Pittenger, *Methods Mol. Biol.*, 2008, **449**, 27–44.
26. H. J. Kim, *BMB Rep.*, 2010, **43**, 524–529.

27. E. Fukada and I. Yasuda, *J. Phys. Soc. Jpn.*, 1957, **12**, 1158–1162.
28. F. R. Baxter, C. R. Bowen, I. G. Turner and A. C. Dent, *Ann. Biomed. Eng.*, 2010, **38**, 2079–2092.
29. S. Tarafder, S. Banerjee, A. Bandyopadhyay and S. Bose, *Langmuir*, 2010, **26**, 16625–16629.
30. S. Bodhak, S. Bose and A. Bandyopadhyay, *Acta Biomater.*, 2009, **5**, 2178–2188.
31. S. Bodhak, S. Bose and A. Bandyopadhyay, *Acta Biomater.*, 2010, **6**, 641–651.
32. S. Tarafder, S. Bodhak, A. Bandyopadhyay and S. Bose, *J. Biomed. Mater. Res. B Appl. Biomater.*, 2010, **97**, 306–314.
33. M. Ohgaki, T. Kizuki, M. Katsura and K. Yamashita, *J. Biomed. Mater. Res.*, 2001, **57**, 366–373.
34. M. Nakamura, A. Nagai, Y. Tanaka, Y. Sekijima and K. Yamashita, *J. Biomed. Mater. Res. A*, 2009, **92**, 783–790.
35. D. Kumar, J. P. Gittings, I. G. Turner, C. R. Bowen, L. A. Hidalgo-Bastida and S. H. Cartmell, *Acta Biomater.*, 2010, **6**, 1549–1554.
36. T. Kizuki, M. Ohgaki, M. Katsura, S. Nakamura, K. Hashimoto, Y. Toda, S. Udagawa and K. Yamashita, *Biomaterials*, 2003, **24**, 941–947.
37. S. Nakamura, T. Kobayashi, M. Nakamura, S. Itoh and K. Yamashita, *J. Biomed. Mater. Res. A*, 2010, **92**, 267–275.
38. S. Itoh, S. Nakamura, M. Nakamura, K. Shinomiya and K. Yamashita, *Biomaterials*, 2006, **27**, 5572–5579.
39. M. Nakamura, Y. Sekijima, S. Nakamura, T. Kobayashi, K. Niwa and K. Yamashita, *J. Biomed. Mater. Res. A*, 2006, **79**, 627–634.

CHAPTER 9

Interactions of Biofilm-forming Bacteria with Abiotic Surfaces

S. ROBIN,*[a] T. SOULIMANE[a] AND S. LAVELLE[b]

[a] University of Limerick, Department of Chemical and Environmental Sciences, National Technology Park, Limerick, Ireland; [b] Cook Ireland Ltd., O'Halloran Road, National Technology Park, Limerick, Ireland

9.1 Introduction

Microorganisms are capable of living in any environment where minimal conditions for life are encountered. One of the factors enabling them to do so is their ability to form biofilms – aggregates of microorganism adhered to each other and/or to the surface. This structure offers unicellular microorganisms a better chance of survival in poorly favourable or detrimental conditions. Formation of bacterial biofilm was reported in the early 1930s when Henrici revealed the existence of a sessile form of microbial development and described the basic behaviour of biofilm-forming bacteria.[1] However, it is only since the 1980s that biofilms have been regarded as a naturally preferable, most common mode of bacterial growth, whereas planktonic life is rather artificial or an intermediary stage in development of bacterial communities. Biofilms are found in all kinds of natural environments such as rocks, sand or even ice and play an important role in the ecology of our planet and the sustainability of life in general. They are also able to colonize biotic surfaces and, hence, biofilms can develop on leaves in plants or contaminate seeds where they can participate in nitrogen fixation in symbiotic relationships or cause crop diseases. In animals, biofilms naturally occur on teeth, forming dental plaques and causing tooth

RSC Nanoscience & Nanotechnology No. 21
Biological Interactions with Surface Charge in Biomaterials
Edited by Syed A. M. Tofail
© Royal Society of Chemistry 2012
Published by the Royal Society of Chemistry, www.rsc.org

decay and gum diseases, or in guts, where they participate in digestion. Biofilms play an important role in human society as they are, for example, widely employed in sewage treatment facilities. However, due to the resistance of biofilms to detergents, disinfectants, antibiotics and also to immunological defence, the ability of microorganisms to form biofilms causes a serious concern in industrial, domestic and hospital environments. Nevertheless, the biggest threat that the formation of biofilms poses is their involvement in 80% of all microbial infections in the body, which makes biofilms a primary health concern.[2] Many of those infections are related to the use of internal medical devices that offer a perfect support for the microbes to attach to and form a biofilm. The success of microorganisms as pathogens relies on their ability to adhere to a host surface, to develop as a community and to remain there by producing extracellular polysaccharides (EPS), a polymeric constituent that forms protective matrix that serves as a 'cement' for the biofilm.

This Chapter will focus on the mechanisms driving bacterial adhesion to abiotic surfaces and the formation of biofilms in bacterial development. In addition, the thermodynamics of bacterial adhesion, especially the importance of double-layer interactions, will be also discussed.

9.2 Biofilms in Bacterial Development

Although it is only now that we have started to understand biofilms at a molecular level, the cycle of their formation and the five major steps associated with it have been well described and are presented in Figure 9.1. The formation of every biofilm starts with the initial attachment of planktonic cells from the bulk environment to a solid surface. The transitionally anchored sessile cells form micro-colonies and secrete an extracellular matrix that will ensure the permanent attachment of the bacteria. Subsequently, the biofilm enters a phase of growth and maturation to finally reach a phase of dispersion, in which some cells return to a planktonic state to further colonize the environment.[3–5]

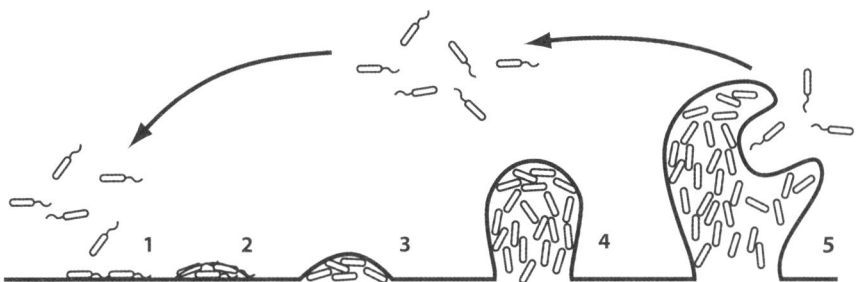

Figure 9.1 Stages of the biofilm development. Stage 1, initial adhesion; stage 2, irreversible attachment; stage 3, biofilm development; stage 4, maturation; stage 5, dispersion.

9.2.1   Adhesion

Microbial adhesion is the process of cell transfer from an unbound state in the bulk phase to a transitional attached state at an interface. It remains an elusive mechanism, dependent on a number of factors; however, it can be divided in two important stages. During the first stage, the bacterial cells can be assimilated to inert colloidal particles. They evolve in the bulk fluid and are subjected to Brownian motion and to physico-chemical mechanisms dictating their reversible adhesion to the abiotic surface. In a second step, bacteria consolidate this adhesion to a permanent attachment by triggering specific biological processes and/or by using specific ligands, such as pili or fimbriae that may complex with the surface. The main event is the production of the EPS matrix that constitutes the cement of the micro-colonies at first and subsequently of the entire biofilm.[6] Surface-related environmental signals and surface contact mediated by specific cell surface structures have been recognized to trigger changes that allow bacteria to undergo stable cell–surface interactions by initiating production of EPS.[7,8] For instance, the *Cpx* system in *Escherichia coli* reacts to stress stimuli caused by contact with hydrophobic surfaces and plays a key role in the adhesion-induced gene expression.[9] In *Pseudomonas aeruginosa*, the transcription of *algC*, a key gene involved in the biosynthesis of alginate required for the synthesis of EPS, is up-regulated 3- to 5-fold in adhered cells compared with their planktonic counterparts.[10] This idealistic view is, nevertheless, too simplistic as bacteria also employ mechanisms to actively promote their adhesion. It is known that bacteria can use their appendages to pierce an unfavourable thermodynamic barrier to promote adhesion.[11] Similarly, bacterial motility plays a role that cannot be underestimated,[12,13] especially for microorganisms showing a great mobility on surfaces, such as *Proteus mirabilis*.[14]

9.2.2   Proliferation of the Colonizers and Maturation of the Biofilm

Detailed observations of biofilms clearly show that bacteria within the biofilm multiply to form micro-colonies. Cells in those colonies are held together by the polysaccharide matrix forming the building blocks of the biofilm. These are subsequently assembled into larger structures that constitute the biofilm.[3] Mature biofilms of many species show structures that can be described as 'mushroom-like', with biofilms of *P. aeruginosa* or *P. mirabilis* being important examples.[15,16] Ultimately, the shape and thickness of the biofilm depend on the availability of nutrients[15] which are distributed into the matrix by diffusion through the water channels that are embedded in the mature biofilm.[3] As previously described for adhesion, a number of changes in the gene expression of microorganisms are triggered during the formation and maturation of biofilms such as, for example, elevated expression of genes involved in secretion of EPS.[17] They can be caused by the establishment of chemical gradients within the biofilm[15] and lead to the alteration of various metabolic pathways that are necessary to fit the requirements of the biofilm. The thick biofilm matrix in the biofilm mode of growth provides microbes with protection to survive hostile environments. This not only offers a shear stress protection, but, in addition, microorganisms embedded in a biofilm can survive concentrations of antiseptics or antibiotics more than 1000 times higher than the concentrations that can kill planktonic cells of the same species.[18]

9.2.3 Dissemination

The last step in the life cycle of the biofilm is the return of some cells to the planktonic state. Non-adhered and some adhered daughter cells escape from the EPS matrix layer, either by switching off the production of EPS or because of exhaustion conditions that support its production. Hence, they are free to drift to new colonization sites to repeat the process and thereby spread the colonization.[19,20] The phenomena involved in the detachment of cells populating the biofilm are poorly understood. It appears that dissemination of the biofilm is triggered by various sequential mechanisms, cross-species cell-to-cell communication and most likely inter-species quorum sensing.[21] However, several studies have shown that molecular mechanisms can be responsible for the weakening of the extracellular matrix that holds sessile cells. For example, *Streptococcus mutans* and *Staphylococcus aureus* produce specific surface proteases that degrades their own surface proteins,[22,23] which promote the release of the cells from the biofilm. Enzymes from other classes such as polysaccharidases, *e.g.* alginate lyase from *P. aeruginosa*,[22] and nucleases, *e.g.* thermonuclease from *S. aureus*,[22] have been also shown to participate in the degradation of the extracellular matrix and in the release of cells. In addition, pathogens that possess fimbriae or pili can regulate micro-colony cohesion and dispersal by structural modulation of those appendages, as found to take place in *E. coli*.[24,25] Finally, other mechanisms can be involved, such as for the very mobile *P. mirabilis* in which swarming-type cells can reach nutrients outside the biofilm.[15]

9.3 The Thermodynamics of Bacterial Adhesion

9.3.1 Extended Derjaguin–Landau–Verwey–Overbeek (DLVO) Theory

Although bacteria are diverse, complex and heterogeneous entities, due to their small size, they can be assimilated to colloidal particles. Therefore, the thermodynamics associated with such particles can be used to describe and understand the behaviour of bacteria towards inert surfaces.[26] In 1971, the DLVO theory of colloid stability was used to explain ionic strength effects on bacterial adhesion.[27] It presented the interaction between cell and surface as the sum of the van der Waals and Coulomb interactions. Later on, additional theories such as the cell surface hydrophobicity (CSH) have been also employed,[28] however, none of those approaches can fully explain bacterial adhesion. Currently, the extended DLVO theory, which considers hydrophobic/hydrophilic and osmotic interactions, as described by Van Oss *et al.*,[29] is regarded as the most comprehensive of all of the available models.

The first stage of the bacterial adhesion, where the cell is trapped in the secondary interaction minimum, is reversible. Then, the deposition develops, eventually, into an irreversible attachment through primary minimum interactions if the macroscopic potential barrier is overcome. According to the extended DLVO theory, the surface tension of colloid-sized particles, including microorganisms, is composed of an apolar (Lifshitz–van der Waals), a polar (Lewis acid–base) and an electrostatic component. Both, Lifshitz-van der Waals and hydrophobic interactions are short ranged, with the latter one becoming operative only if the distance between the interactive surfaces is less than 5 nm.[30] In contrary, double-layer interactions have a long range (>50 nm). The total adhesion energy can be hence expressed in eqn (1):

$$\Delta G^{adh} = \Delta G^{vdW} + \Delta G^{EL} + \Delta G^{LAB} \qquad (1)$$

where ΔG^{adh} is a total free energy of adhesion, ΔG^{vdW} is the Lifshitz–van der Waals component, ΔG^{EL} the double-layer interactions component and ΔG^{LAB} the Lewis acid–base component.

According to the extended DLVO theory, adhesion will occur if ΔG^{adh} is negative as interacting surfaces will tend to reach a state of minimal energy. ΔG^{vdW}, in microbial interactions, is almost always a negative value and therefore Lifshitz–van der Waals interactions are usually of a positive value. Although the hydrophobic interactions are very important in bacterial adhesion and are believed to be more influential than Lifshitz–van der Waals interactions and long-range electrostatic interactions, the importance of the latter should not be underestimated as numerous reports suggest their major role in bacterial adhesion.[31] Indeed, bacteria are first transported to the surface by the long-range interactions and then at closer proximity the short-range interactions become more predominant.[32]

9.3.2   Origins of Cellular Surface Charge

The cellular charge results from the presence of anionic groups (*e.g.* carboxylic, phosphoric and hydroxyl), as well as cationic groups (*e.g.* amine), on the cell surface. These functional groups are associated with peptidoglycan, teichoic acid and teichuronic acid on the surfaces of Gram-positive bacteria and with lipopolysaccharides (LPS), phospholipids and proteins on the surfaces of Gram-negative bacteria.[33,34] The effective net charge of the bacterium is influenced by the pH immediately adjacent to the cell surface which itself is a function of the cell surface electrostatic potential and bulk pH.[34] The latter dictates the protonation state of acidic and basic groups.[35] Table 9.1, compiled

Table 9.1 pK_a of cellular functional groups.

Reaction	Molecules	pK_a	Charge at pH 7
COOH → COO⁻ + H⁺	Polysaccharides	2.8	–
	Proteins, peptidoglycan	$4.0 < pK_a < 5.0$	–
$NH_3^+ \rightarrow NH_2 + H^+$	Proteins, peptidoglycan	$9.0 < pI_a < 11.0$	+
$HPO_4 \rightarrow PO_4^- + H^+$	Teichoic acid	2.1	–
$H_2PO_4 \rightarrow HPO_4^- + H^+$	Phospholipids	2.1	–
$HPO_4^- \rightarrow PO_4^{2-} + H^+$	Phospholipids	7.2	–/+

from Rijnards *et al.*,[36] shows the pKa for the functional groups and the molecules with which they are associated, as well as the net charge resulting from their protonation state at neutral pH.

As a result of the basic nature of most of the surface features, bacteria usually exhibit a negative net charge at neutral pH. Several strains have however been found to be exceptions to this rule and *Stenotrophomonas maltophilia*[35] serves here as an important example. Evidently, the net cellular charge varies not only with the species[37] and strains,[38] but also with the age of bacteria[39] and the growth or environmental conditions.[40]

9.4 Bacterial Surface Determinants of Bacterial Adhesion/Biofilm Formation

The surface of bacterial cells is a complex structure composed of various molecules that will play a particular role in the net charge of cells and therefore adhesion. The most important surface features participating to the cell charge in Gram-positive and -negative bacteria are briefly reviewed below.

9.4.1 Lipopolysaccharides (LPS)

LPS constitute a major component of the outer membrane of Gram-negative bacteria. They protect the membrane from the environment and, hence, contribute to the structural integrity of the cell. They consist of O-polysaccharides, core oligosaccharides and lipid A and they can be considered as charged polymer brushes grafted on to the cellular surface. The major charged groups are carboxylic and phosphate groups carried by the KDO (keto-deoxy-octolusonate) core of the LPS and the entire LPS, respectively, and they contribute mainly to the negative charge of the cell membrane. The variation of the length of LPS, which is modulated by ionic strength,[41,42] also plays an important role in bacterial adhesion.[43] Some organisms have shown to be able to modulate charges of these LPS in an unusual manner. For instance, *P. aeruginosa* produces two LPS forms, a neutral and hydrophobic band A and a negatively charged and hydrophilic band B.[42] The surface hydrophobicity is increased when the charged LPS contract with increasing ionic strength and the hydrophobic LPS are more exposed to the medium.

9.4.2   Peptidoglycan Layer and Lipoteichoic Acid (LTA)

Peptidoglycans are important constituents of the exposed cell wall of Gram-positive bacteria and can represent 20 to 30% of their total dry weight.[44] In contrast, in Gram-negative bacteria, the peptidoglycan is entrapped within the periplasmic space and is therefore not exposed to the extracellular environment. Thus, the existence of peptidoglycans influences the surface electronegativity of Gram-positive bacteria due to the exposed phosphoryl and carboxylic groups,[45] whereas they have little or no effect on the net charge of Gram-negative bacteria. LTA, another major constituent of the cell wall of Gram-positive bacteria, comprises teichoic acids and long chains of ribitol phosphate. It is anchored to the lipid bilayer via a glyceride and it can be up to 80 nm thick.[46] The net negative charge of LTA has been shown to play a key role in adhesion to abiotic surfaces and biofilm formation by *S. aureus*[47] and *Staphylococcus epidermidis*.[48]

9.4.4   Surface Layer (S-Layer)

The surface layer consists of a monomolecular, two-dimensional and 5–25 nm thick layer, comprising identical proteins or glycoproteins,[49] and it is commonly found in bacteria. When present, this usually positively charged protein layer increases the hydrophobicity of the cell surface.[50,51]

9.4.5   Polysaccharide Structures of Bacterial Envelope

More or less organized polysaccharide structures, capsule and slime layer, respectively, are produced by numerous bacteria, such as uropathogenic *E. coli*[52,53] or *S. aureus*.[54] They play a role in bacterial adhesion and irreversible attachment[13] due to their chemical and structural heterogeneity which provides multiple specific interactions with a surface.[55] Electrostatic interactions have been found to dominate the adhesive behavior of bacteria that overproduce EPS, *e.g.* colanic acid,[56] which can modulate bacterial adhesion and attachment by masking the charge/hydrophobicity of the cell surface.[57] However, there are examples where the EPS structures were found to decrease bacterial adhesion. This is due to long-range steric repulsion forces generated in the case of capsular colanic acid of *S. aureus*[54] or due to the blocking of the specific binding between cells and abiotic surfaces as is in the case of uropathogenic *E. coli*.[52]

9.4.6   Outer Membrane Proteins (OMPs)

OMPs are polypeptides embedded in the outer part of the cell wall of Gram-negative bacteria. They are constitutively expressed or are induced under certain growth conditions and can make up approximately half of the total outer membrane of the cell.[58] Their unspecific role in adhesion to abiotic surfaces is yet to be understood in detail. In general, however, they increase the overall hydrophobicity of the cell surface as, for example, in the case of the hydrophobic cell lytic enzyme AltE of *S. epidermis* that has been shown to be involved in initial adhesion to plastic and glass.[59,60] Additionally, members of the autotransporter subfamily of OMPs, including antigen 43,[61,62] the AIDA adhesin associated with some diarrheagenic *E. coli*[63] and the TibA adhesin/invasin from enterotoxigenic *E. coli*[64] play an active and specific role in bacterial adhesion. Recently, auto-agglutinins from *P. mirabilis*, such as PMI2122 and PMI2575, have been shown to promote bacterial deposition and biofilm formation by driving cell–cell aggregation.[65,66]

9.4.7 Bacterial Appendages: Flagella, Fimbriae and Curli

Flagella are rotating protein appendages that are rich in protein-associated carboxylic groups and therefore negatively charged.[40] They provide two of several kinds of bacterial motility,[13,67] allowing the cell to eventually reach a region of the substrate where attachment can occur.[13] Those helical filaments are also believed to increase the forces that hold cells on the surface.[57] The role of flagella in attachment has been demonstrated for *P. aeruginosa* and *Vibrio cholerae*.[17] Fimbriae are hair-like, several micrometers long protein structures with a radius of approximately 3–10 nm.[68] Their presence generally increase cell surface hydrophobicity and bacterial adhesion.[69] Curli are heteropolymeric, proteinaceous, amyloidic fibres[70,71] that are involved in bacterial adhesion to surfaces, cell aggregation and biofilm formation, as suggested by several reports.[71,72] However, other studies clearly indicate that there is no correlation between curli production, cell surface hydrophobicity and bacterial adhesion,[73] suggesting complexity of this issue that awaits to be elucidated.

9.5 Electric Properties of Abiotic Surfaces that Influence Bacterial Adhesion

Bacterial adhesion to inert surfaces constitutes a significant problem in a wide range of environmental, industrial and medical settings. Surface properties such as chemical composition,[74] hydrophobicity[75] and roughness or microtopography[57,75–80] are important factors affecting bacterial adhesion and motility. As mentioned previously, long-range electric double-layer interactions are one of the most predominant and influence the initial phase of bacterial adhesion on to solid surfaces.[81–86] These non-specific interactions usually cause the cell to be captured into the secondary minimum of interaction.[87,88] The occurrence of reversible adhesion[89] generally precedes biomolecular recognition, which leads to attachment of the cell to the surface.

Most of the surfaces, naturally occurring or manufactured, present an overall negative charge.[43,90,91] Glass surfaces have been found to be generally electronegative[35,43,47] with a zeta-potential stretching from –9 mV for a typical borosilicate microscope slide to –56 mV for a tin-modified glass.[43] Similarly, numerous metal oxides present a strong negatively charged surface,[43] whereas metals, such as stainless steel, are weakly electronegative.[90] Finally, synthetic polymers used in the production of medical devices, such as polyurethanes and silicones, in particular, are also electronegative.[91] In addition, most bacteria also show a negative surface charge at neutral pH with only few exceptions from this rule,[35] as described above. Naturally, an increasingly negative charge of the surface would be expected to result in increased repulsion of bacteria. However, the correlation between surface charge and adhesion is a much more complex phenomenon and experimental data often contradict the theory with numerous studies showing that bacterial adhesion to various surfaces is not significantly affected by the relative surface charge of bacteria.[92–94]

The engineering of surfaces in order to challenge and understand the role of electrostatic interactions in bacterial adhesion often shows discrepancies between DLVO or extended DLVO theories and experiments. Nevertheless, in some cases, the experimental results correlate with the thermodynamics theory. Interestingly, in one of these cases, Gottenbos *et al.*[95,96] discovered that, upon electropositive coating, bacterial adhesion was increased but the coating had a bacteriostatic effect by repressing micro-colony formation after adhesion. Also, the adhesive behaviour of *S. maltophilia,* which exhibits a positive zeta-potential over a wide range of pH values, complies with the DLVO theory.[35] The difficulty in correlating the theory and the actual behaviour of microorganisms when in contact with surfaces results from the fact that the thermodynamic models use entities presenting homogeneous surface properties. In reality, abiotic and biotic surfaces are extremely heterogeneous and this greatly contributes to the surface charge. In particular, the parameter used to estimate the surface potential of bacteria and materials, the zeta-potential, does not indicate the surface charges at the microscopic level, while the presence of localized positive charges on the surface of bacteria or abiotic surfaces is most likely a very important factor in cell adhesion. Recent studies focusing on surface charge heterogeneity unveiled the significance of local charges at a nanoscale level. Kalasin *et al.*[97] showed that the adhesion of the electronegative bacterium *S. aureus* to anionic surface can be strictly explained by electrostatic interactions by creating pDMAEMA {poly[(2-dimethylamino)ethyl methacrylate]} cationic nanopatches. Depending on the concentration of this cationic polymer, the zeta-potential of silica can be shifted from –80 mV to +15 mV.[87] At intermediate concentration, from 5% pDMAEMA upwards, although the overall surface charge is negative and unfavourable for bacterial adhesion, the presence of positively charged patches allows bacterial transitional interaction with the substrate.[97]

Similarly, bacterial surfaces are also highly heterogeneous with the components constituting the cell surface being individually influenced by the nature of the species and strains, the age and the physiological state of the cells,[39] as well as the growth conditions. In result, surface heterogeneities extend the matter of bacterial adhesion beyond predictions by thermodynamic theories.[89] Finally, the importance of localized charges has to be also considered in the context with ionic force, which is known to influence the potential of surfaces.[44] At low ionic strengths, the net surface charge prevails, while the localized charges become more influential with higher ionic strength as the Debye length decreases.

9.6 Conclusions

Biofilms are complex dynamic systems that have existed for more than 3 billion years and constitute an integral component of the prokaryotic lifestyle. The biofilm offers a protected mode of growth that not only allows microorganisms to colonize natural environments, but living hosts as well. Bacterial adhesion is the first and crucial step in surface colonization and biofilm formation, and it is mediated by an ensemble of physical and molecular interactions. Although our knowledge about this process is already substantial, existing thermodynamic models are often insufficient to accurately predict the behaviour of cells towards a surface. At present, it seems that a major cause of the failure of those models lies in the heterogeneity of both the cells and abiotic surfaces. Very recent studies on electrostatic interactions, clearly demonstrate that bacterial adhesion could be predicted using existing thermodynamic models only if the surface charge heterogeneity at the microscale level is considered. Furthermore, the available models reduce the microbe/abiotic surface arrangements to static systems, while the dynamic nature of microbial cells has to be accounted for as they can actively modulate their surface properties to allow adherence. Therefore, the development of efficient and versatile anti-biofilm treatments, in particular for medical applications, after decades of research, still poses an extreme challenge.

References

1. A. T. Henrici, *J Bacteriol*, 1933, **25**, 277–287.
2. L. Hall-Stoodley, J. W. Costerton and P. Stoodley, *Nat Rev Microbiol*, 2004, **2**, 95–108.
3. J. W. Costerton, *Int. J. Antimicrob. Agents*, 1999, **11**, 217–221; discussion, 199, **11**, 237–219.
4. M. Habash and G. Reid, *J. Clin. Pharmacol.*, 1999, **39**, 887–898.
5. G. Reid, *Int. J. Antimicrob. Agents*, 1999, **11**, 223–226; discussion, 1999, **11**, 237–229.
6. M. Hermansson, *Colloids Surf., B Biointerfaces*, 1999, **14**, 105–119.
7. C. Prigent-Combaret, O. Vidal, C. Dorel and P. Lejeune, *J Bacteriol*, 1999, **181**, 5993–6002.
8. J. P. Zhang and S. Normark, *Science*, 1996, **273**, 1234–1236.
9. K. Otto and T. J. Silhavy, *Proc. Natl. Acad. Sci. U.S.A.*, 2002, **99**, 2287–2292.
10. D. G. Davies, A. M. Chakrabarty and G. G. Geesey, *Appl. Environ. Microbiol.*, 1993, **59**, 1181–1186.
11. K. Hori and S. Matsumoto, *Biochemical Engineering Journal*, 2010, **48**, 424–434.

12. H. Morisaki, S. Nagai, H. Ohshima, E. Ikemoto and K. Kogure, *Microbiology*, 1999, **145**(10), 2797–2802.
13. R. M. Harshey, *Annu. Rev. Microbiol.*, 2003, **57**, 249–273.
14. M. M. Pearson, M. Sebaihia, C. Churcher, M. A. Quail, A. S. Seshasayee, N. M. Luscombe, Z. Abdellah, C. Arrosmith, B. Atkin, T. Chillingworth, H. Hauser, K. Jagels, S. Moule, K. Mungall, H. Norbertczak, E. Rabbinowitsch, D. Walker, S. Whithead, N. R. Thomson, P. N. Rather, J. Parkhill and H. L. Mobley, *J. Bacteriol.*, 2008, **190**, 4027–4037.
15. S. M. Jones, J. Yerly, Y. Hu, H. Ceri and R. Martinuzzi, *FEMS Microbiol. Lett.*, 2007, **268**, 16–21.
16. M. Klausen, A. Heydorn, P. Ragas, L. Lambertsen, A. Aaes-Jorgensen, S. Molin and T. Tolker-Nielsen, *Mol. Microbiol.*, 2003, **48**, 1511–1524.
17. G. O'Toole, H. B. Kaplan and R. Kolter, *Annu. Rev. Microbiol.*, 2000, **54**, 49–79.
18. J. N. Anderl, M. J. Franklin and P. S. Stewart, *Antimicrob. Agents Chemother.*, 2000, **44**, 1818–1824.
19. P. Tenke, C. R. Riedl, G. L. Jones, G. J. Williams, D. Stickler and E. Nagy, *Int. J. Antimicrob. Agents*, 2004, **23**(Suppl. 1), S67 S74.
20. R. Wang, B. A. Khan, G. Y. Cheung, T. H. Bach, M. Jameson-Lee, K. F. Kong, S. Y. Queck and M. Otto, *J. Clin. Invest.*, 2010, **121**, 238–248.
21. T. Abee, A. T. Kovacs, O. P. Kuipers and S. van der Veen, *Curr. Opin. Biotechnol.*, 2011, **22**, 172–179.
22. A. Boyd and A. M. Chakrabarty, *Appl. Environ. Microbiol.*, 1994, **60**, 2355–2359.
23. S. F. Lee, Y. H. Li and G. H. Bowden, *Infect. Immun.*, 1996, **64**, 1035–1038.
24. J. B. Kaplan, *J. Dent. Res.*, 2010, **89**, 205–218.
25. J. Sheikh, J. R. Czeczulin, S. Harrington, S. Hicks, I. R. Henderson, C. Le Bouguenec, P. Gounon, A. Phillips and J. P. Nataro, *J. Clin. Invest.*, 2002, **110**, 1329–1337.
26. M. N. Bellon-Fontaine, N. Mozes, H. C. van der Mei, J. Sjollema, O. Cerf, P. G. Rouxhet and H. J. Busscher, *Cell Biophys.*, 1990, **17**, 93–106.
27. K. C. Marshall, R. Stout and R. Mitchell, *Can. J. Microbiol.*, 1971, **17**, 1413–1416.
28. K. C. Marshall and R. H. Cruickshank, *Arch. Microbiol.*, 1973, **91**, 29–40.
29. C. J. Van Oss, R. J. Good and M. K. Chaudhury, *J. Colloid Interface Sci.*, 1986, **111**, 378–390.
30. R. Bos, H. C. van der Mei and H. J. Busscher, *FEMS Microbiol. Rev.*, 1999, **23**, 179–230.
31. M. C. M. Loosdrecht, W. Norde, J. Lyklema and A. J. B. Zehnder, *Aquat. Sci.*, 1990, **52**, 103–114.
32. Y. H. An and R. J. Friedman, *J. Biomed. Mater. Res.*, 1998, **43**, 338–348.
33. Y. Hong and D. G. Brown, *Colloids Surf., B: Biointerfaces*, 2006, **50**, 112–119.
34. Y. Hong and D. G. Brown, *Langmuir*, 2008, **24**, 5003–5009.
35. B. A. Jucker, H. Harms and A. J. Zehnder, *J. Bacteriol.*, 1996, **178**, 5472–5479.

36. H. H. Rijnaarts, W. Norde, E. J. Bouwer, J. Lyklema and A. J. Zehnder, *Appl. Environ. Microbiol.*, 1993, **59**, 3255–3265.
37. D. O. Ukuku and W. F. Fett, *J. Food Prot.*, 2002, **65**, 1093–1099.
38. L. Rivas, N. Fegan and G. A. Dykes, *J. Appl. Microbiol.*, 2005, **99**, 716–727.
39. S. L. Walker, J. A. Redman and M. Elimelech, *Environ. Sci. Technol.*, 2005, **39**, 6405–6411.
40. R. Briandet, T. Meylheuc, C. Maher and M. N. Bellon-Fontaine, *Appl. Environ. Microbiol.*, 1999, **65**, 5328–5333.
41. N. I. Abu-Lail and T. A. Camesano, *Environ. Sci. Technol.*, 2003, **37**, 2173–2183.
42. J. J. Shephard, D. M. Savory, P. J. Bremer and A. J. McQuillan, *Langmuir*, 2010, **26**, 8659–8665.
43. B. Li and B. E. Logan, *Colloids Surf., B: Biointerfaces*, 2004, **36**, 81–90.
44. A. van der Wal, W. Norde, A. J. B. Zehnder and J. Lyklema, *Colloids Surf., B: Biointerfaces*, 1997, **9**, 81–100.
45. W. W. Wilson, M. M. Wade, S. C. Holman and F. R. Champlin, *J. Microbiol. Methods*, 2001, **43**, 153–164.
46. C. Weidenmaier and A. Peschel, *Nat. Rev. Microbiol.*, 2008, **6**, 276–287.
47. M. Gross, S. E. Cramton, F. Gotz and A. Peschel, *Infect. Immun.*, 2001, **69**, 3423–3426.
48. M. Hussain, C. Heilmann, G. Peters and M. Herrmann, *Microb. Pathog.*, 2001, **31**, 261–270.
49. U. B. Sleytr, H. Bayley, M. Sara, A. Breitwieser, S. Kupcu, C. Mader, S. Weigert, F. M. Unger, P. Messner, B. Jahn-Schmid, B. Schuster, D. Pum, K. Douglas, N. A. Clark, J. T. Moore, T. A. Winningham, S. Levy, I. Frithsen, J. Pankovc, P. Beale, H. P. Gillis, D. A. Choutov and K. P. Martin, *FEMS Microbiol. Rev.*, 1997, **20**, 151–175.
50. K. Gruber and U. B. Sleytr, *Arch. Microbiol.*, 1991, **156**, 181–185.
51. H. C. van der Mei, B. van de Belt-Gritter, P. H. Pouwels, B. Martinez and H. J. Busscher, *Colloids Surf., B: Biointerfaces*, 2003, **28**, 127–134.
52. A. Hanna, M. Berg, V. Stout and A. Razatos, *Appl. Environ. Microbiol.*, 2003, **69**, 4474–4481.
53. A. N. Hassan and J. F. Frank, *Int. J. Food Microbiol.*, 2004, **96**, 103–109.
54. J. L. Prince and R. B. Dickinson, *Langmuir*, 2002, **19**, 154–159.
55. M. M. Sharma, Y. I. Chang and T. F. Yen, *Colloids Surf.*, 1985, **16**, 193–206.
56. A. Razatos, Y. L. Ong, M. M. Sharma and G. Georgiou, *Proc. Natl. Acad. Sci. U.S.A.*, 1998, **95**, 11059–11064.
57. C. Diaz, P. L. Schilardi, R. C. Salvarezza and M. F. de Mele, *Langmuir*, 2007, **23**, 11206–11210.
58. J. Lin, S. Huang and Q. Zhang, *Microbes Infect.*, 2002, **4**, 325–331.
59. C. Heilmann, C. Gerke, F. Perdreau-Remington and F. Gotz, *Infect. Immun.*, 1996, **64**, 277–282.
60. C. Heilmann, O. Schweitzer, C. Gerke, N. Vanittanakom, D. Mack and F. Gotz, *Mol. Microbiol.*, 1996, **20**, 1083–1091.

61. P. N. Danese, L. A. Pratt, S. L. Dove and R. Kolter, *Mol. Microbiol.*, 2000, **37**, 424–432.

62. K. Kjaergaard, M. A. Schembri, C. Ramos, S. Molin and P. Klemm, *Environ. Microbiol.*, 2000, **2**, 695–702.

63. O. Sherlock, M. A. Schembri, A. Reisner and P. Klemm, *J. Bacteriol.*, 2004, **186**, 8058–8065.

64. O. Sherlock, R. M. Vejborg and P. Klemm, *Infect. Immun.*, 2005, **73**, 1954–1963.

65. P. Alamuri, M. Lower, J. A. Hiss, S. D. Himpsl, G. Schneider and H. L. Mobley, *Infect. Immun.*, 2010, **78**, 4882–4894.

66. S. Robin, T. Soulimane, S. Lavelle and S. Tofail, *unpublished data*.

67. T. L. Jahn and E. C. Bovee, *Annu. Rev. Microbiol.*, 1965, **19**, 21–58.

68. A. H. Nobbs, R. J. Lamont and H. F. Jenkinson, *Microbiol. Mol. Biol. Rev.*, 2009, **73**, 407–450.

69. A. Zita and M. Hermansson, *Appl. Environ. Microbiol.*, 1997, **63**, 1168–1170.

70. J. H. Ryu, H. Kim, J. F. Frank and L. R. Beuchat, *Lett. Appl. Microbiol.*, 2004, **39**, 359–362.

71. M. M. Barnhart and M. R. Chapman, *Annu. Rev. Microbiol.*, 2006, **60**, 131–147.

72. A. L. Cookson, W. A. Cooley and M. J. Woodward, *Int. J. Med. Microbiol.*, 2002, **292**, 195–205.

73. R. M. Goulter, I. R. Gentle and G. A. Dykes, *Curr. Microbiol.* **61**, 157–162.

74. G. Speranza, G. Gottardi, C. Pederzolli, L. Lunelli, R. Canteri, L. Pasquardini, E. Carli, A. Lui, D. Maniglio, M. Brugnara and M. Anderle, *Biomaterials*, 2004, **25**, 2029–2037.

75. C. Faille, C. Jullien, F. Fontaine, M. N. Bellon-Fontaine, C. Slomianny and T. Benezech, *Can. J. Microbiol.*, 2002, **48**, 728–738.

76. J. Palmer, S. Flint and J. Brooks, *J. Ind. Microbiol. Biotechnol.*, 2007, **34**, 577–588.

77. K. Oliveira, T. Oliveira, P. Teixeira, J. Azeredo, M. Henriques and R. Oliveira, *J. Food Prot.*, 2006, **69**, 2352–2356.

78. G. Lerebour, S. Cupferman and M. N. Bellon-Fontaine, *J. Appl. Microbiol.*, 2004, **97**, 7–16.

79. C. G. Kumar and S. K. Anand, *Int. J. Food Microbiol.*, 1998, **42**, 9–27.

80. J. Verran, D. L. Rowe and R. D. Boyd, *J. Food Prot.*, 2001, **64**, 1183–1187.

81. M. C. van Loosdrecht, J. Lyklema, W. Norde, G. Schraa and A. J. Zehnder, *Appl. Environ. Microbiol.*, 1987, **53**, pp. 1898–1901.

82. M. C. van Loosdrecht, J. Lyklema, W. Norde, G. Schraa and A. J. Zehnder, *Appl. Environ. Microbiol.*, 1987, **53**, 1893–1897.

83. A. A. Mafu, D. Roy, J. Goulet and L. Savoie, *Appl. Environ. Microbiol.*, 1991, **57**, 1969–1973.

84. P. Gilbert, D. J. Evans, E. Evans, I. G. Duguid and M. R. Brown, *J. Appl. Bacteriol.*, 1991, **71**, 72–77.

85. J. Gannon, Y. H. Tan, P. Baveye and M. Alexander, *Appl. Environ. Microbiol.*, 1991, **57**, 2497–2501.
86. A. L. Mills, J. S. Herman, G. M. Hornberger and T. H. Dejesus, *Appl. Environ. Microbiol.*, 1994, **60**, 3300–3306.
87. N. Tufenkji and M. Elimelech, *Langmuir*, 2005, **21**, 841–852.
88. J. A. Redman, S. L. Walker and M. Elimelech, *Environ. Sci. Technol.*, 2004, **38**, 1777–1785.
89. S. L. Walker, J. A. Redman and M. Elimelech, *Langmuir*, 2004, **20**, 7736–7746.
90. S. Fukuzaki, H. Urano and K. Nagata, *J. Ferment. Bioengi.*, 1995, **80**, 6–11.
91. M. T. Khorasani, S. Moemenbellah, H. Mirzadeh and B. Sadatnia, *Colloids Surf., B: Biointerfaces*, 2006, **51**, 112–119.
92. A. H. Hogt, J. Dankert and J. Feijen, *J. Gen. Microbiol.*, 1985, **131**, 2485–2491.
93. G. Harkes, J. Feijen and J. Dankert, *Biomaterials*, 1991, **12**, 853–860.
94. A. Abbott, P. R. Rutter and R. C. W. Berkeley, *Journal of General Microbiology*, 1983, **129**, pp. 439–445.
95. B. Gottenbos, H. C. van der Mei, F. Klatter, D. W. Grijpma, J. Feijen, P. Nieuwenhuis and H. J. Busscher, *Biomaterials*, 2003, **24**, 2707–2710.
96. B. Gottenbos, D. W. Grijpma, H. C. van der Mei, J. Feijen and H. J. Busscher, *J. Antimicrob. Chemother.*, 2001, **48**, 7–13.
97. S. Kalasin, J. Dabkowski, K. Nusslein and M. M. Santore, *Colloids Surf., B: Biointerfaces*, 2010, **76**, 489–495.

CHAPTER 10

Endothelial Cells and Smooth Muscle Cells: Interactions at Biomaterials' Surfaces

M. WAWRZYŃSKA,[a] B. SOBIESZCZAŃSKA,[b] D. BIAŁY[a] AND J. ARKOWSKI*[a]

[a] Wrocław Medical University, Department of Cardiology, ul. Pasteura 4, 50-369, Wrocław, Poland; [b] Wrocław Medical University, Department of Microbiology, ul. T. Chałubińskiego 4, 50-368 Wrocław, Poland

10.1 Introduction

Endothelial cells (ECs) and smooth muscle cells (SMCs) are very important cell groups in vascular biology. ECs build the inner layer of the vessel wall. Their surface is the area of contact and barrier between blood and tissues. SMCs are the main cellular components of the mechanically important middle layer of the vessel wall. In small- and medium-sized arteries they build up a major portion of the strength of the vessel wall. Their reactivity is a very important factor in regulating blood flow. The relaxation of SMCs dilates the vessel and increases blood flow. Their contraction narrows the vessel thus decreasing the blood flow. In physiological conditions, SMCs in muscular arteries adjust the blood flow according to the demands of the tissue. If their reactivity is altered or lost, the vessels can no longer adapt to an increased oxygen demand.

RSC Nanoscience & Nanotechnology No. 21
Biological Interactions with Surface Charge in Biomaterials
Edited by Syed A. M. Tofail
© Royal Society of Chemistry 2012
Published by the Royal Society of Chemistry, www.rsc.org

Once a foreign body (a stent or a prosthetic device) is placed inside the blood vessel, it is put into direct contact with blood and vessel wall. ECs have a tendency to cover the foreign body, although the intensity of the process depends strongly on the material used. If the so-called re-endothelialization takes place, blood cells and proteins are no longer in contact with foreign substance.[1] This process strongly diminishes the risk of thrombosis. In physiological conditions chemical signal exchange between ECs and SMCs makes it possible for both tissues to function in a way that characterizes a healthy vessel. In some instances, a malfunction of ECs (for example caused by stent implantation) stimulates SMCs to migrate into the vessel lumen and to form a new layer. This cell layer is called neointima (as it was previously believed to originate from ECs). Unlike true endothelium, its cells are built in many layers and narrow the vessel lumen. The process described above is macroscopically visible as restenosis – a re-narrowing of the vessel wall after stent implantation.[2]

10.2 The Central Role of ECs in Vascular Biology

The vascular endothelium is often considered as an integral organ due to its mass, surface and the ability to act as an endocrine gland. The localization of ECs puts them in the first line of contact between the blood stream and surrounding tissue. ECs take part in the regulation of vasomotor reactions, haemostasis, angiogenesis, inflammatory processes and immunology. The endothelium also reacts to the changes in blood flow and blood pressure, and controls them. Mechanical damage or functional disorder of the endothelium is believed to be the first pathophysiological stage of hypertension, atherosclerosis and thrombosis.

The endothelium is composed of a single layer of cells, approximately 0.2–0.3 μm in thickness. The cells are elongated in the direction parallel to the blood flow. ECs are the only known surface that can support the liquid state of the blood in prolonged contact.[3] This is mainly due to the expression of anti-coagulation molecules such as prostaglandin I2 (PGI2), thrombomodulin, heparin sulfate and tissue plasminogen activator (t-PA) on the surface of the cells. On the luminal surface of endothelium several adhesion molecules are also situated. The most important of them are selectin E, selectin P, intercellular adhesion molecule 1 (ICAM-1), ICAM-2, platelet/endothelial cell adhesion molecule 1 (PECAM-1) and vascular cell adhesion molecule 1 (VCAM-1). These molecules are crucial in the endothelium–neutrophil interactions that initiate an inflammatory reaction.

The ECs secrete vasodilatory substances such as nitric oxide (NO) or PGI2. The most important vasoconstrictive factors are thromboxane A, angiotensin II and endothelin-1. Distorted equilibrium between those two groups of substances may provoke ischaemia in the part of tissue supplied by the vessel.

Healthy endothelium secretes NO, cyclo-oxygenase 2 and various anti-oxidants. Functional disorder of ECs is believed to be the first stage of atherosclerosis. The main factors impairing endothelial function are: the toxic ingredients of cigarette smoke, oxidative stress caused by free oxygen radicals, antibodies and some infections, glycated molecules in the course of diabetes and oxidated low-density lipoprotein (LDL) particles. The role of infection is currently disputed, but an inflammatory reaction in the course of many contagious diseases results in endothelial dysfunction.

NO is the key molecule for the vasodilatory function of ECs. There are several factors initiating NO release. Among the most important ones are shear stress, bradykinin, histamine, adenosine, vascular endothelial growth factor (VEGF), thrombin, substance P, natriuretic factor A and serotonin. Not only does NO provoke the relaxation of vascular wall, it also inhibits SMC proliferation and synthesis of extracellular matrix (ECM). It also decreases the expression of adhesion molecules thus slowing down the inflammatory reaction. Prostacyclin A is another important vasodilatory molecule. It impairs platelet thrombus formation. There are also other, less important, NO-independent vasodilatory molecules such as cytochrome-derived particles or natriuretic peptide C.

When the availability of NO decreases, the endothelium starts to produce vasoconstrictory molecules. Cells in the endothelium began to convert angiotensin I into angiotensin II, a process which leads to vascular contraction, platelet activation and leukocyte adhesion. A healthy endothelium does not allow leukocyte adhesion. The increased expression of adhesive molecules, such as VCAM-1, makes it possible for mononuclear leukocytes to adhere to ECs.

Immunological reactions play a very important role in the development of atheroma. Oxidated LDLs and heat-shock proteins are the most important antigens that stimulate an immunological response. The presentation of these antigens provokes synthesis of pro-inflammatory molecules, adhesion proteins and metalloproteinases.

All the mechanisms mentioned above (vasodilation/vasoconstriction disequilibrium, leukocyte adhesion and altered immunological reactions) are crucial in the development of atherosclerosis. The consequences or late complications of atherosclerosis – such as myocardial infarction or stroke – are among the most frequent causes of death in modern developed societies. In recent years physicians have begun to appreciate that one needs to understand endothelial function in order to successfully treat these life-threatening diseases. It is now believed that almost all of the most important groups of drugs used in the treatment of ischaemic heart diseases have some influence on the endothelial function. For example, most of the so-called 'modern' beta-blockers or angiotensin-converting enzyme (ACE) inhibitors have great effects in improving the survival of the patients, even at doses that provoke only minor changes in heart rate or blood pressure. Similarly, the 'pleiotropic action' of statins (cardiovascular risk reduction and improved patient survival) is independent of the lipid- lowering action of these drugs. Improved endothelial function and the slower progression of atherosclerosis are postulated explanations for these effects.

10.3 Endothelial Cell Culture as a Model for Vascular Function Studies

10.3.1 Endothelial Cell Culture Models

ECs occur as various species of different anatomical origin. In designing an investigation of biomaterial–endothelium interactions, it is of great importance to consider the appropriateness of the methodology for a given application.[3] There are currently four types of ECs commercially available: bovine, human, porcine and rat. Protocols for harvesting canine, sheep and mouse ECs have been reported also. Human umbilical vein endothelial cells (HUVECs) are considered the most useful for vascular implant testing. HUVECs are thought to be less thrombotic than other human ECs (i.e. atrial or omentum) as they do not show surface tissue factor expression. Human saphenous vein endothelial cells (HSVECs) and human omentum microvascular endothelial cells (HOT-MECs) also show a good correlation with canine models to test vascular graft patency. For tissue engineering in cardiac tissue regeneration and neovasculature formation, cardiovascular endothelial progenitor cells (EPCs) are of great interest. Also, EPCs are used in combination with surface-immobilized anti-CD34 antibodies for the study of the cardiovascular stent re-endothelialization process.[4]

In the *in vivo* state, ECs present a heterogeneous group, depending on the place in the vascular tree and surrounding tissue.[5] For instance, large vessel ECs are surrounded by SMCs and the ECs of microvessels are surrounded by pericytes. The growth, differentiation, physiological barrier function and the paracrine modulation of the vascular wall of ECs are all dependent on the surrounding structures. This also applies to cell–cell contacts (see Section 10.4), protein–ECM interactions and the luminal shear forces during blood flow. Shear forces, for example, may alter the endothelial cytoskeleton, gene expression and the production of surface adhesion molecules.[6] Appropriate flow conditions are necessary for any *in vitro* study to enable an adequate evaluation of the functions of ECs.

10.3.2 Endothelial Interactions with ECM Proteins

One of the most important parameters of biomaterial surface–EC interactions is the assessment of proliferation and the rate of adhesion.[7] Various studies of biomaterials' surface covered with ECM proteins (collagen and fibronectin) have provided information on cellular behaviour. Earlier experiments with ECs cultured on polystyrene [tissue culture grade polystyrene (TCPS)] raised the question about distinguishing the influence of ECM proteins from the cellular shapes on cell behaviour. Studies have shown that an optimal cell shape and spreading can be reached by varying the ECM protein concentrations independently of the type of ECM protein.[8] When different concentrations of fibronectin/collagen were used, no differences in cell morphology were observed between different coatings.[9] In further studies with different TCPS coatings the above strategy was also valid.

10.3.3 Fluid Shear Stress

It has been noticed that atherosclerotic lesions occur most frequently at bends, angulations and bifurcation of the arteries. Arteries with few side branches rarely develop atherosclerotic lesions. This observation led to the conclusion that altered blood flow at these locations facilitates the development of atherosclerotic plaques.[10] Laminar blood flow protects from atheroma development; it promotes the expression of superoxide dismutase and NO synthase. Turbulent blood flow creates shear stress that increases the expression of nuclear factor κB factor (NF-κB), which, in turn, increases the production of VCAM-1. It also increases the production of the adhesion molecules and tissue factors, which initiate the coagulation process. High shear stress inhibits the transcription factor Krueppel-like factor 2 (KLF2; a molecule essential for the expression of NO synthase and thrombomodulin).[11] This process also promotes the expression of adhesion molecules. It is now believed that the mechanical stimuli from the shear stress causes endothelial dysfunction in selected areas and is the first step towards atheroma development.[12]

10.4 Molecular Basis of Restenosis

10.4.1 Stent-induced Inflammatory Response

The response of the vascular wall to stent placement is a foreign body reaction initiated by an immediate inflammatory response.[4] The stented section of the artery is characterized by pressure-induced tissue damages, such as the compression of the atherosclerotic plaque (with possible dissections of arterial wall), de-endothelialization and expansion of the vessel diameter. On the de-endothelialized implant surface a fibrin layer is deposited with activated platelets expressing adhesion molecules for inflammatory cells such as leukocytes. Leukocytes adhere to implant surface first through binding receptors such as P-selectin glycoprotein ligand-1 (PSGL-1) and then they form a tight junction to the surface through the leukocyte integrin Mac-1 (CD11b/CD18) attached to platelet receptors. Finally, they transmigrate across surface-adherent platelets to the vessel wall where they induce inflammation. It is generally postulated that Mac-1 is an important signalling molecule in the process of restenosis. In an experimental angioplasty and stenting study, it was shown that inhibiting Mac-1 production reduced neointimal thickening after the procedure.[13] The inflammatory response is accompanied by a significant infiltration of monocytes into arterial tissues surrounding the stent. Activated monocytes secrete various pro-inflammatory chemokines such as interleukin 6 (IL-6), IL-1β, and monocyte chemotactic protein 1 (MCP-1) which stimulate the proliferative response of SMCs.

The chemical properties of stent alloys may increase inflammatory response. The most widely used material is stainless steel 316L. During corrosion of 316L, metal ions are released into the surrounding tissue. This was shown to alter cellular function. Messer *et al.*[14] showed that exposure to stainless steel leads to monocyte activation and promotes pro-inflammatory chemokine secretion.

10.4.2   SMC Proliferation

The migration of cells from tunica media to intima and the proliferation of vascular SMCs constitute the basic processes of in-stent restenosis.[15] Under normal conditions, vascular SMCs are quiescent and exhibit low levels of proliferative activity. Percutaneous coronary intervention (PCI)-induced mechanical injury triggers G1/S transition in the SMC cell cycle.[1] Platelet activation after endothelial denudation and subsequent release of the chemotactic mitogen platelet-derived growth factor (PDGF) are also important factors in promoting inward migration of SMCs. Migration of SMCs is regulated by the cell cycle. SMCs are only able to migrate on mitogenic stimuli in phase G1, not in later phases of the cell cycle. Different phases of the cell cycle are regulated by protein complexes such as cyclin-dependent kinases (CDKs) and their cyclin-dependent inhibitors (CKIs), such as $p27^{Kip1}$ and $p21^{Cip1}$. The up-regulation of inhibitors leads to cell cycle arrest, whereas down-regulation increases proliferation. In early phases of the healing response, $p27^{Kip1}$ is down-regulated, which causes an increase in SMC cell proliferation. In later phases, $p27^{Kip1}$ is up-regulated and a significant decline in cell proliferation is seen.

10.4.3   Vascular Repair after Stent Implantation: Role of Re-endothelialization

Implantation of a stent results in damage to endothelium, which triggers proliferation of SMCs until the endothelium is regenerated. Endothelial inhibition of SMCs is an essential process for maintaining arterial health and its loss is a precursor to restenosis and arterial thickening. If the growth of ECs is slower than that of SMCs, the latter enters a proliferative state that can eventually lead to restenosis. Thus a desirable stent should be a better substrate for ECs than for SMCs. Immediate re-endothelialization is a hallmark of good biocompatibility of an endovascular implant. Recent studies show that some molecules such as heparan sulfate proteoglycans (HSPGs) are potent regulators of vascular remodelling and repair.[2] Heparanase is the major enzyme capable of degrading heparan sulfate in mammalian cells. *In vitro* studies showed that heparanase expression in ECs serves as a negative regulator of endothelial inhibition of vascular SMC proliferation. Arterial structure and remodelling after injury were also modified by heparanase expression. Baker *et al.* have shown that HSPGs are essential for endothelial inhibition of vascular SMC proliferation.[24] In a study on the response to drug-eluting stents, endothelial recovery was observed more rapidly between stent struts. ECs over the stent struts did not reached their mature bioregulatory phenotype, as indicated by increased expression of phospho-S6 ribosomal protein (p-S6RP). In healthy conditions p-S6RP expression on ECs does not occur. Stents struts remain uncovered and in direct contact to blood flow, thus increase the risk of in-stent thrombosis.

Stent-induced vascular injury and inflammation also induces the release of progenitor stem cells from bone marrow into the blood stream. EPCs contribute to stent surface re-endothelialization and vascular repair.[16] On the other hand, smooth muscle progenitor cells (SMPCs) are also postulated to be recruited to stented areas to differentiate into SMCs, proliferate and form intimal hyperplasia. The mediator molecules in both processes are yet to be defined.

10.5   Endothelial Cell Culture on Different Vascular Prosthetic Surfaces

Many studies have demonstrated that titanium and titanium oxides inhibit platelets aggregation and fibrin deposition, as well as promoting and accelerating EC growth. According to Windecker *et al.*[17] titanium–NO coatings reduced neointimal hyperplasia by approximately 50% in comparison with 316L stainless steel. Methe *et al.*[18] demonstrated that levels of endothelialization on TiN and TiO_2 sheets were comparable or even higher to the cells growing on tissue culture-treated polystyrene dishes coated with gelatin.

To improve biocompatibility of biomaterials many strategies are being explored, *i.e.* nanoscale surface modifications. It has been found that nanostructured surfaces greatly improve cellular adhesion. According to Wood *et al.*,[5] sub-micron to nanotopographically patterned surfaces affected gene expression significantly in ECs as compared with flat surfaces. Lu *et al.*[7] showed increased vascular EC adhesion on random-nanostructured titanium surfaces compared with conventional titanium surfaces.

In our study on silica titania sol-gel coatings on 316L stainless steel, partially as pure silica (T13 film) and partially as pure titania material (T15 film) depending on the molar ratios of the substrates, which were non-irradiated or pre-irradiated with blue laser light (λ_{405} CW; light energy, 117.6 mJ cm^{-2}), enhanced EC and SMC attachment was seen in comparison to the control 316L stainless steel surface (Figure 10.1). However, no significant differences in the number of attached ECs and SMCs were observed on the pre-irradiated surfaces, indicating that pre-irradiation had no effect on cell attachment. Similar results were obtained in the proliferation experiments in which EC and SMC proliferation rates on the non-irradiated or pre-irradiated surfaces were comparable.

10.6   SMC Interactions on Different Vascular Prosthetic Surfaces

SMCs play an important role in atheroma development. SMCs migrate from the media into the intima under the influence of PDGF, endothelin 1, thrombin and angiotensin II. Some SMCs can proliferate in the intima. If the process is

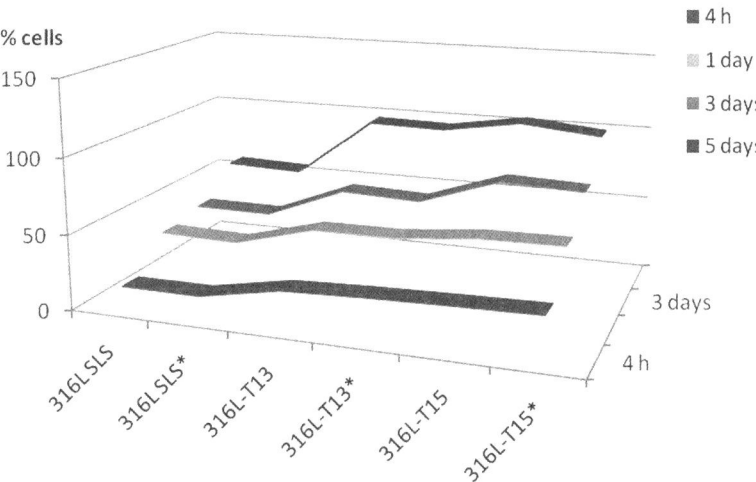

Figure 10.1 Comparison of EC (HUVECs) and SMC (T/GHAUSMC) attachment to and proliferation on the pre-irradiated and non-irradiated with blue laser light silica titania sol-gel pure silica (T13) and pure titania (T15) films on 316L stainless steel (316L SLS) in comparison with 316L SLS surface. *Pre-irradiated surface.

accelerated the atheroma grows. The phenotype of the SMCs situated in intima layer is less mature, which makes them more embryonic-like. The bursts of SMC replication after vessel wall injury are believed to be the main mechanism of atheroma growth. The vessel wall injury during stenting provokes SMC to migrate onto the implant surface and to proliferate.[19] The migration of SMCs on to prosthetic surface is the principal mechanism of restenosis. The restenotic tissue (also called myxomatous tissue) comprises star-like SMCs and a loose highly hydrated ECM. Therapeutical interventions that reduce the number of SMCs and stop their proliferation are efficient in treating restenosis. Such a mechanism of action has been shown for both brachytherapy and *in situ* drug release.

In our study we have investigated the attachment and proliferation of SMCs on two types of films: titania-doped silica (T13) and pure titania (T15) films which were photoactivated with blue laser light (λ_{405} CW; light energy, 117.6 mJ cm^{-2}). Enhanced SMC attachment and proliferation was observed with both of these coatings when compared with the 316L stainless steel surface. There were no differences in the attachment and the proliferation ratio of SMCs after photoactivation of the titania sol–gel coating in comparison with non-photoactivated titania surfaces (Figure 10.2).

10.7   Interactions of ECs with SMCs

The resident blood vessel cell types, ECs and SMCs, are engaged in various interactions. Such cross-talk between cell types is an important factor for vessel

T/GHAUSMCs attachment and proliferation

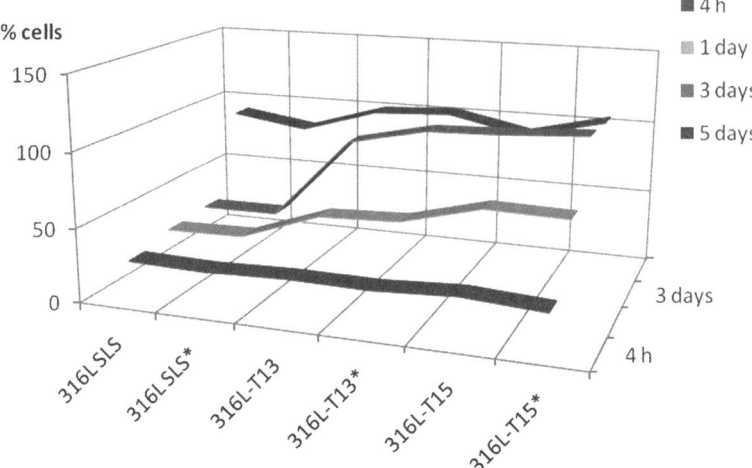

Figure 10.2   SMC (T/GHAUSMC) attachment to and proliferation on the photo-activated and non-photoactivated with blue laser light silica titania sol-gel pure silica (T13) and pure titania (T15) films in comparison with 316L SLS surface. *Photoactivated surface.

wall haemostasis regulation and post-injury repair processing. By paracrine modulation, each cell type may induce or suppress phenotypic transformation or bioregulatory state changes in the other. Some co-culture studies performed to date on mutual interactions shed more light on the pathophysiology of post-angioplasty wound healing in the vessel wall. In healthy conditions, ECs form an internal monolayer of the vessel wall that shields SMCs from forming direct contact with the blood flow. After stent-induced injury, ECs can modulate the growth, migration, proliferation and contractile function of SMCs. On the other hand, SMCs can influence re-endothelialization of the stent surface.

It was found that the cell-to-cell adhesion molecule N-cadherin mediated interactions between ECs and SMCs to promote vascular stability. A hypothesis was raised that ECs may suppress inward migration of SMCs by presenting N-cadherin.[20] Therefore post-injury de-endothelialization may be a trigger for SMC inward migration from the arterial tunica media to the intima. It was shown that smooth muscle migration is initiated with polarized reorganization of the cell cytoskeleton. Moreover, in the same study it was shown that migrating SMCs lose their polarized phenotype when they make contact N-cadherin-expressing ECs. Therefore, loss of ECs may be an important factor leading to inward SMC migration and initiation of processes that lead to neointima formation and restenosis. This finding indicates the importance of endothelial-to-smooth muscle signalling for maintaining vascular wall stability. The above findings lead to the conclusion that N-cadherin-targeted therapy may be a way to combat restenosis by inhibiting SMC recruitment from the tunica media to the stent area.

The p-S6RP/vascular mammalian target of rapamycin (mTOR) pathway is critical for vascular reactivity and metabolism. Thus its function is crucial for endothelial repair after injury. Phosphorylation of the S6RP and mTOR pathways in ECs is activated by luminal flow. It was shown that EC S6RP signalling is mediated by adjacent SMCs, and SMCs inhibit flow-mediated mTOR activation in ECs.[12] *In vitro* observations were confirmed in *in vivo* studies on stented porcine arteries. In bare metal stents, p-S6RP expression and phosphorylation was greatest in ECs farthest from intact SMCs. In drug-eluting stents after sirolimus elution, phosphorylation in ECs was absent. The above observations are in accordance with clinical findings of impaired re-endothalialization of drug-eluting stents.[21,22] Wallace *et al.*[23] reported that ECs cultured on fibronectin-coated plastic plates spread in significantly higher rate compared with EC spreading on SMCs. The study found differences in the focal complex formation. It was observed that the ECs that adhered to fibronectin-coated plastic had formed focal adhesions, whereas the ECs in the co-culture lacked focal adhesions and had a reduced spreading rate when compared with those adhered to polyacrylamide gels of similar stiffness. The ECs in co-culture attached to SMCs through the $\alpha5\beta1$ integrin complex. ECs attached to SMCs can resist shear stress as high as 300 dyn cm^{-2}.

Interestingly, when co-cultured on different sides on the membrane, and not in contact with each other, SMCs were shown to promote endothelial adhesion and increase the proliferation rate by altering EC function. Co-cultured SMCs induced focal EC adhesion through microtubule depolymerization and activation of the paxilin pathway.

Paracrine and direct contact-mediated SMC and EC interactions need to be further defined. These observations have led to the conclusion that the vascular responses to stent-induced injury, as well as drug-eluting and bare metal stents, are regulated in a complex manner. One has to remember that cardiovascular stents are placed in atherosclerotic plaques with ongoing inflammation and an altered vascular geometry. The presence of other cells, as well as the altered pathophysiologycal conditions, may also affect the interactions between ECs and SMCs. Only few of the existing *in vivo* interactions can be reproduced consistently in *in vitro* models.

10.8   Conclusions

ECs and SMCs are the key elements in the pathogenesis and treatment of vascular disease. We are beginning to understand their intricate interactions. Dysfunction of the arterial endothelium is the first stage of atherosclerosis, but it is the migration and proliferation of other cells (including SMCs) that forms the plaque. Our current understanding of in-stent restenosis is based on the concept of vessel wall injury, de-endothelization and exposure of SMCs from tunica media. Endothelialization of the implant ensures healing of the intervention site. The migration of SMCs into the vessel lumen, their proliferation and the secretion of ECM are the key elements of restenosis. It is the delicate play between those two types of cells that needs to be understood and, ideally, also regulated in order to make the effects of intravascular interventions long-lasting.

References

1. A. Shahryari, F. Azari, H. Vali and S. Omanovic, *Acta Biomater.*, 2010, **6**, 695–701.
2. B. Balakrishnan, A. R. Tzafriri, P. Seifert, A. Groothuis, C. Rogers and E. R. Edelman, *Circulation*, 2005, **111**(22), 2958–2965.
3. F. Zhang, Z. Zheng, Y. Chen, X. Liu, A. Chen and Z. Jiang, *J. Biomed. Mater. Res.*, 1998, **42**, 128–133.
4. A. V. Finn, G. Nakazawa, M. Joner, F. D. Kolodgie, E. K. Mont, K. Herman, H. K. Gold and R. Virmani, *Arterioscler. Thromb. Vasc. Biol.*, 2007, **27**, 1500–1510.
5. J. A. Wood, S. J. Liliensiek, P. Russell, D. A. Stephan, P. F. Nealey and J. Murphy, *Materials*, 2010, **3**, 1610–1639.
6. P. F. Davies, *Nat. Clin. Pract. Cardiovasc. Med.*, 2009, **6**(1), 16–26.
7. J. Lu, M. P. Rao, N. C. MacDonald, D. Khang and T. J. Webster, *Acta Biomater.*, 2008, **4**, 192–201.
8. P. A. Underwood, P. A. Bean and J. M. Whitelock, *Atherosclerosis*, 1998, **141**(1), 141–152.
9. D. G. Della Rocca and C. J. Pepine, *Clin. Cardiol.*, 2010, **33**(12), 730–732.
10. S. Xu, Y. He, M. Vokurkova and R. M. Touyz, *Hypertension*, 2009, **54**, 427–433.
11. E. Fine, L. Zhang, H. Fenniri and T. J. Webster, *Int. J. Nanomedicine*, 2009, **4**, 91–97.
12. J. Steffel, R. A. Latini, A. Akhmedov, D. Zimmermann, P. Zimmerling, T. F. Lüscher and F. C. Tanner, *Circulation*, 2005, **112**, 2002–2011.
13. Y. H. Wang, Z. Q. Yan, B. R. Shen, L. Zhang, P. Zang and Z. L. Jiang, *Eur. J. Cell Biol.*, 2009, **88**, 701–709.
14. S. Windecker, R. Simon, M. Lins, V. Klauss, F. R. Eberli, M. Roffi, G. Pedrazzini, T. Moccetti, P. Wenawesser, H. Togui, D. Tuller, R. Zbinden, C. Seiler, J. Mehilli, A. Kastrati, B. Meier and O. M. Hess, *Circulation*, 2005, **111**, 2617–2622.
15. M. Balcells, J. Martorell, C. Olivé, M. Santacana, V. Chitalia, A. A. Cardoso and E. R. Edelman, *Circulation*, 2010, **121**, 2192–2199.
16. Q. Lin, X. Ding, F. Qiu, X. Song, G. Fu and J. Ji, *Biomaterials*, 2010, **31**, 4017–4025.
17. S. Windecker, M. Billinger and O. M. Hess, *EuroIntervention.*, 2006, **2**(2), 146–148.
18. H. Methe, M. Balcells, M. Alegret, M. Santacana, B. Molins, A. Hamik, M. K. Jain and E. R. Edelman, *Am. J. Physiol. Heart Circ. Physiol.*, 2007, **292**, H2167–H2175.
19. T. Inoue and K. Node, *Circ. J.*, 2009, **73**, 615–621.

20. P. J. Sabatini, M. Zhang, R. Silverman-Gavrila, M. P. Bendeck and B. L. Langille, *Circ. Res.*, 2008, **103**(4), 405–412.

21. J. A. S. Muldowney, J. R. Stringham, S. E. Levy, L. A. Gleaves, M. Eren, R. N. Piana and D. E. Vaughan, *Arterioscler. Thromb. Vasc. Biol.*, 2007, **27**, 400–406.

22. G. G. Camici, J. Steffel, I. Amanovic, A. Breitenstein, J. Baldinger, S. Keller, T. F. Luescher and F. C. Tanner, *Eur. Heart J.*, 2010, **31**, 236–242.

23. C. S. Wallace, S. A. Strike and G. A. Truskey, *Am. J. Physiol. Heart Circ. Physiol.*, 2007, **293**(3), H1978–H1986.

24. A. B. Baker, A. Groothuis, M. Jonas, D. S. Ettenson, T. Shazly, E. Zcharia, I. Vlodavsky, P. Seifert and E. R. Edelman, *Circ. Res.*, 2009, **104**, 380–387.

CHAPTER 11

Interactions of Bacteria and Fungi at the Surface

E. DWORNICZEK,* R. FRANICZEK, U. NAWROT AND G. GOŚCINIAK

Wrocław Medical University, Department of Microbiology, Chałubiński Street 4, Wrocław, 50-368, Poland

11.1   Introduction

There is an ever-increasing likelihood of hospital acquired infections due to the modern-day lifestyle with the increasing number of long-term hospitalized, immuno-compromised patients and the usage of medical implants and devices. In many cases they are caused by microorganisms growing on the surface in sessile communities (biofilms) that have been regarded as tissue-associated and medical device-related infections. Examples include periodontitis, osteomyelitis, native heart valve endocarditis, prostatitis and infections that results from colonization of implanted peacemakers, prosthetic joints, voice prosthesis, mechanical heart valves etc.[1,2] Although significant progress has been made in the detection and control of microbial adhesion, and the associated pathology, novel approaches of prevention are crucial. For this, it is important to know how microorganisms behave and grow on solid surfaces.

11.2   Clinical Pathogens and Medical Device-related Infections

Medical devices such as catheters, defibrillators and ventilators have become an essential part of the management in most hospital units. However, their

RSC Nanoscience & Nanotechnology No. 21
Biological Interactions with Surface Charge in Biomaterials
Edited by Syed A. M. Tofail
© Royal Society of Chemistry 2012
Published by the Royal Society of Chemistry, www.rsc.org

long-term usage usually leads to the development of hospital acquired (nosocomial) infections such as urinary tract infections, pneumonia and life-threatening septicemias, particularly in critically ill patients.[2–7] Bacteria originating from the skin or the external environment can easily colonize an artificial surface and promote biofilm formation, which is difficult to eradicate. Such a microbial community localized at the surface of the biomaterials or the tissue becomes a source of infection and may spread to other distant areas of the human body. Microorganisms involved in biofilm formation include Gram-positive and Gram-negative bacteria and yeasts.[3]

Indwelling urinary catheters, typically made of latex and silicone, are standard medical devices in many hospital wards. Inserted through the urethra into the bladder to collect urine and prevent its retention, urinary catheters simultaneously constitute the most important risk factor increasing the prevalence of urinary tract infections (UTIs) in hospitalized patients. Catheter-associated urinary tract infections (CAUTIs) are considered as one of the most common type of hospital-acquired infections (80% of all nosocomial UTIs) and the most significant cause of secondary nosocomial septicemia.[8,9] The majority of Gram-negative uropathogens in both catheterized and non-catheterized patients belongs to the *Enterobacteriaceae* family.[10,11] Among the Gram-positive bacteria, *Staphylococcus epidermidis* and *Enterococcus faecalis* are the dominant organisms.[12] The enterobacteria are mostly fecal contaminants displaying specific properties such as an involvement in the adherence to uroepithelial cells and urinary catheters, evasion of host defence mechanisms and damage of the mucosal cells. *Escherichia coli* is undoubtedly the predominant aetiological agent of both community-acquired and nosocomial UTIs and the most common agent of CAUTIs, followed by *Proteus mirabilis, Pseudomonas aeruginosa* and *Klebsiella pneumoniae*.[10,11,13–16]

Central venous catheters (CVCs) are used for the administration of fluids and antibiotics and for haemodynamic monitoring.[17] Being in direct contact with the blood, CVCs are coated with tissue proteins and plasma that form a 'conditioning film' on the surface.[18] Bacteria with adhesins to specific blood proteins easily attach to the device. They originate from the patient skin insertion site, the hub of the device or the skin of healthcare workers.[18,19] Microorganisms that universally colonize the inner lumen or the outer surface of CVC include Gram-positive, coagulase-negative staphylococci (CNS), *Staphylococcus aureus* and *E. faecalis*, and Gram-negative rods such as *K. pneumoniae, E. coli, Enterobacter cloacae* and *Serratia marcescens*. Non-fermentative rods such as *Stenotrophomonas maltophilia, Pseudomonas* spp. and *Acinetobacter* spp. have also been reported, especially in critically ill patients and cancer patients.[2,4,5,7,20]

Infections associated with the cardiac implantable electronic devices (CIED), including permanent pacemakers and implantable cardioverter-defibrillators, have been recently reported.[21] Although the majority of the CIED-related infections were caused by Gram-positive cocci (primarily staphylococci), Gram-negative rods (*P. aeruginosa, S. marcescens* and *E. cloacae*) as well as *Candida* spp., *Aspergillus niger, Nocardia* spp. and *Mycobacterium fortuitum* have also been isolated.

Ventilator-associated pneumonia (VAP) is one of the most frequent device-related infections in intensive care units.[22–24] The common VAP bacterial pathogens are Gram-negative non-fermentative rods and species of the *Enterobacteriaceae* family. *E. coli* and *K. pneumoniae* are the predominant causative organisms in early-onset VAP, whereas non-fermentative rods including *Pseudomonas* and *Acinetobacter* were significantly associated with late-onset VAP.[25]

Implantations of mechanical devices and tissue heart valves often cause damage to the surrounding tissues and result in the accumulation of fibrin and platelets on the device and suture sites. The conditioning film that forms on such devices and sites constitutes a perfect place for colonization by opportunistic microorganisms. Prosthetic heart valves endocarditis is predominantly caused by CNS, the so-called early colonizers, originating from the contaminated surgical site during implantation. Infection that develops several months after implantation is usually caused by 'late colonizers' such as streptococci, enterococci and fungi.[26]

Tympanosomy tubes (TT) are implanted into patients with a middle ear infection to alleviate hearing loss. One of the main pathogens isolated from these devices are *Pseudomonas* and *Staphylococcus*. *In vivo* studies that involved TTs made of silicone, fluoroplastic, silver oxide-coated Armstrong-style silicone and ion-treated silicone showed that, except for the ion-treated silicone, all of these materials were colonized by bacteria.[27,28] Similar colonization and associated medical problems are also encountered in ophthalmology, gynecology or orthopedics. Mixed bacterial flora have been isolated from various types of contact lenses, intrauterine devices and prosthetic joints.[3,26]

Infections caused by opportunistic fungi constitute another important problem in current day medicine. Fungi are widely distributed and usually harmless microorganisms but can, in certain conditions, induce superficial, skin- or mucosa-associated infections, as well as deep-seated and life-threatening mycoses. Alarmingly, the most serious infections of blood and innate organs usually occur in hospitalized patients. Immuno-suppressed patients (*e.g.* patients with leukaemia, after bone marrow and solid organ transplantation, and patients with congenital or acquired immunodeficiences), patients treated in intensive care units and patients who have received an indwelling medical device are usually at the highest risk of infection.[29–31] The most clinically relevant genera are *Candida* and *Aspergillus*. However, an increased number of infections due to emerging species, *e.g. Fusarium, Scedosporium* and *Zygomycetes*, has been observed in recent years, especially in haematological patients. *Candida*, which cause up to 70% of invasive mycoses, is regarded as the fourth pathogen responsible for septicaemia in hospitalized patients.[32,33] The predominant species is *Candida albicans* (35–70% of candidaemia), although an increased incidence of non-*albicans* species has been observed, especially those with intrinsic resistance to antifungal agents (*Candida glabrata* and *Candida krusei*) and those with enhanced ability to colonize the catheters, such as *Candida parapsilosis*. The source of candidaemia may be endogenous or exogenous. In many cases vascular catheters colonized by yeasts constitutes a source of infection that spread *via* the blood to internal organs.[29] The yeast of *Candida* spp., especially *C. albicans*, exhibits an extraordinary ability to adhere, colonize and form biofilms on many types of host cells, extracellular proteins (*e.g.* collagen, fibronectin, fibrinogen and vitronectin) and medical devices.[34,35] Fungal cells thus can easily settle on endothelium and prosthesis surfaces when spreading with blood, which leads to origin of the invasion and lesions formation. This is a common way that comes to the development of candidal endocarditis, ocular candidosis, hepatic and hepato-splenic candidosis, osteoarticular candidosis, meningitis and others.[36–40] In addition to vascular catheters, candidal infections can be associated with several other medical devices and implants, including haemodialysis and peritoneal dialysis catheters, central nervous system shunts, intracardiac prosthetic devices (heart valves), joint prostheses, stents, breast implants and cardioverter defibrillators, which can either be infected directly during surgery or indirectly such as haematogenously. Treatment of such infections is difficult and, in addition to antifungal therapy, device removal is necessary in most cases, which is often associated with serious medical complications.[41–43]

11.3 Microbial Adhesion

Permanent and irreversible bindings of microbial cells to animate (eukaryotic cells) or inanimate surfaces are a complicated phenomena, which are influenced by different physicochemical characteristics of the cell surface, materials' surface and the environment in which such interactions take place.[44] Microorganisms may adhere specifically to: abiotic surfaces (*e.g.* biomaterials of medical devices), host cell molecules or molecules of extracellular matrix (ECM) and blood plasma.[45] In the first (initial) stage of the adhesion, the cells of the microorganism are brought into the contact with the surface due to Brownian motion, gravitational and hydrodynamic forces. Next, bacteria bind to the surface *via* physico-chemical and molecular interactions (reversible and irreversible, respectively). Different sensors and intracellular signalling pathways allow the cell to modify microbial metabolism and change their phenotype from planktonic to sessile.[46] After successful attachment, the microorganisms proliferate, colonize the surface and form biofilms.

Various theories attempting to explain interactions between a microorganism and a surface have been discussed in Chapter 9, especially in the light of the DLVO theory, extended DLVO and thermodynamic theory. Cellular and molecular interactions is also important as non-specific interactions between a microorganism and a surface can bring the microorganism closer to the surface to a distance at which specific interactions can occur.

It is the irreversible binding (specific adhesion) conditioned by interactions that occur between the ligands (adhesins) of the bacterial/fungal cells and the surface receptors. These interactions activate signal transduction pathways that lead to enhanced adhesion or invasion.[47]

There are two groups of bacteria that are distinguished by Gram staining: Gram-positive (G+) and Gram-negative (G–) bacteria.[48] These two groups considerably differ in the composition and structure of their cell walls. In Gram-negative bacteria, the cell wall consists of a thin peptidoglycan [alternating *N*-acetylglucosamine and *N*-acetylmuramic acid connected by a β-(1,4)-glycosidic bond] layer, approximately 1–2 nm, the inner cytoplasmic membrane and the outer membrane with lipopolysaccharide (LPS), and multiple porin channels responsible for selective transport of molecules. The Gram-positive bacteria cell wall is characterized by a rigid, thick peptidoglycan layer (15–80 nm), covering the phospholipid cytoplasmic membrane.[49]

Ligands for cell attachment of bacteria and fungi to a particular area may be polymers (polysaccharides, glycoproteins, LPS or uronic acids) secreted by the microorganism and are referred to as extracellular polymeric substances (EPS). EPS form capsules or pseudocapsules around the cell, or a slime layer loosely associated with the cell wall.[49,50] Mucinous material surrounding the cell facilitates the microorganism to 'anchor' in a defined area. For example, in staphylococci capsular polysaccharides, PS/A (polysaccharide adhesin) is responsible for adhesion to the surfaces of both animate and inanimate surfaces, whereas PIA (polysaccharide intercellular adhesin) is involved in biofilm formation and aggregation of bacteria.[51,52] There are, however, examples of specific polymer layers that may also impede cell adhesion.[50,53] Surface adhesins, directly associated with the cell wall of Gram-positive bacteria or outer membrane in Gram-negative bacteria, play a substantial role in tissue-specific adhesion. All bacteria produce multiple adhesions. These adhesins may be a single protein, such as TibA and AIDA of *E. coli,* or can be associated with specialized structural surface appendages as short, rod-shaped fimbriae (pili), longer thicker fibrils and flagella involved in cell motility.[54–56]

An important group of adhesins constitutes factors that recognize ECM proteins. The MSCRAMMs (microbial surface components recognizing adhesive matrix molecules) are either the membrane proteins or cell wall-bound proteins that bind, for example, collagen, fibrinogen, fibronectin, laminin and vitronectin. The examples include the SdrG adhesin of *S. epidermidis*, ClfA (clumping factor A) of *S. aureus* and several other proteins of the *Streptococcus bovis* group.[57] A vast array of *S. epidermidis* adhesins interact with matrix proteins covering indwelling medical device surfaces and increase the risk of biofilm-associated diseases.[58] For example, SdrF adhesins, which are commonly present in most strains of *S. epidermidis*, bind to polystyrene and Dacron[R] with high affinity and promote staphylococcal CVC-associated infections, and ventricular assist device driveline-related infections.[59,60] Other bacteria possess similar adhesins.

Fimbriae (pili) are rigid, filamentous structures of Gram-positive and Gram-negative bacteria originating in the cytoplasm of the cell and project through the cell wall. They are composed of protein subunits called pilin.[49,61] The presence of adhesins on fimbriae may be beneficial for microorganism, especially those covered by polysaccharide layers/capsules. Due to the adhesins placed on protruding appendages, the pathogens may still efficiently colonize the surface.[62] Although fimbriae play the same adhesive function, their structures differ in particular species of bacteria. Gram-negative bacteria fimbriae comprise multiple subunits which are non-covalently attached to each other. An adhesin protein is located at the tip of the pilus.[63] The subunits of Gram-positive bacterial pili are covalently linked by transpeptidase.[63]

A large number of Gram-negative and Gram-positive bacterial adhesion organelles have been studied, *e.g.* type 1 pili in *E. coli*, type 3 pili in *K. pneumoniae* and type IV pili found, among others, in *Clostridium perfringens*.[64–66] *E. coli* express a variety of surface adhesions, such as type 1, 2, 3 and P fimbriae, F1C pili, S pili and F9 fimbriae, with the most common being type 1 and type 3 fimbriae.[11] Strains with type 1 (so-called mannose-sensitive) fimbriae are widespread among the *Enterobacteriaceae*.[67] *Via* lectin FimH fimbrial adhesion, they specifically bind to different eukaryotic cells, ECM proteins and also colonize polyethylene surfaces.[68] Therefore, they are believed to play an important role in device-associated infections. Most clinical isolates of *K. pneumoniae* express fimbrial adhesins 1 and 3, and the latter ones strongly adhere to plastic materials.[69] Uropathogenic rods, *P. mirabilis*, tend to attach to different catheter biopolymers, including polystyrene, propylene and silicone.[70] These species may express a wide variety of adhesive factors as fimbriae MR/P, MR/K, NAF, ATF, PMF and UEA. All of them may be involved in adhesion and colonization of catheters and biofilm formation.[11] In *P. aeruginosa* and *S. maltophilia*, strains involved in opportunistic nosocomial infections associated with medical devices, type IV pili and SMF fimbriae, respectively, play an important role in binding to abiotic surfaces.[71,72]

Fibrils, proteinaceous appendages, are more common in Gram-positive than Gram-negative bacteria. In *Streptococcus* for example they are localized or distributed all around the surface (as hair-like structures). Bacterial fibrils mediate adhesion to ECM, host cell receptors and probably strengthen adhesion to abiotic surfaces as well.[73]

Flagella, the organelle of movement, do not play a pivotal role in adhesion but still may be important in binding to abiotic surfaces. Rotation of flagellum enables bacteria to strengthen the adhesion at high flow rates or, on the other hand, detach from the surface when flow rates is low. Its role is rather important at the reversible stage of adhesion by overcoming the repulsive forces.[49,74]

Apart from the case of *C. albicans*, the knowledge of adhesion mechanisms of pathogenic fungi has so far been limited. One thing is certain: the fungal cell surface plays a crucial role for adhesion and colonization. The inner polysaccharide layers of the cell wall of *C. albicans* is formed by β1,3-glucans, β1,6-glucans, small amount of chitin and Pir proteins (internal proteins covalently linked to β1,3-glucans). The outer protein layer of the cell wall is formed mostly by mannoproteins, strongly mannosylated and phosphorylated proteins covalently linked to β1,6–glucans through a glycosylphosphatidylinositol (GPI) anchor.[75] The external proteins are negatively charged due to the presence of phosphodiester bridges and N-linked carbohydrates and are responsible for binding with positively charged ions and proteins.[76,77] The fungal cell wall is a highly dynamic structure and is remodelled depending on the cell cycle, morphological form (yeast/filaments) and environmental condition (*e.g.* oxidative stress, pH, iron restriction and ethanol concentration). These processes are under the control of several signalling pathways affecting protein expression and epigenetic mechanisms, which has been extensively investigated in the last decade.[78,79]

Among the *C. albicans* adhesins, the agglutinin-like sequence (Als) family, Hwp1, Csh1 and Int1p have been indentified to play a major role in the attachment to surfaces.[34,79] Eight Als proteins have been identified so far, and particular family members are differently expressed depending on the cell morphology (yeast/ filaments) and infection site. For example, Alsp5 binds to ECM proteins, Alsp1 and Alsp3 mediate the binding to epithelial and endo-thelial cells, contribute to cell aggregation, the formation of biofilms and they are also involved in the development of oropharyngeal candidiasis. Another important adhesin is Hwp1p (a hyphal wall protein) expressed on the germ tube surface that play crucial role in biofilm formation. The Csh1p is a hydrophobic substrate-binding protein, which has an impact on cell surface hydrophobocity, and the Eap1p adhesion mediates binding to plastic surfaces.[34,76,77,80]

Other factors that influence microbial adhesion include hydrophobicity, chemistry, configuration and roughness of the surface, and the environment in which the interaction between the microorganism and the surface occurs. These factors are discussed in Chapter 9. A number of serum and tissue proteins also affect microbial adhesions either directly or indirectly. Host proteins from tissue, serum or cerebrospinal fluid change the physic-chemical properties of the liquid environment through their presence in the environment and also through direct binding to the biotic or abiotic surface.

A number of investigations have shown that fibronectin, fibrinogen, laminin and thrombin promote adhesion of staphylococci to biomaterials and tissues, and thus contribute to the pathogenesis of device-associated infections.[81,82] On the other hand, albumin is a protein that inhibits microbial binding when deposited on polymers and metal surfaces.[83] Coating the biomaterial surfaces with antimicrobial agents is an effective method, which leads to a reduction or total exclusion of microbial adhesion. For example, α-tropomyosin, a hydro-philic protein from fish muscle, when coated on to glass, polystyrene, PVC or stainless steel surface effectively reduces the binding of bacteria.[84]

In the case of fungi, the ability of *C. albicans* to adhere to many ECM proteins results from the activity of 'integrin-like receptors', sharing of the structure and binding properties with human integrins. For example, laminin, a component of eukaryotic basement membranes, binds to *Candida* germ tubes.[85] Fibronectin, a multifunctional dimeric protein present in plasma in a soluble form and in ECM in an insoluble form, mediates the adherence of yeasts to vaginal epithelial cells.[86] Vitronectin, a constituent of vascular walls and dermis, is highly adsorb to biomaterials. Binding to this glycoprotein promotes candidal infections.[87] Similarly, the development of dental diseases may be associated with the high affinity of *C. albicans* for salivary proteins adsorbed on hydroxyapatite and silicone biomaterials used in dentistry.[88,89]

In other groups of human fungal pathogens (opportunistic molds and dimorphic fungi) different multifunctional adhesive surface components have been identified as mediating the attachment to various surface receptors.[34]

11.4 Interactions of Bacteria and Fungi with Charged Surfaces

The DLVO theory (see Chapter 9) used in interpreting bacterial adhesion assumes that the surface of bacteria has a constant charge or constant surface potential. However, the electrostatic behaviour of bacterial surface may differ depending on factors associated with the environment and the microorganism.[90] Microbial electrokinetic cell surface properties contribute to bacterial attachment and biofilm formation. This is due to charged molecules associated with the bacterial cell wall. In Gram-positive bacteria, they are the carboxylic, phosphoric, hydroxyl and amine groups associated with peptidoglycan, teichoic and teichuronic acids of the cell wall. In Gram-negative bacteria, the above functional groups are associated with LPS, phospholipids and different proteins of the cell wall.[90]

Under physiological conditions most microorganisms possess a net negative electrostatic charge. This is due to a larger number of carboxylic and phosphate groups compared with the amino groups. Despite the presence of positively charged surface structures, the number of anionic molecules exceeds the number of cationic molecules conferring negative charge of the whole bacterial cell.[91] In Gram-positive cocci of *Staphylococcus aureus*, the presence of teichoic acids, highly charged cell wall polymers, play a pivotal role in adhesion. A low number of positively charged D-alanine residues and the domination of negatively charged phosphate groups probably confer the moderately negative net charge of *S. aureus* at physiological pH.[92] In aqueous environment the surface negative charge attracts a layer of mobile ions forming an 'electric double layer' with an external cloud of positively charged ions. The electric double layer interactions may also be affected by the charged groups inside the bacterial cells.[56]

There are some bacteria that demonstrate positive surface charge. Strains of *S. maltophilia* and *Streptococcus thermophilus*, which are positively charged at the neutral pH, have been isolated from clinical specimens.[93,94] The LPS of *S. maltophilia* is uncharged and not able to shield charges present in the outer membrane. This strain has been reported to bind efficiently to negatively charged Teflon and glass.[94]

The surface charge of microorganisms varies among different species, depending on the surface structures of the microorganism in question and the pH and the ionic strength of the medium.[95–97] The pH of the surrounding medium strongly affects the number of charged acidic or basic groups of microbial surface that, upon pH changes, may associate or dissociate, contributing to conformational changes in bacterial surface structures.[94] It has been shown, for example, that endospores of *Bacillus cereus* efficiently attached to the surface when the pH of the growth medium was 3. An increase in the pH led to a decrease in endospores attachment which probably results from repulsive interactions between both negatively charged surface and spores.[97] As it has been discussed in Chapter 9, a high ionic strength of the medium can suppress the natural surface charge of microorganisms and the subsequent electrostatic interactions between the bacteria and the surface.[94,98]

It was observed that the so-called heterogeneous populations of bacteria adhered better to the surface in comparison with a homogeneous population of bacteria. The heterogeneous population consisted of subpopulations of cells that differed in their negative charge. This fact, during the initial step of adhesion, enabled more efficient binding and retention of heterogeneous population on the surface.[99]

Some of bacterial species, *e.g. P.s aeruginosa*, adhere more rapidly than others (*e.g. S. aureus*). Depending on the surface of the bacterial and conducting/semiconducting material, the cells may accept or donate electrons.[100,101] By donating the negative charge from the region of contact of two surfaces, the cells may decrease electrostatic repulsion.[101] Bacteria that donated more electrons adhere more strongly to the surface.[102]

Adhesion and surface growth of microorganisms may be oppositely affected by the charge of the surface substratum. A negatively charged biomaterial surface reduces attachment of most bacteria and delays development of biofilm. In turn, a positively charged surface promotes adhesion of negatively charged bacteria but inhibits their proliferation.[103] Positively charged surfaces are more adhesive and attract more bacterial cells, but this phenomenon is counterbalanced by the absence of the microbial growth.[103] It was demonstrated for the Gram-negative *P. aeruginosa* that the consequence of increasing binding strength to the positively charged surface was reduced surface multiplication of these bacteria.[104] Several other reports confirm this observation.[105,106] The proliferation of Gram-positive bacteria is less affected by the substratum charge. Their thick layer of peptidoglycan protects, to some extent, the cytoplasmic membrane from contact with the surface. This ultimately protects the cell from destruction.[103] Therefore, the positively charged surfaces can be considered as potentially antimicrobial for Gram-negative bacteria. The use of such biomaterials in medicine can prevent microbial proliferation, especially on implants contaminated with pathogens during surgery.[103]

11.5   Conclusions

The prevention and efficient eradication of pathogenic bacteria and fungi adhering to implants or tissue surfaces are an important pursuit in current medicine. A fundamental understanding of how microbes interact with biotic and abiotic surface is crucial in finding new methods for the control and treatment of microbial infections. Thus, therapeutic strategies based on the inhibition and the modulation of microbial binding to the surfaces are very promising for a successful control and containment of implant-related infections, chronic diseases and nosocomial infections.

References

1. W. J. Marone, W. Jarwis, D. Culver and R. Haley, in *Hospital Infection*, ed. J. V Bennet and P. S. Brachman, Little Brown and Company, Boston, 1992, p. 577.
2. L. S. Elting and G. P. Bodey, *Medicine*, 1990, **69**, 196.
3. J. Jass, S. Surman and J. Walker, *Medical Biofilms*, John Wiley & Sons Ltd, Chichester, 2003.
4. H. Hanna, C. Afif, B. Alakech, M. Boktour, J. Tarrand, R. Hachem and I. Raad, *Infect. Control Hosp. Epidemiol.*, 2004, **25**, 646.
5. E. L. Larson, J. P. Cimiotti, J. Haas, M. Nesin, A. Allen, P. Della-Latta and L. Saiman, *Pediatr. Crit. Care Med.*, 2005, **6**, 457.
6. I. Raad, H. Hanna and D. Maki, *Lancet Infect. Dis.*, 2007, **7**, 645.
7. M. Yeshurun, A. Gafter-Gvili, M. Thaler, N. Keller, A. Nagler and A. Shimoni, *Infection*, 2010, **38**, 211.
8. I. Hartstein, S. B. Garber, T. T. Ward, S. R. Jones and V. H. Morthland, *Infect. Control*, 1981, **2**, 380.
9. V. Gould, C. A. Umscheid, R. K. Agarwal, G. Kuntz and D. A. Pegues, *Infect. Control Hosp. Epidemiol.*, 2010, **31**, 319.
10. L. E. Nicolle, *Drugs Aging*, 2005, **22**, 627.
11. S. M. Jacobsen, D. J. Stickler, H. L. T. Mobley and M. E. Shirtliff, *Clin. Microbiol. Rev.*, 2008, **21**, 26.
12. D. J. Stickler, *Biofouling*, 1996, **94**, 293.
13. M. Potic and M. I. Ignjatovic, *Int. Urol. Nephrol.*, 2009, **41**, 461.
14. W. R. Jarvis and W. J. Martone, *J. Antimicrob. Chemother.*, 1992, **29**, 19.
15. C. Fluit, F. J. Schmitz and J. Verhoef, *Eur. J. Clin. Microbiol. Infect. Dis.*, 2001, **20**, 188.
16. R. Mittal, S. Aggarwal, S. Sharma, S. Chhibber and K. Harjai, *J. Infect. Public Health*, 2009, **2**, 101.
17. R. H. Flowers, K. J. Schwenzer, R. F. Kopel, M. J. Fisch, S. J. Tucker and B. M. Farr, *JAMA*, 1989, **261**, 878.
18. I. Raad, *Lancet*, 1998, **351**, 893.
19. T. S. J. Elliott, H. A. Moss, S. E. Tebbs, I. C. Wilson, R. S. Bonser, T. R. Graham, L. P. Burke and M. H. Faroqui, *Eur. J. Clin. Microbiol. Infect. Dis.*, 1997, **16**, 210.
20. I. Raad, W. Costerton, U. Sabharwal, M. Sacilowski, W. Anaissie and G. P. Bodey, *J. Infect. Dis.*, 1993, **168**, 400.
21. G. M. Viola, L. L. Awan and R. O. Darouiche, *Circulation*, 2010, **121**, 2085.
22. K. A. Davis, *J. Intensive Care Med.*, 2006, **21**, 211.
23. S. M. Koenig and J. D. Truwit, *Clin. Microbiol. Rev.*, 2006, **19**, 637.
24. L. Berra, J. Sampson and J. Wiener-Kronish, *Minerva Anestesiol.*, 2010, **76**, 824.
25. N. M. Joseph, S. Sistla, T. K. Dutta, A. S. Badhe, D. Rasitha and S. C. Parija, *J. Infect. Dev. Ctries*, 2010, **4**, 218.
26. R. M. Donlan and J. W. Costerton, *Clin. Microbiol. Rev.*, 2002, **15**, 167.

27. S. Saidi, J. F. Biedlingmaier and P. Whelan, *Otolaryngol. Head Neck Surg.*, 1999, **120**, 621.
28. J. Biedlingmaier, R. Samaranayke and P. Whelen, *Otolaryngol. Head Neck Surg.*, 1998, **118**, 444.
29. M. C. Arendrup, *Curr. Opin. Crit. Care*, 2010, **16**, 445.
30. Z. Bhatti, A. Shaukat, N. G. Almyroudis and B. H. Segal, *Mycopathologia*, 2006, **162**, 1.
31. B. P. Guery, M. C. Arendrup, G. Auzinger, E. Azoulay, S. M. Borges, E. M. Johnson, E. Müller, C. Putensen, C. Rotstein, G. Sganga, M. Venditti, R. Zaragoza Crespo and B. J. Kullberg, *Intensive Care Med.*, 2009, **35**, 55.
32. H. Wisplinghoff, H. Seifert, S. M. Tallent, T. Bischoff, R. P. Wenzel and M. B. Edmond, *Pediatr. Infect. Dis. J.*, 2003, **22**, 686.
33. H. Wisplinghoff, T. Bischoff, S. M. Tallent, H. Seifert, R. P. Wenzel and M. B. Edmond, *Clin. Infect. Dis.*, 2004, **39**, 309.
34. G. Tronchin, M. Pihet, L. M. Lopes-Bezerra and J. P. Bouchara, *Med. Mycol.*, 2008, **46**, 749.
35. M. K. Hostetter, *Clin. Microbiol. Rev.*, 1994, **7**, 29.
36. S. A. Klotz, *Clin. Infect. Dis.*, 1992, **14**, 340.
37. W. J. Steinbach, J. R. Perfect, C. H. Cabell, V. G. Fowler, G. R. Corey, J. S. Li, A. K. Zaas and D. K. Benjamin, Jr, *J. Infect.*, 2005, **51**, 230.
38. S. S. Feman, J. C. Nichols, S. M. Chung and T. A. Theobald, *Trans. Am. Ophthalmol. Soc.*, 2002, **100**, 67.
39. R. Kohli and S. Hadley, *Infect. Dis. Clin. North. Am.*, 2005, **19**, 831.
40. J. Sánchez-Portocarrero, E. Pérez-Cecilia, O. Corral, J. Romero-Vivas and J. J. Picazo, *Diagn. Microbiol. Infect. Dis.*, 2000, **37**, 169, 41.
41. G. Ramage, J. P. Martínez and J. L. López-Ribot, *FEMS Yeast Res.*, 2006, **6**, 979.
42. E. M. Kojic and R. O. Darouiche, *Clin. Microbiol. Rev.*, 2004, **17**, 255.
43. H. H. Wolf, M. Leithäuser, G. Maschmeyer, H. Salwender, U. Klein, I. Chaberny, F. Weissinger, D. Buchheidt, M. Ruhnke, G. Egerer, O. Cornely, G. Fätkenheuer and S. Mousset, *Ann. Hematol.*, 2008, **87**, 863.
44. M. Katsikogianni and Y. F. Missrilis, *Eur. Cell Mater.*, 2004, **8**, 37.
45. J. Palmer, S. Flint and J. Brooks, *J. Ind. Microbiol. Biotechnol.*, 2007, **34**, 577.
46. K. Anselme, P. Davidson, A. M. Popa, M. Giazzon, M. Liley and L. Ploux, *Acta Biomater.*, 2010, **6**, 3824.
47. H. M. Dalton and P. E. March, *Curr. Opin. Biotechnol.*, 1998, **9**, 252.
48. S. M. Hammond, P. A. Lambert and A. N. Rycroft, *The Bacterial Cell Surface*, Croom Helm, London, 1984.
49. P. R. Murray, K. S. Rosenthal and M. A. Pfaller, *Medical Microbiology*, Elsevier Mosby, Philadelphia, 2005.
50. J. Azeredo, J. Visser and R. Oliviera, *Colloids Surf.*, 1999, **14**, 141.
51. M. Bartoszewicz-Potyrała and A. Przondo-Mordarska, *Post. Mikrobiol.*, 2002, **41**, 351.
52. P. Vasudevan, M. Nair, T. Annamalai and K. S. Venkitanarayanan, *Vet. Microbiol.*, 2003, **92**, 179.

53. S. Tsuneda, H. Aikawa, H. Hayasi, A. Yuasa and A. Hirata, *FEMS Microbiol. Lett.*, 2003, **223**, 287.

54. O. Sherlock, M. A. Schembri, A. Reisner and P. Klemm, *J. Bacteriol.*, 2004, **186**, 8058.

55. O. Sherlock, R. M. Vejborg and P. Klemm, *Infect. Immun.*, 2005, **73**, 1954.

56. A. T. Poortinga, R. Bos, W. Norde and H. J. Busscher, *Surf. Sci. Rep.*, 2002, **47**, 1.

57. J. Sillanpää, S. R. Nallapareddy, X. Qin, K. V. Singh, D. M. Muzny, C. L. Kovar, L. V. Nazareth, R. A. Gibbs, M. J. Ferraro, J. M. Steckelberg, G. M. Weinstock and B. E. Murray, *J. Bacteriol.*, 2009, **191**, 6643.

58. M. Otto, *Nat. Rev. Microbiol.*, 2009, **7**, 555.

59. C. Arrecubieta, F. A. Toba, M. von Bayern, H. Akashi, M. C. Deng, Y. Naka and F. D. Lowy, *PloS Pathog.*, 2009, **5**, 1.

60. B. Guo, X. Zhao, Y. Shi, D. Zhu and Y. Zhang, *Infect. Immun.*, 2007, **75**, 2991.

61. J. C. G. Ottow, *Annu. Rev. Microbiol.*, 1975, **29**, 79.

62. P. Klemm, R. Munk Vejborg and V. Hancock, *Appl. Microbiol. Biotechnol.*, 2010, **88**, 451.

63. J. R. Scott and D. Zähner, *Mol. Microbiol.*, 2006, **62**, 320.

64. L. A. Pratt and R. Kolter, *Mol. Microbiol.*, 1998, **30**, 285.

65. P. Di Martino, N. Cafferini, B. Joly and A. Darfeuille-Michaud, *Res. Microbiol.*, 2003, **154**, 9.

66. J. J. Varga, B. Therit and S. B. Melville, *Infect. Immun.*, 2008, **76**, 4944.

67. M. A. Schembri and P. Klemm, *Infect. Immun.*, 2001, **69**, 1322.

68. T. J. Marrie and J. W. Costerton, *Appl. Environ. Microbiol.*, 1983, **45**, 1018.

69. C. Schroll, K. B. Barken, K. A. Krogfelt and C. Struve, *BMC Microbiol.*, 2010, **10**, 179.

70. J. A. Roberts, E. N. Fussell and M. B. Kaack, *J. Urol.*, 1990, **144**, 264.

71. G. A. O'Toole and R. Kolter, *Mol. Microbiol.*, 1998, **30**, 295.

72. D. de Oliveira-Garcia, M. Dall'Agnol, M. Rosales, A. C. Azzuz, M. B. Martinez and J. A. Girón, *Emerg. Infect. Dis.*, 2002, **8**, 918.

73. A. H. Nobbs, R. J. Lamont and H. F. Jenkinson, *Microbiol. Mol. Biol. Rev.*, 2009, **73**, 407.

74. J. W. McClaine and R. M. Ford, *Biotechnol. Bioeng.*, 2002, **78**, 179.

75. S. M. Bowman and S. J. Free, *Bioessays*, 2006, **28**, 799–808.

76. F. M. Klis, G. J. Sosinska, P. W. de Groot and S. Brul, *FEMS Yeast Res.*, 2009, **9**, 1013.

77. W. L. Chaffin, *Microbiol. Mol. Biol. Rev.*, 2008, **72**, 495.

78. A. Halme, S. Bumgarner, C. Styles and G. R. Fink, *Cell*, 2004, **116**, 405.

79. K. J. Verstrepen and F. M. Klis, *Mol. Microbiol.*, 2006, **60**, 5.

80. L. L. Hoyer, C. B. Green, S. H. Oh and X. Zhao, *Med. Mycol.*, 2008, **46**, 1.

81. P. Vaudaux, R. Suzuki, F. A. Waldvogel, J. J. Morgenthaler and U. E. Nydegger, *J. Infect. Dis.*, 1984, **150**, 546.

82. M. Herrmann, P. E. Vaudaux, D. Pittet, R. Auckenthaler, P.D. Lew, F. Schumacher Perdreau, G. Peters and F. A. Waldvogel, *J. Infect. Dis.*, 1988, **158**, 693.

83. Y. H. An and R. J. Friedman, *Appl. Biomater.*, 1998, **43**, 338.

84. R. M. Vejborg, N. Bernbom, L. Gram and P. Klem, *J. Appl. Microbiol.*, 2008, **105**, 141.

85. J. B. Bouchara, G. Tronchin, V. Annaix, R. Robert and J. M. Senet, *Infect. Immun.*, 1990, **58**, 48.

86. A. Kalo, E. Segal, E. Sahar and D. Dayan, *J. Infect. Dis.*, 1998, **157**, 1253.

87. E. Jakab, M. Paulsson, F. Ascencio and A. Ljungh, *APMIS*, 1993, **101**, 187.

88. R. D. Cannon, K. Nand and H. F. Jenkis, *Microbiology*, 1995, **141**, 213.

89. A. R. Holmes, P. van der Wielen and R. D. Cannon, *Oral Surg. Oral Med. Oral Pathol. Oral Radiol. Endod.*, 2006, **102**, 488.

90. Y. Hong and D. G. Brown, *Langmuir*, 2008, **19**, 5003.

91. D. S. Jones, C. G. Adair, M. W. Mawhinney and S. P. Gorman, *Int. J. Pharm.*, 1996, **131**, 83.

92. M. Gross, S. E. Cramton, F. Gotz and A. Peschel, *Infect. Immun.*, 2001, **69**, 3423.

93. H. J. Busscher, M. N. Bellon-Fontaine, N. Mozes, H. C. van der Mei, J. Sjollema, O. Cerf and P. G. Rouxhet, *Biofouling*, 1990, **2**, 55.

94. B. A. Jucker, H. Harms and A. J. B. Zehnder, *J. Bacteriol.*, 1996, **178**, 5472.

95. J. Dankert, A. H. Hogt and J. Feijen, *CRC Crit. Rev. Biocompat.*, 1986, **2**, 219.

96. N. Dan, *J. Colloid Interface Sci.*, 2003, **27**, 41.

97. U. Husmark and U. Rönner, *J. Appl. Bacteriol.*, 1990, **69**, 557.

98. A. A. Mafu, D. Roy, J. Foulet and P. Magny, *J. Food. Prot.*, 1990, **53**, 742.

99. A. E. J. van Merode, H. C. van der Mei, H. J. Busscher and B. P. Krom, *J. Bacteriol.*, 2006, **188**, 2421.

100. A. T. Poortinga, R. Bos and H. J. Busscher, *J. Microbiol. Methods*, 1999, **38**, 183.

101. A. T. Poortinga, R. Bos and H. J. Busscher, *Biophys. Chem.*, 2001, **19**, 273.

102. S. Bayoudh, A. Othmane, L. Ponsonnet and H. B. Ouada, *Colloids Surfaces A*, 2008, **318**, 291.

103. B. Gottenbos, D. W. Grijpma, H. C. van der Mei, J. Feijen and H. J. Busscher, *J. Antimicrob. Chemother.*, 2001, **48**, 7.

104. B. Gottenbos, H. C. van der Mei and H. J. Busscher, *J. Biomed. Mater. Res.*, 2000, **50**, 208.

105. J. L. Speier and J. R. Malek, *J. Colloid. Interface Sci.*, 1982, **89**, 68.

106. E. R. Kenawy, H. F. Abdel, A. El-Raheem, R. El-Shanshoury and M. H. El-Newehy, *J. Control. Release*, 1998, **50**, 145.

CHAPTER 12

Immunological Response of Electrostatic Charge at the Surface of Biomaterials

B. SOBIESZCZAŃSKA,* M. WAWRZYŃSKA AND D. BIAŁY

Wroclaw Medical University, Poland

12.1 Introduction

Vascular endothelial cells (ECs) play an important role in the regulation of immune and inflammatory responses by the secretion and expression of a variety of adhesive molecules, inflammatory mediators, antithrombic and procoagulant factors. While implantation of a stent into a narrowed, diseased artery helps to restore normal blood flow, it also causes an extensive endothelial damage, which is a strong inflammatory stimulus. The foreign body reaction between the immune system, ECs and the surface of the stent influences the outcome of an implanted device. Immediately after the stent implantation, acute inflammatory response to injured tissue is mediated by a variety of cell types such as ECs, platelets, neutrophils and fibroblasts. In contrast, during the wound healing and the reconstruction of damaged tissue at the implant site, the predominant cell types at the surface of a long-term implant are lymphocytes and monocyte-derived macrophages. The ensuing chronic inflammation impairs vascular healing, may result in neointimal hyperplasia, thrombosis,

RSC Nanoscience & Nanotechnology No. 21
Biological Interactions with Surface Charge in Biomaterials
Edited by Syed A. M. Tofail
Published by the Royal Society of Chemistry, www.rsc.org

vascular remodelling and a recurrence the narrowing of the artery, commonly known as restenosis.

12.2 ECs as an Active Component of Innate Immune Response

In physiological conditions, the vascular endothelium is immunologically inert. However, the vascular endothelium actively participates in the immune and inflammatory responses through a constitutive or an induced expression of a great variety of molecules and through the synthesis and secretion of soluble mediators released or tethered on the surface of ECs.

12.2.1 Mediators Involved in the Regulation of Immune and Inflammatory Responses Secreted by ECs

Vascular endothelial damage is associated with an increase in inflammatory cytokine/chemokine release.[1] ECs synthesize several mediators that participate in the immune response and are important for tissue repair after endothelial injury, *e.g.* cytokines [interleukin-1 (IL-1) and IL-6] and chemoattractant molecules [IL-8 is chemotactic for neutrophils, and monocyte chemotactic protein (MCP-1) attracts monocytes and leukocytes].[2,3] Constitutively released nitric oxide (NO) by the vascular endothelium maintains its surface antithrombogenic properties by inhibiting platelet adhesion, activation, secretion and aggregation. Endothelial-derived NO also inhibits leukocyte adhesion and vascular smooth muscle cells (SMCs) migration and proliferation.[4] The synthesis of NO can be modulated by a number of exogenous chemical and physical stimuli, *e.g.* shear stress. However, an excessive amount of NO produced during the secretion of pro-inflammatory mediators by activated ECs, as well as neutrophils and monocytes, may generate NO radicals, especially peroxynitrite ($ONOO^-$), which are implicated in the promotion of vascular injury.[1] ECs are also a source of growth factors that display a variety of effects. For example, vascular endothelial growth factor (VEGF) is one of the most potent endogenous regulators of endothelial integrity after injury and protects the artery from the progression of neointimal proliferation.[5]

In animal models, VEGF-eluting stents accelerated re-endothelialization.[6] On the other hand, increased expression of VEGF has been reported in artherosclerotic and restenotic lesions.[7,8] Through the induction of monocyte migration and activation, VEGF is also an active component of inflammatory responses.[9] Transforming growth factor-β (TGF-β) synthesized by ECs, T-lymphocytes and macrophages has an important role in the inflammatory response as a chemoattractant for leukocytes and fibroblasts. It is also a factor relevant in wound healing due to its stimulatory effects on the synthesis of extracellular matrix (ECM) proteins and increasing levels of protease inhibitors. Platelet-derived growth factor (PDGF) is produced by several cell types, including ECs, and is a potent mitogen for SMCs and fibroblasts. PDGF is also chemotactic for monocytes and neutrophils and enhances ECM protein production. Collectively, all these factors are important in immune and inflammatory responses, and may be implicated in the overproduction of the ECM and inadequate SMC proliferation that can lead to fibrosis and, subsequently, vessel narrowing.[2]

12.2.2   Mediators Involved in Activation of ECs

A number of pro-inflammatory cytokines, *e.g.* tumor necrosis factor α (TNFα), released by inflammatory cells that are activated on contact with biomaterials, can activate ECs.[10] TNFα, a prototype cytokine synthesized by leukocytes, induces the expression of leukocyte adhesion molecules and chemokines in ECs, as well as vascular leakiness.[11] Xue *et al.*[12] demonstrated that polyurethane (PU) and polytetrafluoroethylene (PTFE), two commonly used blood-contacting biomaterials, induced monocyte activation with the subsequent release of cytokines, such as TNFα, IL-1β and IL-6, which contribute to the inflammatory activation in ECs. Titanium dioxide, in its three common nano-architectures comprising anatase, rutile and nanotube, induces dendritic cells to secrete cytokines.[13] Martinesi *et al.*[14] demonstrated that a plasma-treated titanium alloy (Ti-6Al-4V), a widely employed implant material, stimulated a remarkable increase in TNFα release on contact with mononuclear leukocytes, as well as increased expression of adhesion molecules on cultured ECs. Moreover, in an animal study, Omar *et al.*[15] showed that surface properties of anodically oxidized and machined titanium implants modulated the expression of chemokine and integrin receptors important for the recruitment and adhesion of cells which are crucial for the inflammatory processes.

12.3   EC Adhesion Molecules Involved in Inflammation

12.3.1   EC Adhesion Molecules

The ECs of blood vessels provide the primary physical barrier between blood and tissue. Expression of cell-surface molecules [cell adhesion molecules (CAMs)] on ECs orchestrates the trafficking of circulating blood cells and helps to direct leukocyte transendothelial migration (TEM), a process that occurs during immune and inflammatory responses. Endothelial CAMs are involved in the circulation and activation of leukocytes and platelets at the sites of biomaterial implantation. The CAMs involved in EC interactions with leukocytes comprise: (i) selectins, which support the rolling of leukocytes along the surface of ECs through low-affinity interactions between leukocytes and endothelium, and (ii) immunoglobulin superfamily-specific receptors [*e.g.* intercellular adhesion molecule-1 (ICAM-1) and ICAM-2, vascular cell adhesion protein-1 (VCAM-1) and platelet/endothelial cell adhesion molecule-1 (PECAM-1)], which govern high-affinity interactions between ECs and integrins [*e.g.* very late antigen-4 (VLA-4), lymphocyte function-associated antigen-1 (LFA-1) and Mac-1] on the leukocyte surface, promoting the firm adhesion of leukocytes to endothelium followed by leukocyte TEM into sub-endothelial intima.[2,3,16] The importance of these integrin-mediated events in leukocyte trafficking has been observed through a reduction in neointimal hyperplasia in Mac-1-deficient mice.[3]

Endothelial CAMs can influence many aspects of vascular biology, although one of their key functions is the recruitment of leukocytes from blood into tissues.[17] An increased expression of the CAMs, *e.g.* E-selectin, VCAM-1 and ICAM-1, which promote leukocyte adhesive interactions, is stimulated by inflammatory cytokines [*e.g.* IL-1, TNFα and interferon γ (IFNγ)] secreted at the site of EC injury.[18] CAMs are usually expressed on the inner wall of ECs in response to cytokines and inflammatory mediators, but can also also be expressed upon a mechanical injury and/or EC contact with biomaterials' surfaces. CAMs thus play a critical role in EC adhesion, and in conditioning their regeneration and revascularization of biomaterials' surfaces later after implantation.[19]

12.3.2 Interactions of EC Adhesion Molecules with Biomaterials

Distinct patterns of adhesion molecules expressed by activated ECs contribute to the selectivity of leukocyte recruitment and migration, *e.g.* VCAM-1 binds with monocytes, but not with neutrophils. On the contrary, E-selectin binds neutrophils, but not monocytes. Overexpression of E-selectin and neutrophil infiltration are characteristics of acute inflammation and appear early after stenting, whereas the expression of ICAM-1 and VCAM-1 is most commonly associated with chronic conditions such as atherosclerosis, in which monocytes play a more prominent role in cellular infiltration.[3,20–23] Several studies have shown that the chemistry and structure of the stent materials have an impact on expression of EC adhesion molecules. For example, Nitinol, an equi-atomic metal alloy of nickel and titanium, has many applications in medical devices, including endovascular stents. Depending on the concentration and exposure time, nickel ions can significantly up-regulate the expression levels of ICAM-1, VCAM-1 and E-selectin in ECs, thus inducing an inflammatory response.[24,25] Similarly, other metal ions, *e.g.* zinc and cobalt ions, even at the micromolecular concentrations released from commonly utilized prosthetic alloys during the processes of metal corrosion have been found to up-regulate the expression of CAMs.[26,27]

On the other hand, by co-culturing ECs with functionalized biomaterials [poly(ethylene glycol) with either hydroxyl- or amine-terminal groups, PS-PEG-OH and PS-PEG-NH$_2$, respectively] and mononuclear leukocytes, Lester *et al.*[28] have shown that the expression of CAMs on EC surfaces is not dependent on the surface chemistry of the material but predominantly on adherent cells. Apart from these chemical factors, blood flow through the branched network of the vasculature generates biomechanical forces (*e.g.* hydrostatic pressure, stretch, laminar and turbulent wall shear stresses) modulating the expression of CAMs. Quiescent ECs express VCAM-1 and selectins at low or undetectable levels, whereas ICAM-1 is expressed at modest level.[17] Shear stress, on the other hand, regulates VCAM-1 expression negatively in contrast with the ICAM-1 expression.[29] In addition to the effect of the fluid shear stress on ECs, the stress caused by stent struts may also play an important role in vessel injury and remodeling.[30] According to Binns *et al.*,[31] wall shear stress determined by the graft diameter have been implicated in the initiation of neointimal hyperplasia and EC healing after stent implantation. Furthermore, these haemodynamic forces can stimulate leukocyte and platelet interactions with ECs.[32]

12.4 Leukocyte-derived Cells in Innate Immune Response to Prosthetic Surfaces

12.4.1 Role of Platelets in Innate Immune Response to Biomaterials' Surfaces

In normal conditions, intact ECs secrete mediators (*e.g.* NO and the prostaglandin PGI_2), while maintaining circulating platelets in a quiescent state. However, the injury to ECs that occurs during the implantation results in platelet activation. Furthermore, depending on the chemistry and structure, biomaterials' surfaces can also activate platelets: (i) directly, via platelet membrane integrin and interactions of protein receptors with the specific groups present on the biomaterial interface, or (ii) indirectly, via the interactions of platelet receptors with plasma proteins (*e.g.* fibrinogen) adsorbed on the biomaterial's surface or proteins (*e.g.* collagen) exposed by endothelial damage.

Owing to the secretion and expression of various molecules, activated platelets are highly active components of inflammatory processes and thrombogenesis at the site of stenting.[33] Platelets produce the CD40L ligand which engages CD40 on the EC surface, leading to immune activation and inflammation, *i.e.* up-regulation of endothelial CAMs, chemokine secretion and leukocyte recruitment.[34] Activated platelets produce and secrete a variety of pro-inflammatory mediators, which in turn activate many different cells. Upon activation by contact with a biomaterial's surface, platelets synthesize and excrete several inflammatory modulators. Among these are PDGF, a potent mitogen promoting tissue remodelling and cell differentiation in angiogenesis and atherosclerosis, VEGF, a potent angiogenic cytokine inducing mitosis and permeability of ECs, and TGF-β, involved in proliferative responses of SMCs and the development of fibrotic organ failure.[35]

Titanium and its alloys are considered important biomaterials due to their biocompatibility, which originates from the presence of a stable titania surface layer. On the other hand, titania (TiO_2) in its nanoparticulate form can decrease platelet adhesion. For example, we observed a 10% decrease in the platelet adhesion in a nano-titania coating on stainless steel in comparison with

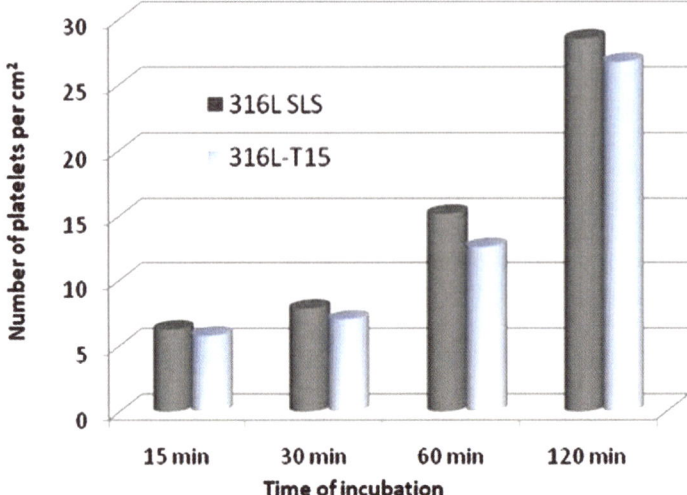

Figure 12.1   The contact of platelets with an artificial biomaterial surface stimulates their activation, resulting in a characteristic change of platelet shape from a globular to a hedgehog-like appearance and the release of factors that enhance their aggregation. Phalloidin–FITC-stained actin in activated star-like platelets on a titania-based coating has been used.

an uncoated steel (Figure 12.1), indicating that even biocompatible biomaterials such as titania, in its nanocrystalline form, can exert deleterious effects on some components of the inflammatory responses (Figure 12.1).

12.4.2   Neutrophils in Inflammation and Restenosis

Neutrophils [polymorphonuclear leukocytes (PMNs)] play an important role in immune responses and wound healing, although, paradoxically, factors released by activated neutrophils mediate local tissue injury. Proteases and reactive oxygen species (ROS), such as hydrogen peroxide, NO and NO-derived radicals, released by neutrophils are well-known factors mediating ECs cytotoxicity.[1] Neutrophil infiltration occurs early after tissue injury and precedes the recruitment of monocytes, which is dictated by chemotactic factors activated early in the inflammatory response.[36,37] All clinically used biomaterials, stimulate a foreign body reaction associated with activation of many types of cells, including neutrophils.[38] At the site of tissue injury, spontaneous adsorption of plasma proteins, *e.g.* fibrinogen, and complement components on the implant surface plays a crucial role in the adherence to and activation of neutrophils. The major role of neutrophils in inflammatory responses is phagocytosis of tissue debris, pathogens or foreign materials. However, because of the size disparity of biomaterials' surfaces, neutrophils are unable to phagocytose such a big object, and therefore release their products in an attempt to degrade the biomaterial as foreign body. ROS released by neutrophils cause self-destruction accompanied by the release of enzymes that damage surrounding tissue, thus potentiating the inflammatory response to the biomaterial's surface.[39]

Several studies have shown that the structure of biomaterials influences neutrophil behaviour. Young and Copper[40] reported that when PUs, which are chemically inert and very commonly used in medical devices, contain a phosphorylcholine moiety they can effectively reduce neutrophil adherence. Chang *et al.*[41] demonstrated that neutrophils adhering to rough biomaterial surfaces produced more oxygen intermediates in contrast with neutrophils attached to smooth biomaterial surfaces. Tomczok *et al.*[42] have shown that hydrophilic, in contrast with hydrophobic, biomaterials' surfaces stimulated significant neutrophil adherence. Apart from biomaterial surface chemistry, various heavy metal ions released from implants during corrosion processes can also stimulate leukocyte chemoattraction and activation, as shown by the study of Hujanen *et al.*[43] Metal ions such as chromium and iron had no influence on the migratory activity of neutrophils in contrast with gold ions that caused the inhibition in neutrophil migration. On the contrary, zinc, copper and nickel ions were found to enhance neutrophil attraction.

12.5 Monocyte/Macrophage Interactions in the Immunological Response to Biomaterials' Surfaces

Monocytes/macrophages also play an important role in inflammatory processes as active producers of potent inflammatory mediators (Figure 12.2). These include: TNFα, macrophage inflammatory protein 1α (MIP-1α), MCP-1, VEGF, IL-1β, IL-6 and IL-8, which plays a role in recruitment of leukocytes to areas of vascular injury.[44] Furthermore, there is an increasing evidence that monocytes play also a central role in in-stent restenosis, as monocyte numbers after implantation have been directly correlated with later neointimal hyperplasia.[45–47] Although these cells appear early in the tissue response to biomaterials, they are the predominant cell type at the surface of a long-term implants.[48,49] Monocytes may be recruited and activated at the site of biomaterial implantation due to tissue injury and upon contact with the biomaterial surface or proteins adsorbed on the implant surface, *e.g.* fibrinogen. A key mediator of monocyte adhesion is the Mac-1 receptor and its various ligands. It has been shown that specific anti-Mac-1 antibodies inhibit monocyte adhesion on stent surfaces.[46] Activated adherent monocytes secrete numerous growth factors, cytokines, tissue factor and metalloproteinases and generate ROS, which have an impact on the surrounding tissues, *e.g.* promoting stent thrombosis as well as migration and proliferation of SMCs to the subendothelial space.[45,50] Several studies have shown that the expression levels of cytokines secreted from adhered monocytes are greatly influenced by the surface chemistry that can modulate surface-adherent monocyte/macrophage activation.[51] These differences have been illustrated in the study of Chang *et al.*[47] in which it was

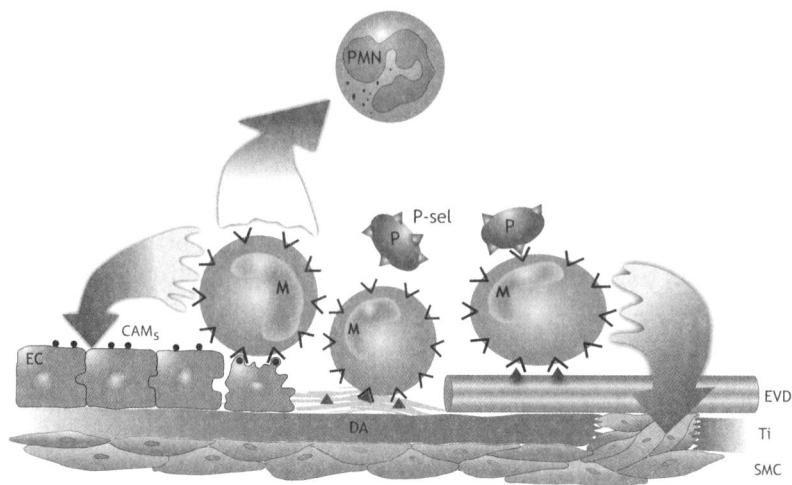

Figure 12.2   Involvement of monocyte/macrophage in the inflammatory response after biomaterial implantation. Pro-inflammatory mediators released by monocytes/macrophages infiltrate the subendothelial space, attract PMNs, promote expression of CAMs on ECs, and influence cellular proliferation, differentiation and apoptosis. CAM, cell adhesion molecule; DA, denudated area; EC, endothelial cell; EDV, endovascular stent; M, monocyte; P, platelet; PMN, polymorphonuclear lymphocyte; P-sel, P-selectin; SMC, smooth muscle cells; Ti, tunica initima.

demonstrated that hydrophilic/neutral and anionic surfaces promoted a pro-inflammatory response (*e.g.* increased levels of IL-8 and TNFα), whereas hydrophilic/cationic surfaces induced an anti-inflammatory response (*e.g.* increased levels of IL-10).

12.6   Conclusions

Complex interactions between ECs and blood-contacting biomaterials, as well as mediators released by inflammatory cells, play a fundamental role in processes accompanying tissue injury after implantation. Depending on the chemical and physical properties of a biomaterial these interactions can lead to a wound healing response or unwanted chronic inflammation associated with neointimal hyperplasia. Biocompatibility, concerning the interaction that take place between the biomaterial and the tissue of the body, is an interdisciplinary field for engineering and the processing of biomaterials for medical applications. Currently, a truly biocompatible implant seems unattainable due to the diversity of components, including ECs, participating in a foreign body reaction to blood-contacting biomaterials. However, a greater understanding of the role of ECs as a major factor orchestrating the foreign body reaction at the site of biomaterial implantation is needed to improve the modification techniques of biomaterial properties.

References

1. B. Tesfamarian and A. F. DeFelice, *Vascul. Pharmacol.*, 2007, **46**, 229–237.
2. D. Carvalho and C. Savage, *Cardiovasc. Pathol.*, 1997, **6**, 61–78.
3. C. Davis, J. Fischer, K. Ley and I. J. Sarembock, *J. Thromb. Haemost.*, 2003, **1**, 1699–1709.
4. D. B. Cines, E. S. Pollak, C. A. Buck, J. Loscalzo, G. A. Zimmerman, R. P. McEver, J. S. Pober, T. M. Wick, B. A. Konkle, B. S. Schwartz, E. S. Barnathan, K. R. McCrae, B. A. Hug, A. M. Schmidt and D. M. Stern, *Blood*, 1998, **91**, 3527–3561.
5. E. Van Belle, F. O. Tio, D. Chen, L. Maillard, M. Kearney and J. M. Isner, *J. Am. Coll. Cardiol.*, 1997, **29**, 1371–1379.
6. D. H. Walter, M. Cejna, L. Diaz-Sandoval, S. Willis, L. Kirkwood, P. W. Stratford, A. B. Tietz, R. Kirchmair, M. Silver, C. Curry, A. Wecker, Y. S. Yoon, R. Heidenreich, A. Hanley, M. Kearney, F. O. Tio, P. Kuenzler, J. M. Isner and D. W. Losordo, *Circulation*, 2004, **110**, 36–45.
7. I. P. Kay, J. M. Ligthart, R. Virmani, vH. M. van Beusekom, K. Kozuma, A. J. Carter, G. Sianos, W. J. van der Giessen, A. J. Wardeh, P. J. de Feyter and P. W. Serruys, *Int. J. Cardiovasc. Intervent.*, 2003, **5**, 137–142.
8. M. Inoue, H. Itoh, M. Ueda, T. Naruko, A. Kojima, R. Komatsu, K. Doi, Y. Ogawa, N. Tamura, K. Takaya, T. Igaki, J. Yamashita, T. H. Chun, K. Masatsugu, A. E. Becker and K. Nakao, *Circulation*, 1998, **98**, 2108–2116.
9. M. A. Costa and D. I. Simon, *Circulation*, 2005, **111**, 2257–2273.
10. J. M. Kułdo, K. I. Ogawara, N. Werner, S. A. Asgeirsdottis, J. A. Kamps, R. J. Kok and G. Molema, *Curr. Vasc. Pharmacol.*, 2005, **3**, 11–39.
11. J. S. Pober, in *Endothelial Biomedicine*, ed. W. C. Aird, Cambridge University Press, New York, 2007, p. 261.
12. Y. Xue, X. Liu and J. Sun, *Toxicol. In Vitro.*, 2010, **24**, 404–410.
13. B. C. Schanen, A. S. Karakoti, S. Seals, D. R. Drake, W. L. Warren and W. T. Self, *ACS Nano.*, 2009, **3**, 2523–2532.
14. M. Martinesi, S. Bruni, M. Stio, C. Treves and F. Borgioli, *J. Biomed. Mater. Res.*, 2005, **74A**, 197–207.
15. O. Omar, M. Lennerås, S. Svensson, F. Suska, L. Emanuelsson, J. Hall, U. Nannmark and P. Thomsen, *J. Mater. Sci. Mater. Med.*, 2010, **21**, 969–80.
16. E. S. Wittchen, *Front. Biosci.*, 2009, **14**, 2522–2545.
17. S. Muro, in *Endothelial Biomedicine*, ed. W. C. Aird, Cambridge University Press, New York, 2007, p. 1058.
18. A. Mantovani and E. Dejana, *Trends Immunol.*, 1989, **10**, 370–375.
19. E. Cenni, D. Granchi, G. Ciapetti, E. Verri E, D. Cavedagna, S. Gamberini, M. Cervellati, A. Di Leo and A. Pizzoferrato, *Biomaterials*, 1997, **1312**, 489–494.
20. C. G. Gurtner, V. Davis, H. Li, M. J. McCoy, A. Sharpe and M. I. Cybulsky, *Genes Dev.*, 1995, **9**, 1–14.
21. Y. Abe, K. Sugisaki and A. M. Dannenberg, *J. Leukoc. Biol.*, 1996, **60**, 692–703.

22. R. De Caterina, G. Basta, G. Lazzerini, G. Dell'Omo, R. Petrucci, M. Morale, F. Carmassi and R. Pedrinelli, *Arterioscler. Thromb. Vasc. Biol.*, 1997, **17**, 2646–2654.

23. L. Feng, D. M. Stern and J. Pile-Spellman, *Radiology*, 1999, **212**, 655–664.

24. E. McLucas, Y. Rocher, W. M. Carroll and T. J. Smith, *J. Mater. Sci. Mater. Med.*, 2008, **19**, 975–980.

25. R. L. Messer, J. C. Wataha, J. B. Lewis, P. E. Lockwood, G. B. Caughman and W. Y. Tseng, *J. Long Term Eff. Med. Implants*, 2005, **15**, 39–47.

26. C. L. Klein, P. Nieder, M. Wagner, H. Köhler, F. Bittinger, C. J. Kirkpatrick and J. C. Lewis, *J. Mater. Sci. Mater. Med.*, 1994, **5**, 798–807.

27. A. Kastrati, A. Dibra, C. Spaulding, G. J. Laarman, M. Menichelli, M. Valgimigli, E. Di Lorenzo, C. Kaiser, I. Tierala, J. Mehilli, M. Seyfarth, O. Varenne, M. T. Dirksen, G. Percoco, A. Varricchio, U. Pittl, M. Syvänne, M. J. Suttorp, R. Violini and A. Schömig, *Eur. Heart J.*, 2007, **28**, 2706–2713.

28. E. A. Lester and J. E. Babense, *J. Biomed. Mater. Res.*, 2003, **64A**, 397–410.

29. M. Braddock, J. L. Schwachtgen, P. Houston, M. C. Dickson, M. J. Lee and C. J. Campbell, *News Physiol. Sci.*, 1998, **1312**, 241–246.

30. H. Y. Chen, J. Hermiller, A. K. Sinha, M. Sturek, L. Zhu and G. S. Kassab, *J. Appl. Physiol.*, 2009, **106**, 1686–1691.

31. R. L. Binns, D. N. Ku, M. T. Stewart, J. P. Ansley and K. A. Coyle, *J. Vasc. Surg.*, 1989, **10**, 326–337.

32. J. A. Adams, in *Endothelial Biomedicine*, ed. W. C. Aird, Cambridge University Press, New York, 2007, p. 1690.

33. A. Zarbok, R. K. Polanowska-Grabowska and K. Ley, *Blood Rev.*, 2007, **21**, 99–111.

34. S. Danese, E. Dejana and C. Fiocchi, *J. Immunol.*, 2007, **178**, 6017–6022.

35. A. Shahryari, F. Azari, H. Vali and S. Omanovic, *Acta Biomaterial.*, 2010, **6**, 695–701.

36. F. G. Welt, E. R. Edelman, D. I. Simon and C. Rogers, *Arterioscler. Thromb. Vasc. Biol.*, 2000, **20**, 2553–2558.

37. J. M. Anderson, in *Biomaterials Science*, ed. B. D. Ratner, A. S. Hoffman, F. J. Schoen and J. E. Lemons, Elsevier Academic Press, London, 2004, p. 296.

38. S. S. Kaplan, R. E. Basford, M. H. Jeong and R. L. Simmons, *J. Biomed. Mater. Res.*, 1992, **26**, 1039–1051.

39. K. Irani, in *Biomaterials Science*, ed. B. D. Ratner, A. S. Hoffman, F. J. Schoen and J. E. Lemons, Elsevier Academic Press, London, 2004, p. 375.

40. L. Z. Young and S. L. Cooper, *Biomaterials*, 1998, **19**, 31–40.

41. S. Chang, J. Popowicz, R. S. Greco and B. Haimowicz, *J. Vasc. Surg.*, 2003, **37**, 1082–1090.

42. J. Tomczok, W. Sliwa-Tomczok, C. L. Klein, T. G. Van Kooten and C. J. Kirkpatrick, *Biomaterials*, 1996, **17**, 1359–1367.

43. E. S. Hujanen, S. T. Seppä and K. Virtanen, *Biochim. Biophys. Acta*, 1995, **1245**, 145–152.

44. V. Vielhauer, X. Cullere and T. Magadas, in *Endothelial Biomedicine*, ed. W. C. Aird, Cambridge University Press, New York, 2007, p. 576.

45. D. Fakuda, K. Shimada, A. Tanaka, T. Kawarabayashi, M. Yoshiyama and J. Yoshikawa, *J. Am. Coll. Cardiol.*, 2004, **43**, 18–23.

46. P. Schuler, D. Assefa, J. Ylanne, N. Blaser, M. Olschewski, I. Ahrens, T. Nordt, C. Bode and K. Peter, *Cell Commun. Adhes.*, 2003, **10**, 17–26.

47. D. T. Chang, J. A. Jones, H. Meyerson, E. Colton, I. K. Kwon, T. Matsuda and J. M. Anderson, *J. Biomed. Mater. Res.*, 2008, **87A**, 676–687.

48. M. Ball, K. O'Brien, F. Dolan, G. Abbas and J. A. McLaughlin, *J. Biomed. Mater. Res.*, 2004, **70A**, 380–390.

49. E. A. Spraque and J. C. Palmaz, *J. Endovasc. Ther.*, 2005, **12**, 594–604.

50. T. Palmerini, B. Coller, V. Cervi, L. Tomasi, A. Marzocchi, C. Marrozzini, O. Leone, M. Piccioli and A. Branzi, *J. Am. Coll. Cardiol.*, 2004, **44**, 1570–1577.

51. R. J. Schutte, A. Parisi-Amon and W. T. Reichert, *J. Biomed. Mater. Res.*, 2009, **88A**, 128–139.

Part IV
Applications

CHAPTER 13

Community- and Hospital-acquired Staphylococcal Infections

R. THORNTON AND J. COONEY*

Deptartment of Life Sciences and Materials and Surface Science Institute, University of Limerick, Ireland

13.1 Introduction

Staphylococci are Gram-positive, non-sporulating, non-motile, facultative anaerobic cocci. There are 49 named species in this genus and almost 1000 other unnamed staphylococcal species. From the point of view of human health, the most significant are *Staphylococcus aureus* and *Staphylococcus epidermidis* as these species cause human infections. While both *S. aureus* and *S. epidermidis* are catalase positive, the species are distinguishable on the basis of coagulase activity. This is the ability of the organism to clot blood plasma. *S. aureus* is coagulase positive, and in many cases this phenotype was used to identify *S. aureus*. In addition, *S. aureus* produces golden-coloured colonies, a characteristic feature from which it derives its name, while those of *S. epidermidis* are white. When these bacteria are grown on blood agar, only *S. aureus* produces zones of haemolysis.

The cell wall of staphylococci is formed of a thick layer outside the plasma membrane composed of a glycine-rich peptidoglycan, teichoic acid and proteins. Enzymes involved in the biosynthesis of peptidoglycan are the targets site

RSC Nanoscienc & Nanotechnology No. 21
Biological Interactions with Surface Charge in Biomaterials
Edited by Syed A. M. Tofail
Published by the Royal Society of Chemistry, www.rsc.org

for action of the β-lactam antibiotics including methicillin. Vancomycin and other glycopeptide antibiotics also target peptidoglycan biosynthesis. Protruding through the peptidoglycan layer, the cell surface is decorated with surface proteins. In addition to the peptidoglycan layer, the bacteria can also produce an exopolysaccharide layer which is known as the capsule or glycocalyx. The capsule can surround an individual cell and make it less susceptible to phagocytosis.[1] Exopolysaccharides are an important component of the slime which is involved in the matrix surrounding a biofilm produced by *S. aureus*.

S. aureus and *S. epidermidis* are amongst the most prevalent organisms associated with infections due to in-dwelling devices, and a contributing factor in colonization is the adhesive properties of a polysaccharide-containing slime and surface proteins they produce. Interestingly ClfA, a *S. aureus* fibrinogen-binding protein, can also associate with materials used in implants, for example, polyvinyl compounds,[2,3] and this property may contribute to the ability of *S. aureus* to rapidly colonize in-dwelling devices.

13.2   Staphylococcal Infections

S. aureus is the aetiological agent for a variety of infections of the skin and soft tissue, as well as deep-seated, localized and systemic infections. Furthermore, toxins produced by *S. aureus* are responsible for staphylococcal food poisoning, scalded-skin syndrome and toxic shock syndrome (TSS). *S. aureus* is carried transiently by between 30 and 50% of the normal population, with 10 to 20% of people being permanent carriers. The most common site of carriage is the anterior nares, or nasal passage, and the skin.[4] There is evidence to suggest that carriage is a risk factor for infection.

13.2.1   Soft Tissue Infections

S. aureus is a significant cause of minor skin infections.[5] These infections are generally localized, and take the form of boils, furuncles and carbuncles. A boil or furuncle is usually an infection of a single hair follicle, with multiple infected follicles forming a carbuncle. Diffuse infections include impetigo and cellulitis.[6,7] Bullous impetigo results in large (2–5 cm), thin-walled blisters containing yellow fluid, which rupture, leaving an area of desquamation. The more common non-bullous form results in a cluster of eruptions with a honey-yellow crust. Cellulitis occurs in the dermis and subcutaneous tissues, most frequently of the legs and digits, and can result in abscesses and local necrosis. The more severe infection necrotizing fasciitis involves the superficial fascia, the connective tissue surrounding the muscles,[8] and the incidence is increasing as community-associated staphylococcal infection increases.[9]

13.2.2   Deep-seated Infection

Although *S. aureus* can cause meningitis, the most significant deep-seated infection caused by *S. aureus* is pneumonia, representing 15% of invasive staphylococcal infections and causing 50 000 infections annually in the US.[10,11] It is a common cause of nosocomial-acquired pneumonia associated with the use of ventilators, and also causes community-acquired infections of previously healthy people.[12] Approximately 50% of these infections are methicillin resistant, resulting in high mortality rates (> 50%).[11,13]

13.2.3 Localized Infections

S. aureus can also cause localized infections, including bone diseases and endocarditis, which can subsequently lead to systemic infections.

Endocarditis is an infection of the heart valves and chambers. The bacteria form a biofilm-like vegetation on the heart valves, with alternating layers of fibrin and platelets, with bacterial colonies interleaved.[14] The incidence rate of endocarditis is 1.7–6.2 cases per 100 000, with mortality rates of up to 40%.[15] The symptoms include chills, fever, fatigue, sweating, aches and pains in joints and muscles. Infections are associated with intravenous drug use, catheter or other device implantation, valve damage, surgery or even recent dental work. The condition can lead to severe damage to heart valves, stroke and dissemination of the infection to other tissues and organs. Treatment involves hospitalization of up to 6 weeks and long-term antibiotic therapy, with severe cases requiring surgery.[16]

S. aureus also causes two bone diseases, septic arthritis and osteomyelitis (reviewed by Wright and Nair[17]). Septic arthritis results in bacterial growth in the synovial fluid, with severe disruption of the cartilage and destruction of collagen. Osteomyelitis causes inflammatory responses resulting in bone distruction.[17] Septic arthritis occurs at a rate of 20–100 cases per million of the population. Rates are higher where a prosthetic joint implant has been made, at 300–700 per million of implants.[18–20] Osteomyelitis occurs at a rate of 10–20 per million of the population and is thought to be decreasing.[18–20] However, infections are becoming more difficult to treat because of the rise of antibiotic resistance, augmented by an increase in the number of joint replacement surgeries.

13.2.4 Systemic Infections

S. aureus causes systemic blood infections, or bacteraemia. These may be a sequelae of other *S. aureus* infections, *e.g.* endocarditis, or they may occur independently. Bacteraemia can also give rise to metastatic infections such as endocarditis.[21] The mortality rates are high, at approximately 30%.[22] Studies have shown that in the UK between 1997 and 2004 the rates of nosocomial bacteraemia have increased from 0.2% to 0.4% of admissions.[23] There are also increasing incidence of community-acquired bacteraemia. The general increase in *S. aureus* resistance to antibiotics has resulted in more difficulty in curing bacteraemia.[24–26]

13.2.5 Toxin-induced Diseases

There are three main toxin-induced diseases, these are TSS, staphylococcal food poisoning and staphylococcal scalded-skin syndrome.

TSS was first identified in 1978, and these outbreaks occurred in menstruating women, predominantly in the US, and were associated with use of high-absorbent tampons.[27] Removal of these products, and other corrective measures, resulted in a significant decline in the incidence from 6 to 1 case per 100 000.[27] Non-menstrual-related TSS accounts for approximately 40% of TSS cases currently reported in the US.[28] Although generally associated with infection of the lower genital tract, it can occur from any case of *S. aureus* infection, *e.g.* abscesses or through nosocomial infection. TSS results in multi-organ failure, combined with vomiting and diarrhoea, and a low-grade fever.[27] The disease is caused by the action of the TSS toxin-1 (TSST-1) and other toxins of the superantigen type (the enterotoxins).[27]

Staphylococcal food poisoning results from the growth of *S. aureus* on foodstuffs, with the production of enterotoxins.[27] The illness presents as a rapid onset of nausea and vomiting which may be accompanied by diarrhoea and fever. The disease is rarely fatal, and generally resolves in 24–48 hours.[27,29] The heat-stable enterotoxins produced by toxinogenic strains of *S. aureus* cause the disease.[29] As little as 0.1 µg is sufficient to cause illness in humans,[30] and evidence suggests that the enterotoxins are more heat resistant in a food matrix than in a standard laboratory setting, compounding the issue of their heat stability.[31]

Outbreaks of scalded-skin syndrome are rare (0.09–0.13 cases per million of the population) and usually confined to neonates. The disease is caused by the epidermolytic toxins (ETs; also known as exfoliative toxins), which have proteolytic activity.[32–34] These enzymes target the desmosomes between the keratinocytes.[35] The activity of the toxins ultimately results in the formation of fluid-filled spaces at the epidermal interface, which leads to splitting and separation at the granular layer. The resultant thin-roofed blisters rupture easily, resulting in desquamation.[36]

13.3   *S. aureus* Virulence and Antibiotic Resistance

S. aureus virulence factors can be divided into three groups: surface-associated factors, extracellular factors and toxins (Figure 13.1). Staphylococcal pathogenicity is the result of the coordinated activity of numerous digestive enzymes and secreted toxins, as well as outer membrane proteins that bind extracellular matrix and plasma proteins.

13.3.1   Surface Proteins of *S. aureus* Implicated in Disease

S. aureus produces many surface-bound proteins which have been implicated by gene-deletion experiments in virulence.[37,38] Several of these bind host proteins. Protein A can bind immunoglobulin through the Fc region, thus helping *S. aureus* evade phagocytosis.[39,40] Protein A also binds to platelets[41] and the von Willebrand factor (VWF).[42] FnBPA and FnBPB are fibronectin-binding proteins which enable invasion of endothelial and epithelial cells.[43] By virtue of

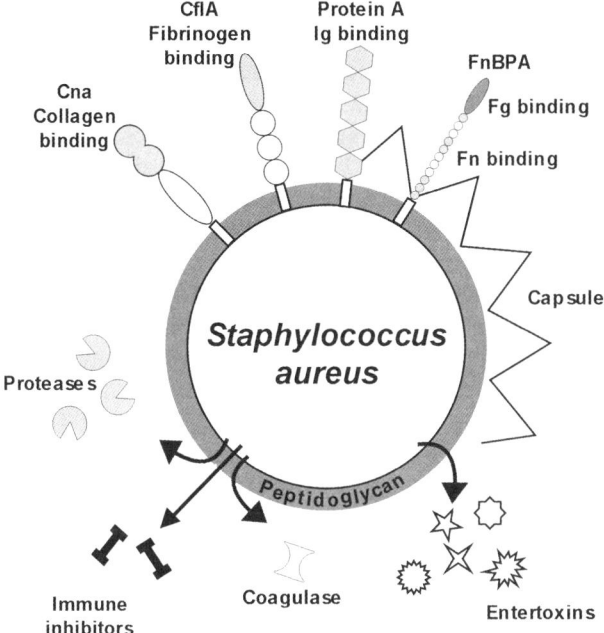

Figure 13.1   Virulence factors of *Staphylococcus aureus*. The diagram shows representations of the major surface-associated and exported virulence factors produced by *S. aureus*. Surface bound proteins are indicated in grey and white shapes, where the grey regions represent domains are thought to be actively involved in the biological function of the protein.

a series of tandem-repeat sequences, the protein binds multiple copies of fibronectin. The proteins can also bind fibrinogen and elastase, and have been implicated in biofilm formation.[44] ClfA and ClfB are fibrinogen-binding proteins.[45-47] These proteins can activate platelets and this activity contributes to endocarditis.[48-50] *S. aureus* also produces a collagen-binding protein, CNA, which has been implicated in *S. aureus* deep-tissue infections such as septic arthritis,[51] osteomyelitis[52] and keratitis.[53] *S. aureus* also binds laminin[54] and vitronectin.[55] Together, these surface proteins are thought to contribute to the establishment of *S. aureus* infection following invasion into the subcutaneous tissue.

13.3.2   Extracellular Proteins of *S. aureus* Implicated in Disease

Pathogenicity of *S. aureus* is also associated with an array of secreted extracellular enzymes and toxins. These extracellular proteins target the host systems in a number of different ways.

Two distinct staphylococcal entities, coagulase and staphylokinase target the blood clotting cascade. They have functionally opposing activities. Coagulase activity results in the deposition of fibrin and clotting of blood.[56,57]

Staphylokinase activity results in the dissolution of the fibrin net around the infection site, and allows staphylococci entry into the deeper host tissues.[58]

Other proteins contribute to the resistance of *S. aureus* to the host innate immunity. Chemotaxis inhibiting protein of *S. aureus* (CHIPS) inhibits chemotaxis by binding to and blocking the C5a receptor and the formyl peptide receptor (FRP) on neutrophils.[59,60] Staphylococcal complement inhibitor (SCIN) inhibits complement activation by the three main pathways (classical, alternative and lectin). It does this by binding to the C3 convertase complex and prevents the production and deposition on the bacterial surface of C3b and, in this manner, helps *S. aureus* circumvent host defences.[61]

S. aureus expresses four characterized extracellular proteases. The V8 serine protease (SspA), two cysteine proteases staphopains A and B and a metallo-protease aureolysin (Aur).[62] Amongst other activities, aureolysin can active pro-urokinase and cleave plasminogen activator inhibitor 1, suggesting that aureolysin augments the activity of staphylokinase in the dissolution of clots.[63] Both staphopains cleave fibrinogen in a destructive manner and also cleave collagen.[64] In addition, they both target H-kininogen and release bradykinin, which causes increased vascular permeability,[65] and target the mammalian cystatins, which are cysteine protease inhibitors.[66]

Staphopain A has elastinolytic activity and can also inactivate α1-proteinase inhibitor, whereas Staphopain B has gelatinase activity.[67] In addition to these proteases, the epidermolytic toxins ETA and ETB are trypsin-like serine proteases.[33,68]

S. aureus produces a large family of structurally related, emetic enterotoxins,[29,69] which cause inflammation by non-specific activation of T-cells. These enterotoxins cause food poisoning and TSS.

The cytolytic toxins of *S. aureus* include the α-haemolysin (or α-toxin), β-haemolysin, δ-haemolysin and the two-component toxins γ-haemolysin and Panton–Valentine leukocidin (PVL), which lyse a range of host cells including white and red blood cells.[70] The most significant of these are the α-toxin, γ-haemolysin and PVL. These pore-forming toxins form multimeric mushroom-shaped structures on the host cell membrane and generates a pore through which cellular material is lost.[71–73]

13.3.3 The Role of Genetic Plasticity in *S. aureus* in Virulence and Drug Resistance

S. aureus carries various types of mobile genetic elements (MGEs), including transposons (Tn), insertion sequences, pathogenicity islands, prophages, plasmids and cassettes. As many as 46 virulence determinants have been documented to be located on MGEs,[74] indicating that these elements are not only contributing to genetic plasticity but also to staphylococcal pathogenicity. Additionally, MGEs are frequently associated with drug resistance and this will be more fully discussed below.

The first reported methicillin-resistant *S. aureus* (MRSA) isolate, designated COL, was isolated from a patient in Colindale, UK, in the 1960s.[75] A more recently isolated MRSA strain, USA300, which is associated with the community-associated MRSA (CA-MRSA) appeared between 1999–2001 and there is increasing evidence for global dissemination of this isolate.[76–79] Genome sequence comparison of COL and USA300 revealed that both strains harbour genes involved in antibiotic resistance and virulence on plasmids, prophages, pathogenicity islands and Tn. However, prophages ΦSA2 and ΦSA3 were unique to the USA300 strain, suggesting that these prophage could contribute to the variation in virulence seen for the two strains.[80]

S. aureus harbours many plasmids and conjugative plasmids, both of which can carry multiple resistance markers.[81] Over 90% of human isolates are now resistant to penicillin.[82] Resistance is conferred by a β-lactamase, encoded by *blaZ*. This gene is frequently carried by a plasmid. Staphylococcal plasmids pS194 and pI258 encode chloramphenicol- and erythromycin-resistance determinants respectively. It has been shown that these plasmids can be mobilized by members of the u11 family of bacteriophage.[83] Vancomycin resistance is carried on a 57.9 kb plasmid that includes an insertion of a Tn*1546*-like element carrying the *vanA* operon.[84,85] Vancomycin resistance was first reported in 1998 in *Enterococcus faecium*.[86] Vancomycin-resistant *S. aureus* (VRSA) are thought to have acquired the *vanA* operon from enterococci.

The *mecA* gene, which encodes the penicillin-binding protein PbP2a, confers wide spread β-lactam and methicillin resistance in MRSA, and is located on a staphylococcal cassette chromosome *mec* (SCC*mec*).[87] At present, eight SSC*mec* allotypes have been described, SSC*mec*I–SSC*mec*VIII.[88–93] The *S. aureus* pathogenicity islands (SaPIs) make up a family of 15–27 kb genetic elements that encode a range of superantigen toxins.[94–96] SaPI1, the archetype member of SaPIs, encodes genes for TSST-1 and enterotoxins K and Q.[95,97]

Tn carried by *S. aureus* frequently encode antibiotic resistance genes.[74] Two important members of this group include Tn*555* and Tn*554*. These harbour genes that encode resistance to spectinomycin or penicillin and MLSB antibiotics (macrolide, lincosamide and streptogramin B) respectively.[98,99]

13.4 MRSA in the Hospital Setting

The emergence of nosocomial infections in hospital may be regarded rather as a realization of a pre-existing problem. Significant monitoring and intervention was implemented in the middle of the 20th century, however, this was as a response to an already widespread problem of *S. aureus* infections in the hospital context. The advent of the widespread use of penicillin and its derivatives in the 1950s reduced the problems of nosocomial *S. aureus*, resulting in a temporary reduction of infection levels. The first significant hospital-associated methicillin resistance in *S. aureus* was reported in the 1960s. It is not conclusive whether the resistance was a direct response to the use of therapeutic methicillin or occurred naturally in the bacterial population in response to environmental penicillins. Regardless, the emergence of these resistant bacteria posed and continues to pose a significant risk of infection in the hospital context (see 87,100–103 and references therein).

Resistance to methicillin is coupled to resistance to a range of cephalosporin drugs, and the strains are also resistant to other classes of drugs, such as members of the macrolide and aminoglycoside groups, so methicillin-resistant *S. aureus* (MRSA) are also multi-resistant. The designation MRSA, although initially associated with methicillin resistance, is now widely regarded as associated with multi-resistant strains.

Extensive studies on the carriage and transmission of MRSA have been carried out over many decades. An excellent and comprehensive summary of 169 individual studies on MRSA between 1980 and 2006 is presented by Albrich and Harbarth.[101] The main findings of their analysis are presented herein and give a robust insight into factors influencing the spread of MSRA in the hospital context. The reader is recommended to consult the extensive bibliography in that review for further reading.

One of the main points of the study by Albrich and Harbarth[101] was to determine the carriage-levels of MRSA in healthcare workers and to examine the influence of this carriage on the transmission rates to patients. In the 169 studies examined over 33 000 healthcare workers were tested. The main site of colonization by *S. aureus* was the anterior nares or nasal passage, with a carriage rate of 4.1%. MRSA were also detected on hands (6.4%), perineum (1.6%) and pharynx (0.3%). Worryingly, data were also presented which suggested over 5% of healthcare workers had an active infection. These were predominantly skin or soft tissue infections and were often sub-clinical. The carriage of any *S. aureus* is a risk factor for transmission and these data suggests that approximately 1 in 20 hospital workers are a potential source of infection.

The transmission of infections from patients to healthcare staff and from staff to patients is a significant factor in the spread of disease. Although MRSA can be detected on fomites, the major reservoir for this organism is the human. Data from the review by Albrich and Harbarth[101] provide evidence for the transmission of MRSA to the patient from healthcare staff. This was supported in 27 studies by molecular analysis, in which strain typing showed identical strains present in the patient as in the carrier. A further 52 studies showed that transmission from the healthcare worker was the most likely route of infection. Specific studies cited cases where the index case could be identified as asymptomatic staff. In a third of all studies the MRSA present in healthcare workers and patients were clonally related, and thus likely to be cross-contamination in either direction. Studies also found a high level (29%) of carriage of MRSA in family members of healthcare workers. This is a very troubling result and shows a conduit for the spread of bacterial strains into the community and back into the hospital setting.

Although airborne transmission of MRSA is possible, the most likely route is through direct contact. Studies showed an extremely high rate (80%) of MRSA contamination of healthcare workers' hands after dressing an infected wound, and a 17% transmission rate from dealing with an infected patient. These bacteria were readily removed by standard hand-washing procedures. However, the rate of hand-washing in the healthcare setting is lower than perceived by the healthcare staff, leading to a risk potential for transmission.

There are two aspects that are considered when dealing with the elimination of MRSA from the hospital community: firstly, the healthcare worker, who is generally asymptomatic, and, secondly, the infected patient. The procedures and treatments available for dealing with infected patients will be outlined separately below in Section 13.6. The healthcare worker may be an asymptomatic MRSA carrier or have a mild skin infection. Routine testing in hospitals of healthcare staff allows identification of potential reservoirs of MRSA. The recommendations provided by Albrich and Harbarth[101] suggest removal of the carrier from the workplace while a decolonization therapy is performed, with nasal mupirocin treatment being the most efficacious. These measures reduce the MRSA load in the hospital and should be combined with good healthcare practices such as hand-washing and use of gloves.

13.5 CA-MRSA

Although there is potential for the spread of MRSA out of the hospital environment into the community, it was not until the 1990s that the first cases of CA-MRSA were identified (for reviews, see 104,105). There were a number of distinguishing features between nosocomial infections and these community infections. The most obvious one was the target population for the infection. Unlike nosocomial MRSA cases, where the patient had some underlying condition, *e.g.* device implantation, the CA-MRSA cases had no apparent underlying predisposition and were healthy members of the normal population.[106] Additionally, the majority of the CA-MRSA strains were not multi-drug resistant.[89]

Since their initial detection in the 1980s CA-MRSA has, similar to hospital-associated MRSA (HA-MRSA), become a major health problem across the developed world. The incidence of infection is increasing and there is some correlation with the increased asymptomatic carriage of CA-MRSA strains in the general population.[104] A study in the US determined the carriage rate in the normal sample population to be less than 1% in 2001, which had jumped to over 9% 4 years later.[107] Although limited in scope, as it was restricted to a single city, it indicated a trend towards increasing carriage, thus increasing the risk of infection. Furthermore, there is a suggestion that CA-MRSA is more readily transmitted than other types of *S. aureus*.[108] This latter factor, combined with higher general carriage rates, may have contributed to the rapid dissemination of CA-MRSA over the last three decades. More recently, outbreaks of CA-MRSA have been associated with situations of high person-to-person contact, *e.g.* day-care, military and prison facilities, and in team situations.[109–111] In general, CA-MRSA cause soft tissue infections, abscesses or cellulites, and respond to non-β-lactam antibiotics. However, there has been an increased occurrence of more invasive infections such as necrotizing fasciitis.[112]

The ability of CA-MRSA to produce such aggressive diseases has focused interest on the virulence factors produced by these strains. Much of this attention has focussed on PVL, a bacteriophage encoded pore-forming toxin. Although inconclusive, a significant association of PVL with aggressive CA-MRSA has been reported.[113] However, this is controversial and other studies have questioned the role of PVL in CA-MRSA.[114] *S. aureus* has a diverse library of potent virulence factors to use to aid the infectious process. Analysis of the CA-MRSA strain USA300 showed exceptionally high levels of expression of α-toxin, phenol-soluble modulins (PSMs) and several other virulence factors.[105,115] Significantly, genome sequence analysis of two outbreak strains, USA300 and USA400, showed that 20% of the genome was unique in these strains.[115] These differences were as a result of MGEs, and these MGEs encoded enterotoxins. Interesting, recent work has identified a novel PSM on a cassette with a *mec* determinant,[116] which is the first example of an association of a virulence factor with a resistance mechanism on the same MGE.

The US Center for Disease Control put forward the five C's, which defined the risk factors for the spread of CA-MRSA. These are: Crowding, frequent skin-to-skin Contact, Compromised skin integrity, Contaminated fomites and lack of Cleanliness (see www.cdc.gov/niosh/topics/mrsa). These points were put forward as a means of devising control regimes against CA-MRSA spread.

13.6 Current Treatment for *S. aureus*

13.6.1 Antibiotic Therapies

β-Lactams were traditionally the drug of choice for staphylococcal infections due to efficacy and limited side-effects. However, in recent years, this drug has lost its value as a treatment due to the emergence of β-lactam-resistance MRSA. Alternative medications have slowly been in development.

Tetracyclines such as tigecycline and minocycline, linezolid and clindamycin are used for the treatment of skin and soft tissue infections (SSTIs) caused by *S. aureus*.[117] Clindamycin is active *in vitro* against over 80% of CA-MRSA strains, and has been successfully used in treating CA-MRSA infections.[118] Linezoid has *in vitro* activity against VRSA and vancomycin-intermediate *S. aureus* (VISA). Linezolid, tigecycline (a tetracycline), daptomycin (a natural antibiotic) and quinupristin-dalfopristin (streptogramin antibiotics) are used for complicated SSTIs (cSSTIs), deeper soft tissue infections, major abscesses, surgical wound infection, cellulitis, burns and infected ulcers.[119–125]

Vancomycin is a glycopeptide antibiotic and is the most frequent intravenous drug used today for severe CA-MRSA and HA-MRSA infections. However, problems exist in the use of vancomycin as resistance has emerged,[126] and also the compound is nephrotoxic at 15–20 μg mL^{-1}, the levels required to combat severe infections.[127] Derivatives of vancomycin are being developed, and these include telavancin, dalbavancin and oritavancin.[128]

13.6.2   Coatings

Colonization of medical devices by biofilm-forming bacteria plays a significant role in healthcare-associated infections. One strategy for combating this problem has been the direct incorporation of antibiotics and ions such as zinc and silver into the surface coating of devices. A study where silver and zinc were incorporated into novel glass polyalkenoate cement (GPC) coatings showed that eluted metal ions can inhibit MRSA *in vivo*.[129,130]

13.6.3   Phage Therapy and Lysins

Phage therapy has commonly been used in veterinary medicine and some of the most convincing support for the efficacy of phage therapy in *S. aureus* infections has come from animal studies. Phage φMR11 and M[Sa] have exerted a protective effect in mice against lethal infection by MRSA.[131]

A more modern approach to phage therapy involves the use of bacteriolytic agents produced by the virus. The CHAP domain of LysK has recently been shown to contain the anti-staphylococcal activity.[132,133] A second lysin, LysGH15, can protect mice against *S. aureus* bacteraemia.[134] Another bacteriolytic entity lysostaphin has been shown to disrupted both *S. aureus* and *S. epidermidis* biofilms *in vitro*.[135] In combination with Ranalexin, an antimicrobial peptide (AMP) from a bull frog, lysostaphin is bacteriostatic against MRSA,[136] and a lysostaphin-impregnated hydrogel, called CCHL, has been developed for MRSA-infected burn wounds.[137]

13.6.4   Vaccines

Vaccination is a standard therapy in the fight against infective agents. Initial attempts used whole *S. aureus* cells but no protection could be observed in several conditions including peritonitis and catheter-associated infections.[138] However, making a successful anti-staphylococcal component vaccine is complicated by the fact that MRSA infections generally involve several virulence determinants, therefore a successful staphylococcal vaccine would combine multiple antigens. The selected antigens should be expressed by the majority of MRSA strains, be surface exposed, elicit opsonophagocytic killing *in vivo* by human neutrophils and promote antibodies which block MRSA adherence factors. An effective vaccine should also inhibit biofilm formation and neutralize toxic *S. aureus* exoproteins. Several companies have vaccines in pre-clinical development but, at present, they have yet to be proven effective.[139]

StaphVAX from Nabi pharmaceutical targeted capsular polysaccharide,[140] but gave no significant protection against bacteraemia in comparison to the placebo control in a phase III clinical trial. Merck are developing a vaccine based on IsdB, a cell wall-anchored protein. Mice immunized with recombinant IsdB showed improved survival following intravenous challenge with *S. aureus*.[141] Other active vaccines are being developed by several other companies including Pfizer (SA3Ag), Integrated Biotherapeutics (STEBVax),

Sanofi and Syntiron (Staph-SRP), NovaDigm Therapeutics (NDV3) and Novartis (SA-Combo).

Several strategies have targeted ClfA. Administration of pre-formed anti-ClfA antibodies to mice reduced mortality, however, this protection was strain-specific.[142] Combining ClfA antibodies and vancomycin in rabbits with catheter-induced MRSA improved the outcome over vancomycin alone.[143] A murine monoclonal antibody 12-9, from Inhibitex, which binds ClfA preventing fibrinogen binding, showed protection of mice in a lethal sepsis model.[144] Tefibazumab (Aurexis®) is the humanized version of 12-9 and has been used in patients with documented *S. aureus* bacteraemia.[145]

13.7 Conclusions

S. aureus remains one of the dominant causes of hospital-acquired infections, and is an increasingly prevalent cause of community-associated infections. This organism harbours the capability to produce a diverse range of virulence factors to aid it in causing disease. Standard treatment regimes involve use of antibiotics. However, the spectrum of antibiotics effective against *S. aureus* is decreasing with increasing drug resistance. New strategies and therapies for the treatment of *S. aureus* need to be developed and need to encompass both traditional pharmaceutical approaches and new interventions such as phage therapy, anti-bacterial enzymes and novel anti-bacterial materials.

References

1. M. Thakker, J. S. Park, V. Carey and J. C. Lee, *Infect. Immun.*, 1998, **66**, 5183.
2. L. Hall-Stoodley, J. W. Costerton and P. Stoodley, *Nat. Rev. Microbiol.*, 2004, **2**, 95.
3. P. E. Vaudaux, P. Francois, R. A. Proctor, D. McDevitt, T. J. Foster, R. M. Albrecht, D. P. Lew, H. Wabers and S. L. Cooper, *Infect. Immun.*, 1995, **63**, 585.
4. F. Götz, F. T. Bannerman and T. K-H. Schleifer, *The Prokaryotes*, Springer, New York, 3rd edn, 2006, **vol. 4**.
5. F. A. Lopez and S. Lartchenko, *Infect. Dis. Clin. North Am.*, 2006, **20**, 759.
6. A. D. Morris, *Clin. Evid. (Online)*, 2008, **2008**, 1708.
7. D. L. Stulberg, M. A. Penrod and R. A. Blatny, *Am. Fam. Physician*, 2002, **66**, 119.
8. W. R. Morgan, M. D. Caldwell, J. M. Brady, M. E. Stemper, K. D. Reed and S. K. Shukla, *J. Clin. Microbiol.*, 2007, **45**, 668.
9. M. E. Stryjewski and H. F. Chambers, *Clin. Infect. Dis.*, 2008, **46**(Suppl. 5), S368.
10. R. M. Klevens, M. A. Morrison, J. Nadle, S. Petit, K. Gershman, S. Ray, L. H. Harrison, R. Lynfield, G. Dumyati, J. M. Townes, A. S. Craig,

E. R. Zell, G. E. Fosheim, L. K. McDougal, R. B. Carey and S. K. Fridkin, *JAMA*, 2007, **298**, 1763.

11. M. J. Kuehnert, H. A. Hill, B. A. Kupronis, J. I. Tokars, S. L. Solomon and D. B. Jernigan, *Emerg. Infect. Dis.*, 2005, **11**, 868.

12. B. E. Ragle and J. Bubeck Wardenburg, *Infect. Immun.*, 2009, **77**, 2712.

13. Z. Athanassa, I. I. Siempos and M. E. Falagas, *Eur. Respir. J.*, 2008, **31**, 625.

14. D. T. Durack, *J. Pathol.*, 1975, **115**, 81.

15. E. Mylonakis and S. B. Calderwood, *N. Engl. J. Med.*, 2001, **345**, 1318.

16. V. G. Fowler, W. M. Scheld and A. S. Bayer, *Endocarditis and Intravascular Infections*, Elsevier Churchill Livingston, Philadelphia, 7th edn, 2009.

17. J. A. Wright and S. P. Nair, *Int. J. Med. Microbiol.*, 2010, **300**, 193.

18. D. L. Goldenberg, *Lancet*, 1998, **351**, 197.

19. S. Nade, *Best Pract. Res. Clin. Rheumatol.*, 2003, **17**, 183.

20. N. S. Stott, *J. Orthop. Surg. (Hong Kong)*, 2001, **9**, 83.

21. G. R. Corey, *Clin. Infect. Dis.*, 2009, **48**(Suppl. 4), S254.

22. G. E. Thwaites, *PLoS One*, 2010, **5**, e14170.

23. D. H. Wyllie, D. W. Crook and T. E. Peto, *BMJ*, 2006, **333**, 281.

24. P. Collignon, G. R. Nimmo, T. Gottlieb and I. B. Gosbell, *Emerg. Infect. Dis.*, 2005, **11**, 554.

25. A. P. Johnson, A. Pearson and G. Duckworth, *J. Antimicrob. Chemother.*, 2005, **56**, 455.

26. S. E. Cosgrove, V. G. Fowler and Jr , *Clin. Infect. Dis.*, 2008, **46**(Suppl. 5), S386.

27. R. J. Murray, *Intern. Med. J.*, 2005, **35**(Suppl. 2), S106.

28. R. A. Hajjeh, A. Reingold, A. Weil, K. Shutt, A. Schuchat and B. A. Perkins, *Emerg. Infect. Dis.*, 1999, **5**, 807.

29. Y. Le Loir, F. Baron and M. Gautier, *Genet. Mol. Res.*, 2003, **2**, 63.

30. M. L. Evenson, M. W. Hinds, R. S. Bernstein and M. S. Bergdoll, *Int. J. Food Microbiol.*, 1988, **7**, 311.

31. M. S. Bergdoll, in *Staphylococci and Staphylococcal Infections*, ed. C. S. F. Easman and C. Adlam, Academic Press, London, 1983, p. 463.

32. C. J. Bailey and T. P. Smith, *Biochem. J.*, 1990, **269**, 535.

33. J. Cavarelli, G. Prevost, W. Bourguet, L. Moulinier, B. Chevrier, B. Delagoutte, A. Bilwes, L. Mourey, S. Rifai, Y. Piemont and D. Moras, *Structure*, 1997, **5**, 813.

34. J. A. Barbosa, J. W. Saldanha and R. C. Garratt, *Protein Eng.*, 1996, **9**, 591.

35. S. Ladhani, *FEMS Immunol. Med. Microbiol.*, 2003, **39**, 181.

36. M. Mockenhaupt, M. Idzko, M. Grosber, E. Schopf and J. Norgauer, *J. Invest. Dermatol.*, 2005, **124**, 700.

37. A. H. Patel, P. Nowlan, E. D. Weavers and T. Foster, *Infect. Immun.*, 1987, **55**, 3103.

38. N. Palmqvist, T. Foster, A. Tarkowski and E. Josefsson, *Microb. Pathog.*, 2002, **33**, 239.

39. J. Sjodahl, *Eur. J. Biochem.*, 1977, **78**, 471.
40. A. H. Patel, J. Kornblum, B. Kreiswirth, R. Novick and T. J. Foster, *Gene*, 1992, **114**, 25.
41. T. Nguyen, B. Ghebrehiwet and E. I. B. Peerschke, *Infect. Immun.*, 2000, **68**, 2061.
42. J. Hartleib, N. Kohler, R. B. Dickinson, G. S. Chhatwal, J. J. Sixma, O. M. Hartford, T. J. Foster, G. Peters, B. E. Kehrel and M. Herrmann, *Blood*, 2000, **96**, 2149.
43. B. Sinha, P. P. Francois, O. Nusse, M. Foti, O. M. Hartford, P. Vaudaux, T. J. Foster, D. P. Lew, M. Herrmann and K. H. Krause, *Cell. Microbiol.*, 1999, **1**, 101.
44. C. Bisognano, P. Vaudaux, P. Rohner, D. P. Lew and D. C. Hooper, *Antimicrob. Agents Chemother.*, 2000, **44**, 1428.
45. J. Hawiger, S. Timmons, D. D. Strong, B. A. Cottrell, M. Riley and R. F. Doolittle, *Biochemistry*, 1982, **21**, 1407.
46. D. Ni Eidhin, S. Perkins, P. Francois, P. Vaudaux, M. Hook and T. J. Foster, *Mol. Microbiol*, 1998, **30**, 245.
47. D. McDevitt, P. Francois, P. Vaudaux and T. J. Foster, *Mol. Microbiol.*, 1994, **11**, 237.
48. I. R. Siboo, A. L. Cheung, A. S. Bayer and P. M. Sullam, *Infect. Immun.*, 2001, **69**, 3120.
49. P. Pawar, P. K. Shin, S. A. Mousa, J. M. Ross and K. Konstantopoulos, *J. Immunol.*, 2004, **173**, 1258.
50. A. Loughman, J. R. Fitzgerald, M. P. Brennan, J. Higgins, R. Downer, D. Cox and T. J. Foster, *Mol. Microbiol.*, 2005, **57**, 804.
51. J. M.Patti, T. Bremell, D. Krajewska-Pietrasik, A. Abdelnour, A. Tarkowski, C. Ryden and M. Hook, *Infect. Immun.*, 1994, **62**, 152.
52. M. O. Elasri, J. R. Thomas, R. A. Skinner, J. S. Blevins, K. E. Beenken, C. L. Nelson and M. S. Smeltzer, *Bone*, 2002, **30**, 275.
53. M. N. Rhem, E. M. Lech, J. M. Patti, D. McDevitt, M. Hook, D. B. Jones and K. R. Wilhelmus, *Infect. Immun.*, 2000, **68**, 3776.
54. C. R. Carneiro, E. Postol, R. Nomizo, L. F. Reis and R. R. Brentani, *Microbes Infect.*, 2004, **6**, 604.
55. G. S. Chhatwal, K. T. Preissner, G. Muller-Berghaus and H. Blobel, *Infect. Immun.*, 1987, **55**, 1878.
56. A. Hijikata-Okunomiya and N. Kataoka, *J. Thromb. Haemost.*, 2003, **1**, 2060.
57. S. Kawabata, T. Morita, T. Miyata, S. Iwanaga and H. Igarashi, *J. Biol. Chem.*, 1986, **261**, 1427.
58. M. I. Bokarewa, T. Jin and A. Tarkowski, *Int. J. Biochem. Cell Biol.*, 2006, **38**, 504.
59. E. Gustafsson, C. Forsberg, K. Haraldsson, S. Lindman, L. Ljung and C. Furebring, *Protein Expr. Purif.*, 2009, **63**, 95.
60. C. Prat, P. J. Haas, J. Bestebroer, C. de Haas, J. A. van Strijp and K. P. van Kessel, *J. Immunol.*, 2009, **183**, 6569.
61. S. H. Rooijakkers, M. Ruyken, J. van Roon, K. P. van Kessel, J. A. van Strijp and W. J. van Wamel, *Cell. Microbiol.*, 2006, **8**, 1282.

62. J. Potempa, E. Golonka, R. Filipek and L. N. Shaw, *Mol. Microbiol.*, 2005, **57**, 605.
63. N. Beaufort, P. Wojciechowski, C. P. Sommerhoff, G. Szmyd, G. Dubin, S. Eick, J. Kellermann, M. Schmitt, J. Potempa and V. Magdolen, *Biochem. J.*, 2008, **410**, 157.
64. T. Ohbayashi, A. Irie, Y. Murakami, M. Nowak, J. Potempa, Y. Nishimura, M. Shinohara and T. Imamura, *Microbiology*, 2011, **157**, 786.
65. T. Imamura, S. Tanase, G. Szmyd, A. Kozik, J. Travis and J. Potempa, *J. Exp. Med.*, 2005, **201**, 1669.
66. B. Vincents, P. Onnerfjord, M. Gruca, J. Potempa and M. Abrahamson, *Biol. Chem.*, 2007, **388**, 437.
67. K. Rice, R. Peralta, D. Bast, J. de Azavedo and M. J. McGavin, *Infect. Immun.*, 2001, **69**, 159.
68. A. C. Papageorgiou, L. R. Plano, C. M. Collins and K. R. Acharya, *Protein Sci.*, 2000, **9**, 610.
69. H. Li, A. Llera, E. L. Malchiodi and R. A. Mariuzza, *Annu. Rev. Immunol.*, 1999, **17**, 435.
70. L. Song, M. R. Hobaugh, C. Shustak, S. Cheley, H. Bayley and J. E. Gouaux, *Science*, 1996, **274**, 1859.
71. M. Mueller, U. Grauschopf, T. Maier, R. Glockshuber and N. Ban, *Nature*, 2009, **459**, 726.
72. O. Joubert, G. Viero, D. Keller, E. Martinez, D. A. Colin, H. Monteil, L. Mourey, M. Dalla Serra and G. Prevost, *Biochem. J.*, 2006, **396**, 381.
73. M. J. Aman, H. Karauzum, M. G. Bowden and T. L. Nguyen, *J. Biomol. Struct. Dyn.*, 2010, **28**, 1.
74. N. Malachowa and F. R. DeLeo, *Cell Mol. Life Sci.*, 2010, **67**, 3057.
75. K. R. Eriksen, *Ugeskr. Laeg.*, 1961, **123**, 384.
76. W. J. Wannet, E. Spalburg, M. E. Heck, G. N. Pluister, E. Tiemersma, R. J. Willems, X. W. Huijsdens, A. J. de Neeling and J. Etienne, *J. Clin. Microbiol.*, 2005, **43**, 3341.
77. G. R. Nimmo and G. W. Coombs, *Int. J. Antimicrob. Agents*, 2008, **31**, 401.
78. A. Tristan, M. Bes, H. Meugnier, G. Lina, B. Bozdogan, P. Courvalin, M. E. Reverdy, M. C. Enright, F. Vandenesch and J. Etienne, *Emerg. Infect. Dis.*, 2007, **13**, 594.
79. A. R. Larsen, M. Stegger, S. Bocher, M. Sorum, D. L. Monnet and R. L. Skov, *J. Clin. Microbiol.*, 2009, **47**, 73.
80. F. R. DeLeo and H. F. Chambers, *J. Clin. Invest.*, 2009, **119**, 2464.
81. T. Berg, N. Firth, S. Apisiridej, A. Hettiaratchi, A. Leelaporn and R. A. Skurray, *J. Bacteriol.*, 1998, **180**, 4350.
82. J. E. Olsen, H. Christensen and F. M. Aarestrup, *J. Antimicrob. Chemother.*, 2006, **57**, 450.
83. S. O. Jensen and B. R. Lyon, *Future Microbiol.*, 2009, **4**, 565.
84. L. M. Weigel, D. B. Clewell, S. R. Gill, N. C. Clark, L. K. McDougal, S. E. Flannagan, J. F. Kolonay, J. Shetty, G. E. Killgore and F. C. Tenover, *Science*, 2003, **302**, 1569.

85. B. Saha, A. K. Singh, A. Ghosh and M. Bal, *J. Med. Microbiol.*, 2008, **57**, 72.

86. R. Leclercq, E. Derlot, M. Weber, J. Duval and P. Courvalin, *Antimicrob. Agents Chemother.*, 1989, **33**, 10.

87. H. F. Chambers and F. R. Deleo, *Nat. Rev. Microbiol.*, 2009, **7**, 629.

88. W. Higuchi, T. Takano, L. J. Teng and T. Yamamoto, *Biochem. Biophys. Res. Commun.*, 2008, **377**, 752.

89. X. X. Ma, T. Ito, C. Tiensasitorn, M. Jamklang, P. Chongtrakool, S. Boyle-Vavra, R. S. Daum and K. Hiramatsu, *Antimicrob. Agents Chemother.*, 2002, **46**, 1147.

90. R. H. Deurenberg and E. E. Stobberingh, *Infect. Genet. Evol.*, 2008, **8**, 747.

91. T. Ito, Y. Katayama, K. Asada, N. Mori, K. Tsutsumimoto, C. Tiensasitor and K. Hiramatsu, *Antimicrob. Agents Chemother.*, 2001, **45**, 1323.

92. D. C. Oliveira, C. Milheirico and H. de Lencastre, *Antimicrob. Agents Chemother.*, 2006, **50**, 3457.

93. G. Arakere, S. Nadig, T. Ito, X. X. Ma and K. Hiramatsu, *FEMS Microbiol. Lett.*, 2009, **292**, 141.

94. T. J. Foster, *Nat. Rev. Microbiol.*, 2005, **3**, 948.

95. R. P. Novick, *Plasmid*, 2003, **49**, 93.

96. J. E. Alouf and H. Muller-Alouf, *Int. J. Med. Microbiol.*, 2003, **292**, 429.

97. A. Ruzin, J. Lindsay and R. P. Novick, *Mol. Microbiol.*, 2001, **41**, 365.

98. T. Ito, K. Okuma, X. X. Ma, H. Yuzawa and K. Hiramatsu, *Drug Resist.Updat.*, 2003, **6**, 41.

99. S. Phillips and R. P. Novick, *Nature*, 1979, **278**, 476.

100. M. H. Lee, C. Arrecubieta, F. J. Martin, A. Prince, A. C. Borczuk and F. D. Lowy, *J. Infect. Dis.*, 2010, **201**, 508.

101. W. C. Albrich and S. Harbarth, *Lancet Infect. Dis.*, 2008, **8**, 289.

102. G. A. Ayliffe, *Clin. Infect. Dis.*, 1997, **24**(Suppl. 1), S74.

103. M. R. Mulvey and A. E. Simor, *CMAJ*, 2009, **180**, 408.

104. F. R. Deleo, M. Otto, B. N. Kreiswirth and H. F. Chambers, *Lancet*, 2010, **375**, 1557.

105. M. Otto, *Annu. Rev. Microbiol.*, 2010, **64**, 143.

106. B. C. Herold, L. C. Immergluck, M. C. Maranan, D. S. Lauderdale, R. E. Gaskin, S. Boyle-Vavra, C. D. Leitch and R. S. Daum, *JAMA*, 1998, **279**, 593.

107. C. B. Creech 2nd, D. S. Kernodle, A. Alsentzer, C. Wilson and K. M. Edwards, *Pediatr. Infect. Dis. J.*, 2005, **24**, 617.

108. N. F. Crum, R. U. Lee, S. A. Thornton, O. C. Stine, M. R. Wallace, C. Barrozo, A. Keefer-Norris, S. Judd and K. L. Russell, *Am. J. Med.*, 2006, **119**, 943.

109. A. E. Aiello, F. D. Lowy, L. N. Wright and E. L. Larson, *Lancet Infect. Dis.*, 2006, **6**, 335.

110. P. M. Adcock, P. Pastor, F. Medley, J. E. Patterson and T. V. Murphy, *J. Infect. Dis.*, 1998, **178**, 577.

111. E. M. Begier, K. Frenette, N. L. Barrett, P. Mshar, S. Petit, D. J. Boxrud, K. Watkins-Colwell, S. Wheeler, E. A. Cebelinski, A. Glennen, D. Nguyen and J. L. Hadler, *Clin. Infect. Dis.*, 2004, **39**, 1446.

112. L. G. Miller, F. Perdreau-Remington, G. Rieg, S. Mehdi, J. Perlroth, A. S. Bayer, A. W. Tang, T. O. Phung and B. Spellberg, *N. Engl. J. Med.*, 2005, **352**, 1445.

113. M. Labandeira-Rey, F. Couzon, S. Boisset, E. L. Brown, M. Bes, Y. Benito, E. M. Barbu, V. Vazquez, M. Hook, J. Etienne, F. Vandenesch and M. G. Bowden, *Science*, 2007, **315**, 1130.

114. M. Otto, *Nat. Med.*, 2011, **17**, 169.

115. T. Baba, F. Takeuchi, M. Kuroda, H. Yuzawa, K. Aoki, A. Oguchi, Y. Nagai, N. Iwama, K. Asano, T. Naimi, H. Kuroda, L. Cui, K. Yamamoto and K. Hiramatsu, *Lancet*, 2002, **359**, 1819.

116. S. Y. Queck, B. A. Khan, R. Wang, T. H. Bach, D. Kretschmer, L. Chen, B. N. Kreiswirth, A. Peschel, F. R. Deleo and M. Otto, *PLoS Pathog.*, 2009, **5**, e1000533.

117. J. J. Ruhe and A. Menon, *Antimicrob. Agents Chemother.*, 2007, **51**, 3298.

118. G. Martinez-Aguilar, W. A. Hammerman, E. O. Mason, Jr and S. L. Kaplan, *Pediatr. Infect. Dis. J.*, 2003, **22**, 593.

119. K. A. Rodvold, M. H. Gotfried, M. Cwik, J. M. Korth-Bradley, G. Dukart and E. J. Ellis-Grosse, *J. Antimicrob. Chemother.*, 2006, **58**, 1221.

120. R. H. Drew, J. R. Perfect, L. Srinath, E. Kurkimilis, M. Dowzicky and G. H. Talbot, *J. Antimicrob. Chemother.*, 2000, **46**, 775.

121. C. Liu, A. Bayer, S. E. Cosgrove, R. S. Daum, S. K. Fridkin, R. J. Gorwitz, S. L. Kaplan, A. W. Karchmer, D. P. Levine, B. E. Murray, J. M. Rybak, D. A. Talan and H. F. Chambers, *Clin. Infect. Dis.*, 2011, **52**, 285.

122. J. Weigelt, K. Itani, D. Stevens, W. Lau, M. Dryden and C. Knirsch, *Antimicrob. Agents Chemother.*, 2005, **49**, 2260.

123. E. Rubinstein, S. Cammarata, T. Oliphant and R. Wunderink, *Clin. Infect. Dis.*, 2001, **32**, 402.

124. A. A. Apodaca and R. M. Rakita, *N. Engl. J. Med.*, 2003, **348**, 86.

125. F. M. Marty, W. W. Yeh, C. B. Wennersten, L. Venkataraman, E. Albano, E. P. Alyea, H. S. Gold, L. R. Baden and S. K. Pillai, *J. Clin. Microbiol.*, 2006, **44**, 595.

126. C. Hawkins, J. Huang, N. Jin, G. A. Noskin, T. R. Zembower and M. Bolon, *Arch. Intern. Med.*, 2007, **167**, 1861.

127. M. J. Rybak, L. M. Albrecht, S. C. Boike and P. H. Chandrasekar, *J. Antimicrob. Chemother.*, 1990, **25**, 679.

128. S. T. Micek, *Clin. Infect. Dis.*, 2007, **45**(Suppl. 3), S184.

129. S. S. Talsma, *Home Healthc. Nurse*, 2007, **25**, 589.

130. a. Coughlan, K. Scanlon, B. P. Mahon and M. R. Towler, *Biomed. Mater. Eng.*, 2010, **20**, 99.

131. N. H. Mann, *Res. Microbiol.*, 2008, **159**, 400.

132. S. O'Flaherty, A. Coffey, W. Meaney, G. F. Fitzgerald and R. P. Ross, *J. Bacteriol.*, 2005, **187**, 7161.

133. M. Horgan, G. O'Flynn, J. Garry, J. Cooney, A. Coffey, G. F. Fitzgerald, R. P. Ross and O. McAuliffe, *Appl. Environ. Microbiol.*, 2009, **75**, 872.

134. J. Gu, W. Xu, L. Lei, J. Huang, X. Feng, C. Sun, C. Du, J. Zuo, Y. Li, T. Du, L. Li and W. Han, *J. Clin. Microbiol.*, 2011, **49**, 111.

135. J. A. Wu, C. Kusuma, J. J. Mond and J. F. Kokai-Kun, *Antimicrob. Agents Chemother.*, 2003, **47**, 3407.

136. D. P. Clark, S. Durell, W. L. Maloy and M. Zasloff, *J. Biol. Chem.*, 1994, **269**, 10849.

137. F. Cui, G. Li, J. Huang, J. Zhang, M. Lu, W. Lu, J. Huan and Q. Huang, *Drug Deliv.*, 2011, **18**, 173.

138. A. C. Schaffer and J. C. Lee, *Infect. Dis. Clin. North Am.*, 2009, **23**, 153.

139. A. C. Schaffer and J. C. Lee, *Int. J. Antimicrob. Agents*, 2008, **32**(Suppl. 1), S71.

140. H. Shinefield, S. Black, A. Fattom, G. Horwith, S. Rasgon, J. Ordonez, H. Yeoh, D. Law, J. B. Robbins, R. Schneerson, L. Muenz, S. Fuller, J. Johnson, B. Fireman, H. Alcorn and R. Naso, *N. Engl. J. Med.*, 2002, **346**, 491.

141. N. A. Kuklin, D. J. Clark, S. Secore, J. Cook, L. D. Cope, T. McNeely, L. Noble, M. J. Brown, J. K. Zorman, X. M. Wang, G. Pancari, H. Fan, K. Isett, B. Burgess, J. Bryan, M. Brownlow, H. George, M. Meinz, M. E. Liddell, R. Kelly, L. Schultz, D. Montgomery, J. Onishi, M. Losada, M. Martin, T. Ebert, C. Y. Tan, T. L. Schofield, E. Nagy, A. Meineke, J. G. Joyce, M. B. Kurtz, M. J. Caulfield, K. U. Jansen, W. McClements and A. S. Anderson, *Infect. Immun.*, 2006, **74**, 2215.

142. E. Josefsson, O. Hartford, L. O'Brien, J. M. Patti and T. Foster, *J. Infect. Dis.*, 2001, **184**, 1572.

143. J. Vernachio, A. S. Bayer, T. Le, Y. L. Chai, B. Prater, A. Schneider, B. Ames, P. Syribeys, J. Robbins and J. M. Patti, *Antimicrob. Agents Chemother.*, 2003, **47**, 3400.

144. A. E. Hall, P. J. Domanski, P. R. Patel, J. H. Vernachio, P. J. Syribeys, E. L. Gorovits, M. A. Johnson, J. M. Ross, J. T. Hutchins and J. M. Patti, *Infect. Immun.*, 2003, **71**, 6864.

145. J. J. Weems, Jr , J. P. Steinberg, S. Filler, J. W. Baddley, G. R. Corey, P. Sampathkumar, L. Winston, J. F. John, C. J. Kubin, R. Talwani, T. Moore, J. M. Patti, S. Hetherington, M. Texter, E. Wenzel, V. A. Kelley, V. G. Fowler and Jr , *Antimicrob. Agents Chemother.*, 2006, **50**, 2751.

CHAPTER 14

MRSA-resistant Textiles

J. BAUER,[*a] K. KOWAL,[a] S. A. M. TOFAIL[b] AND H. PODBIELSKA[a]

[a] Wroclaw University of Technology, Institute of Biomedical Engineering and Instrumentation I-21, Wybrzeze Wyspianskiego 27, 50-370, Wroclaw, Poland; [b] University Limerick, Materials and Surface Science Institute, Ireland

14.1 Introduction

The textile industry represents one of the most mature technologies of the human race and still marvels with innovation, charm and utility. Almost every society in this world possesses some form of indigenous and/or advanced textile technology. The textile technology, however, has highly benefited from technological advances in the human race and, in return, has given rise to a number of important economic activities. The transition from hand looms to power looms is one such example that brought to fore the Industrial Revolution in the West. This industry has strong backward links with textile machinery and chemical industries, and forward links with apparel manufacture, and the packaging and transport industry. Because of the presence of strong commercial demand, the textile industry represents one of the platform technologies that have hosted innovative ideas, ranging from geotextiles to medical textiles to nanotechnology-enabled textiles.

The application of nanomaterials in the textile industry has increased in the last few decades. At the nanoscale level, many of the changes in the chemical, physical or/and biological properties of materials result in novel, significant functionalities that can have useful applications which are not possible if

RSC Nanoscience & Nanotechnology No. 21
Biological Interactions with Surface Charge in Biomaterials
Edited by Syed A. M. Tofail
Published by the Royal Society of Chemistry, www.rsc.org

conventional materials are used. Bringing nanotechnology to the textile industry thus offers the possibility of creating new types of textile products with novel functions and characteristics.

One of the first applications of nanotechnology in the textile industry was undertaken by Nano-Tex in 2002.[1] Since then, many textile companies, such as Schoeller, Toray, Kanebo, Kuraray, Unitika, Zeneca *etc.*, have been conducting active research on incorporating nano-agents into/onto fabrics to produce innovative fibres and textile finishes. Particular attention had been paid to enable the chemical finishing to being more thorough, controllable and safe. Figure 14.1 shows some of the textile properties that can be achieved by using nanotechnology.

The application of nanotechnology in the textile industry can be broadly classified into two categories. The first category involves technologies that aim to improve current properties and functions of textile materials. The second category involves the use of nanotechnology for the development of smart and intelligent textiles with completely novel functions. Such novel functionalities include properties such as self-cleaning and self-healing, antimicrobial, enabling energy storage capacity, fragrance release, wound healing, human body nourishment, multiple protection and detection, information acquisition and transfer. This work focuses on antimicrobial textiles, especially those resistant to the bacteria known as methicillin-resistant *Staphylococcus aureus* (MRSA).

Medicinal applications of textiles date back to antiquity. A famous example is the use of linen clothes containing salts to wrap the mummies in ancient Egypt. The main reason for the growing interest in the manufacturing of antimicrobial textiles, especially in biomedical field, is mainly functional. Textiles that can kill, halt or slow down the growth of microorganisms may act as protection against bacteria, viruses and fungi, as well as protect fabrics against bad odour and loss of some functional and qualitative properties such as elasticity and/or tensile strength, colour caused by these microbes.

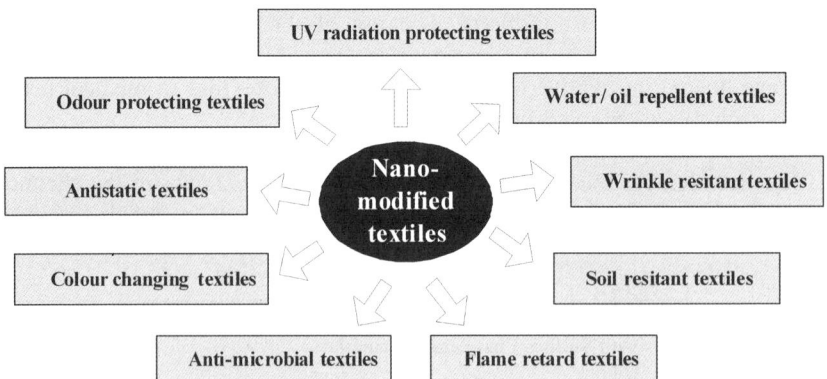

Figure 14.1   Selected properties of fabrics modified by using nanotechnology.

14.2 Textiles and Microbes

The need to preserve textiles against bacteria, viruses, fungi, moulds and mildews has been an area of special scientific interest for a long time. Microbes can cause danger to living organisms, but also destroy non-living matter as fabrics, yarns and threads. They cause degradation of the quality, colour, texture and functionality of fabrics. Microorganisms also produce characteristic odours and a fusty smell. Damage to textiles can occur at almost all stages of manufacturing and storage.

There are a growing number of reports demonstrating the role of microorganisms in causing the health risk and describing the survival and growth of microorganisms in/on textiles.[2–6] Some of these microorgansims are listed in Figure 14.2.

A problem of growing significance in the last few decades is the increasing incidence of infections caused by antibiotic-resistant bacteria, *e.g.* Gram-positive bacteria such as vancomycin-resistant *Enterococci* (VRE) and MRSA,[7,8] and Gram-negative bacteria such as certain strains of *Escherichia coli*. These infections can affect hospitalized patients, especially patients who are immuno-suppressed or are suffering from immunodeficiency disorders, as well as hospital staff members and their families.[9,10] Due to high level of awareness of hospital superbugs, the prevalence of hospital-acquired MRSA-related infections is decreasing in the Western world, however, community-associated MRSA-related infections in crechés, nurseries and nursing homes is on the rise.

One significant factor that contributes to the spread of such pathological microorganisms is the ability of these microorganisms to survive and colonize on various surfaces. Many reports that have been published in last years are

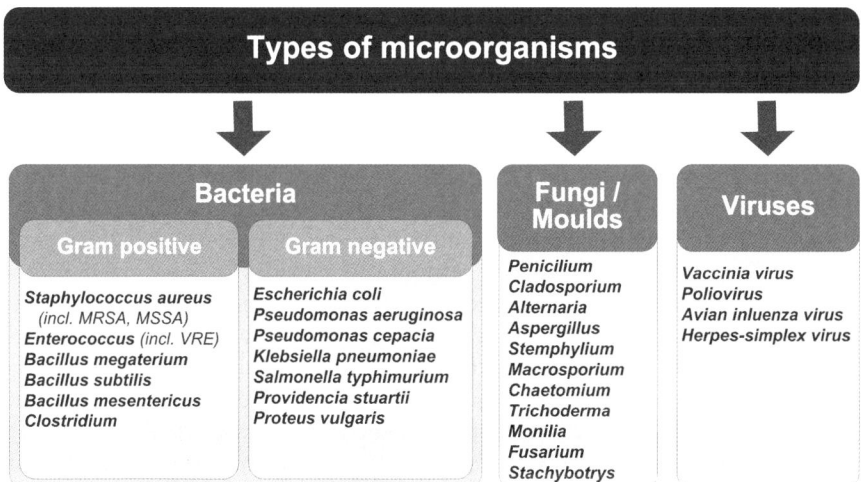

Figure 14.2   Common microorganisms that can cause health problems and textile damage.

devoted to the examination of the survival of Gram-positive bacteria, especially MRSA and VRE on various surfaces as stretchers,[11] stethoscopes and bed rails,[12] bed linens,[13] ear-probe thermometers,[14] room door handles,[15] countertops,[16] dry mops[17] and many others.[18–21] It was found that the spread of microorganisms is strictly connected with hand-to-hand contact and pathogens, when detected in the palms of the patients or hospital staff, pose a higher probability of contamination of the surrounding environment and surfaces.[22,23] In the case of MRSA it was stated that the level of contamination of surrounding surfaces is lower in case of seronegative patients (healthy) than in case of patients that are MRSA seropositive (vectors).[13,15]

Commercially available antimicrobial textiles can be divided into two types: passive and active. Passive materials use anti-adhesive properties to produce the negative effect on living microorganisms. The interaction with microbes can be based, for example, on superhydrophobicity as in the case of lotus leaves or some insects.[24] Superhydrophobic surfaces are achieved mainly by chemical composition modifications along with some topological adaptation. These are the surfaces with a surface contact angle higher than $150°$ and a roll-off angle lower than $5°$.[25] The superhydrophobic effect usually occurs through the combination of high roughness and a low surface energy, which results in the material being water repellent.[26]

Synthetic hydrophobic surfaces with a water contact angle above $140°$ were prepared in the mid-1990s by Onda *et al.*,[27] who anodically oxidized an aluminum surface and treated it with a fluorosilane. In this work,[27] the authors suggested also that the hydrophobicity is strictly connected with the fractal nature of the rough surface. Since then, many reports have been published on superhydrophobic surfaces that combine highly rough fractal surfaces with special hydrophobic chemical coatings.[28–30]

In contrast to passive antimicrobial textiles, active antimicrobial textiles contain specific agents that affect microorganisms in an adverse way. Antimicrobial agents can either kill microorganisms or inhibit microbial growth by damaging cell membranes, affecting their permeability, deactivation of certain proteins and enzymes or inhibition of the synthesis of certain proteins and cell organelles vital for the survival and/or growth of such microorganisms. The first activity indicates the destruction of microorganism and is defined as *cidal* (e.g. bacteriocidal, virucidal, fungicidal *etc.*), whereas the second is known as *static* activity (e.g. bacteriostatic, mildstatic *etc.*).

Generally, the kind of bacterial response to biocides is determined by the type of microorganism and the specificity of used antimicrobial agents. There are two types of bacterial resistance mechanisms to biocides: *intrinsic*, which depends on innate bacteria properties, and *acquired*, which is a consequence of genetic mutations. In the case of bacterial spores, mycobacteria and Gram-negative bacteria the predominant mechanism is intrinsic. Generally, Gram-positive bacteria (*e.g. E. coli*) are less resistant to biocides than Gram-negative. Such factors as environmental pH, temperature, atmospheric humidity, moisture content, and the chemical and physical nature of the host material can significantly amplify or weaken the activity of microorganisms.[31]

In antimicrobial textiles, antimicrobial agents are incorporated either into yarns during processing or in the fabric during finishing either physically or chemically or through a combination of physical and chemical means.[32] These antimicrobial agents can be attached for temporary or permanent actions. In the first case, they are slowly released from the surface to which they were incorporated in the first place. These types of finishes are known as *leaching* type finishes, because they lose their antimicrobial properties over time due to constant leaching of the antimicrobial agents out of the surface. Leaching finishes are usually easy to apply, but are not durable, as the amount of antimicrobial is reduced by laundering.[33] Moreover, the migration of agents can cause environment hazards and health problems.[32] In some cases it is possible to regenerate the antimicrobial functions of the fabric (*e.g.* some halamine-modified textiles).[34,35]

In the case of *non-leaching* finishes, the antimicrobial agents are strongly bound, primarily chemically, to the yarn or fabric. They show high durability but are affective only against microorganisms which come in direct contact with the textile surface. These finishes do not cause risk of environmental contamination and are much more safe for users.[33,34]

14.3   Nanoparticles with Antimicrobial Properties

A number of nanoparticles have shown toxicity that can potentially be exploited for microbiocidal purposes. Examples include carbon nanotubes (CNTs), especially with some residue of the metal catalysts used to prepare CNTs, CdS quantum dots, nanoscale titania (titanium dioxide; TiO_2) in rutile, anatase or a mixed phase, ZnO, gold and silver nanoparticles, clay, copper oxide and iron oxide. The antimicrobial action originates from the shape and morphology (*e.g.* in CNTs), photocatalytic properties giving rise to reactive oxidative species (TiO_2 and ZnO) or chemical super action due to higher surface area and activity at the nanometre scale. They are usually broad spectrum anti-microbial agents, although the effectiveness may vary depending on the type of bacteria and fungi.

Nanoparticles of titania and its doped variants have received considerable attention in recent years and have been actively investigated by researchers as well. Titania is known as a stable, long-lasting, safe and broad spectrum antibiotic.[36] Its capability to perform photo-catalysis is the reason for the authors using it in textiles for photo-sterilization applications and has been discussed in Chapter 2. The antimicrobial activity of TiO_2 can be increased by doping with metals, *e.g.* silver (Ag). TiO_2 nanoparticles containing Ag^+ have been widely used as a filler in the manufacture of antibacterial plastics, coatings, functional fibres, dishware and medical facilities, because Ag^+ has a strong antibacterial activity against many kinds of bacteria, even at lower concentrations.[37]

ZnO nanoparticles have been extensively used in functional devices, catalysts, pigments, optical materials, cosmetics, UV absorbers and as additives in many industrial products. ZnO nanoparticles have some advantages compared to silver nanoparticles, such as lower cost, a white appearance and UV blocking properties.[38] A negative surface charge has been deduced for ZnO nano-particles and illumination increased anti-bacterial performance compared to normal conditions.[39] ZnO-doped polyurethane (PU) films exhibited excellent antibacterial activity, especially for *E. coli*, and that the antibacterial activity of ZnO is considered to be due to the generation of hydrogen peroxide (H_2O_2) from their surfaces.[40]

Due to their unique physical, chemical, thermal and mechanical properties, iron oxide nanoparticles offer a high potential for several biomedical applications such as: cellular therapies research, tissue repair, drug delivery, magnetic resonance imaging (MRI) and hyperthermia. Sub-10 nm Fe_3O_4 nanoparticles have been particularly useful as a super-paramagnetic MRI probe that can be made to target specific cells and tissues inside the body. In the past, iron oxide nanoparticles have been revealed to be very efficient at the decontamination of various organic pollutants[41] and can potentially be used as an antimicrobial agent.

There is now a body of literature available on the antimicrobial activity of silver nanoparticles. In fact one of the most lauded commercial successes of nanotechnology came from silver nanoparticle-loaded wound dressings. Wound dressings are manufactured by means of a bi-layer of silver-coated, high-density polyethylene mesh with a rayon adsorptive polyester core. The dressing delivers nano-crystalline silver from a non-adherent, non-abrasive surface.[42] In vitro studies have shown that the sustained release of this ionized nano-crystalline silver maintains an effective anti-bacterial and fungicidal activity.[43] Silver nanoparticles have been previously identified to have both inhibitory and bactericidal effects. Kim *et al.*[44] suggested that the anti-micro-bial mechanism of silver nanoparticles is related to the formation of free radicals and subsequent free radical induced membrane damages.

Another material which may be used to combat/reduce bacteria is clay. Clay's anti-bacterial efficiency, sterilizing effect, adsorption of toxins and membrane coating indicate the potential biomedical applications of nano-clay.[45] The physical absorption of water, toxic bacteria, viruses and organic matter are among the reasons for the healing properties of clay on wound plasters. In addition, metal elements such as silver, copper, manganese, zinc, iron and metal oxides, such as TiO_2, can also be associated with anti-bacterial properties of mineral clay, however, the mechanism of chemical interaction between clay and bacteria is not clearly known.[38]

Zhang *et al.*[46] have proposed the inhibition of growth and multiplication of different microbes with gold nanoparticles that work efficiently against Gram-positive and Gram-negative bacteria and fungi. Nirmala Grace and Pandian[47] used gold nanoparticles as a carrier core coated with antibiotics, such as streptomycin, and proved that gold nano-composites have an intense anti-bacterial efficiency against various Gram-negative and Gram-positive bacteria, *e.g. E. coli* and *Pseudomonas aeruginosa*, and *S. aureus* and *Micrococcus luteus*. These authors[47] suggested that metal nanoparticles may change the metabolite

Table 14.1 A list of common nanoparticles used for antimicrobial activity.

Nanomaterial	Characteristics
TiO$_2$	Anti-bacterial, antifungal, photocatalyst, self-cleaning, UV-protecting, water/air purifier
Silver	Anti-microbial, disinfectant, anti-fungal, UV-protecting
Zinc Oxide	Anti-bacterial, UV-protecting, photo-catalyst
Copper	Anti-bacterial, UV-protecting
Clay	Anti-bacterial, UV-absorber, flame retardant
Gold	Anti-bacterial, anti-fungal

pathway and the release mechanism of bacterial cells, therefore a better anti-bacterial efficiency can be obtained as a result of the strong efficiency of the gold–drug nano-composites. A list of common antimicrobial nanomaterials is given in Table 14.1.

14.4 MRSA-resistant Textiles and a Few Case Studies

MRSA bacteria are one of the most important healthcare problems nowadays. According to Trautmann et al.,[48] the rate of identified MRSA in hospitals has been rising since 1990. *S. aureus* was one of the major reasons for death before antibiotics were invented. Over the past years, it became resistant to several common antibiotics. MRSA affects the convalescence and harms the healing process. Many strategies have been taken to decrease MRSA-related infections. The social awareness of the problem has increased, primarily due to the undertaking of steps such as increased frequency of hand-washing in communities exposed to the harmful function of MRSA.

The idea of antibacterial textiles has been investigated in the recent years. For example, Borkow and Gabbay[49] have highlighted biocidal textiles as a potential antibacterial barrier to combat nosocomial infections. The main steps to achieve the goal of antibacterial textiles are listed as follows:

- the selection of an appropriate antibacterial agent that can potentially be used in contact with human body and
- the development of a suitable procedure for incorporating this agent in the textiles.

Many types of bacteriocidal nanomaterials incorporated into textiles have been investigated and an excellent review on this topic has recently been published by Dastjesrdi and Montazer.[38] They discussed the use of photocatalytic nanomaterials, *i.e.* TiO$_2$,[50] biocidal Ag[51,52] and CuO[53] in textiles. Some of these compounds match the needs of antimicrobial textiles market.

Kangwansupamonkon et al.[54] have investigated antimicrobial textiles loaded with apatite-coated TiO$_2$ as a potentially good material for sterilization purpose. They have incorporated these particles into textiles by a pad-dry-cure method, in which textile samples were immersed for 1–2 min in a colloidal solution containing apatite-coated TiO_2, then put through padding mangle. The textiles were dried at 100°C for 5 min and cured at 150°C for 3 min. The whole impregnation process is simple and quick. The apatite-coated TiO_2 solution was prepared by immersing TiO_2 powder in distilled water by 5% wt., adding acrylic binding agents and mixing for 30 min. The size of the apatite-coated TiO_2 particles was determined to be 390 ± 60 nm using a dynamic light scattering technique. Various bacteria were tested to investigate the anti-microbial effectiveness of the textile: E. coli (ATCC 25922), M. luteus (ATCC 9341), S. aureus (ATCC 6538) and MRSA (DMST 20627).

The antibacterial activity of the particle incorporated textile was bench-marked against the antibacterial activity of the particles on their own. Also, the influence of irradiation was examined using 20 W black-light lamps (for UV-A radiation), fluorescent-light lamps for visible light and darkness. Irradiation times varied from 0–24 h. Cotton fabrics impregnated with apatite-coated TiO_2 were examined using American Association of Textile Chemists and Colorists (AATCC) 100-1998 method.[55] Fabrics were cut into pieces (diameter of approx. 5cm), autoclaved and exposed to 1 ml of suspension containing 10^5 colony-forming units (CFU) ml^{-1} in tryptic soy broth (TSB). Textiles were also irradiated for 24 h using the same source of light as in case of apatite-coated TiO_2, which showed very good antibacterial activity. For S. aureus, MRSA and E. coli the minimum inhibitory concentration (MIC) of apatite-coated TiO_2 was equal 0.31% (w/v). On the other hand, minimal biocidal concentration (MBC) for these bacteria strains mentioned above was more than 5% (w/v). A reduction in bacteria was visible in the case of all irradiated samples compared with the bacteria suspension without TiO_2 as a control. The black-light irradiation was the most effective, and was equivalent to UV-A radiation. The difference between MIC and MBC values indicates that not only does TiO_2 kill microorganisms, but there is contribution also from the reactive oxygen species (ROS) of the suspension. Apatite-coated TiO_2 textiles showed antibacterial activity, with a 15% bacteria reduction in MRSA species.

In 2009, O'Hanlon and Enright have reported a novel bacteriocidal coating active against MRSA.[56] The experiment compared commercially available products Cliniweave[R] coated textile and 'see it SAFE[R]'. Both products have been claimed as being very efficient in bacteria reduction. For example, anti-bacterial, antifungal, anti-odour, anti-static and thermodynamic properties of 'see it SAFE[R]' have been well publicized.[57] According to the product's website, 'see it SAFE[R]' is a textile covered with a silver layer. Silver fibres are woven into fabrics, suggesting that the product is reusable. Cliniweave[R], on the other hand, is a textile that was produced especially for hospital curtains[58] and does not contain silver.

O'Hanlon and Enright[56] treated polyester fabrics with 3% Cliniweave[R]. Antibacterial analysis was performed on MRSA clone 15 and the methicillin-sensitive S. aureus (MSSA) (NCTC6871). Textiles were autoclaved and 5 cm^2 pieces were immersed in 10 ml of 10^5 CFU and 5×10^5 CFU solutions. After 1 h of incubation of textiles in bacteria solutions, there was no bacteria growth observed on the 3% Cliniweave[R]-treated polyester samples.

In the next step, the researchers tested the antibacterial effect of Cliniweave[R] in comparison to antibiotics. Results showed that Cliniweave[R] has a very good MIC against MRSA and this value is much better than some antibiotics. O'Hanlon and Enright also studied the antibacterial effect of Cliniweave[R] - and 'see it SAFE[R]'-treated fabrics in accordance to the time required to kill MRSA. The aim of the experiment was to compare textiles with and without silver. Again, Cliniweave[R] appeared to be a better product, as, after 60 min, there was no MRSA growth, and in case of 'see it SAFE[R]' there was a continuous bacteria growth. The activity of Cliniweave[R] is very rapid, much better than other commercially available products containing silver.

Parthasarathi and Thilagavati[59] found that nanoparticles of titania have excellent antibacterial properties against *S. aureus* and *Klebsiella pneumonia*. While untreated textile samples showed no reductions in the *S. aureus* population, woven 100% cotton, woven 45/55 polycotton, knitted 100% cotton and knitted 45/55 polycotton exhibited 85%, 93%, 75% and 79% reductions.

We have recently investigated the biocidal effects of textiles with nanoparticles incorporated by ultrasonic impregnation and dip-coating methods.[60] Nanopowders of TiO_2 doped with silver (Ag–TiO_2) were used. Textiles loaded with nanomaterials were tested against the bacteria MRSA and *E. coli*, and the fungi *Candida albicans*. These species were chosen with an intention to explore the utility of such textiles in hospital upholstery and bed linings, as MRSA is one of the major causes of hospital-acquired infections, *E. coli* is omnipresent in hospital areas and hospital garments are ideal places for fungi to grow. The results showed that textiles impregnated with Ag–TiO_2 have excellent antibacterial activity against MRSA and *E. coli*, with almost 100% reduction in 20 min. These textiles showed good antifungal properties as well, with 80% reduction in 20 min.

14.5 Key Issues Associated with MRSA-resistant Textiles

Although recent results have shown that textiles can be modified using nanotechnology to successfully resist MRSA, the technological development to a commercial product needs to consider of issues pertaining to both textile engineering and nanotechnology. This makes the use of nanoparticles in antimicrobial textiles quite a challenging field, starting from the selection of an appropriate nanomaterial to the incorporation into the fabric, to the process of reliability of antimicrobial functions with time and durability of textiles. The technology has to be optimized so as to pose minimal hazard to the environment during its complete lifecycle. To do this, textile engineers, materials scientists, microbiologists and toxicologists need to work together to identify and address the key issues related to successful commercialization of MRSA-resistant textiles.

There are many ways of preparing antimicrobial textiles but the key consideration is to whether to incorporate the antimicrobial agent in or on the yarn, or to incorporate it in the fabric during weaving or knitting or during the finish process. The yarn and the fabric will then need to go through a number of standardized tests specified by American Society of Testing of Materials (ASTM) or AATCC.

To achieve yarn strength, natural staple fibres must be twisted, entangled or bonded together. For continuous filaments this is not essential, although restricted interlacing or twisting can be incorporated to offer some cohesion to the yarns. If a rougher and more natural fabric is required for an application, then coarse tows are prepared and incised into staple fibres for spinning. Texturing processes can provide additional stretch and bulk properties to constant filament yarns and can be built-in to lay down filaments in helical-shaped paths. The majority of textiles are usually planar sheet materials, however, other patterns also occur such as:

- Woven fabrics: warp yarns pass under and over weft yarns
- Braided fabrics: interlacing occurs between diagonal sets of yarns
- Knitted fabrics: yarns are looped collectively in nearby rows, either across the length in weft-knitting or along the length in warp-knitting
- Non-woven fabrics: these are essentially random sheets of fibres, which are held together by adhesive bonding, entanglement or stitching
- One-dimensional fabric assemblies: these assemblies are generally used in cords and ropes
- Three-dimensional fabric assemblies: these assemblies are generally used as composite pre-forms; both in the appearance of shaped sheets or thick structures, with the addition of other applications, such as knitted garments and conveyor belts.[61]

The development of a robust and precisely controllable methodology for durable attachment of antimicrobial agents to these fibres, yarns, fabrics and textiles, while still keeping the desired antimicrobial properties of the agent, is a major concern. Today, the nanoparticles are incorporated into textiles mainly in two ways: (i) a direct addition of the antimicrobial agents to the polymer precursor before extrusion into a fibre or yarn, or (ii) fixing during the finishing stages of fabric preparation, mainly by coating. The choice of method depends on the functionality desired and the durability aimed for. The most frequently used method of coating is padding, in which the nanoagents are attached to the fabrics using padders that are adjusted to appropriate speed and pressure. Alternative ways can be spraying, transfer printing, washing and rinsing.[62] Furthermore, there is a need to control the fixation of nanoparticles on or into textiles to ensure the controllable release of the agents.

Another issue is the availability of nanoparticles with controlled size and proper degree of segregation as nanoparticles have a great tendency to agglomerate, which reduces the effective surface area and reduces antimicrobial activity. Recent production methods of nanoparticles can be broadly divided into three categories: milling, wet and dry synthesis. In the milling approach the nanoparticles are produced by a mechanical breakdown of larger particles.

This method is time consuming, and usually results in a much broader size range of particles. In wet and dry synthesis, nanoparticles are produced from precursors. Sol-gel and precipitation methods are commonly used wet synthesis methods. Dry synthesis usually involves some kind of plasma or evaporation. Both wet and dry processes can be controlled to yield a very small size range of particles. Scalability is also an important problem. Many attractive processes of nanoparticles production have been developed at the laboratory scale, but they are not yet commercialized, because of constraints, including scalability considerations and precursor costs.[63]

While MRSA-resistant textiles offers a very promising solution for contaminated hospital environment, there are some crucial and actual scientific problems connected with the application of such textiles. The selection of the antibacterial agent is the most prominent of all. Current organic-based antibacterial coatings used in reusable hospital garments suffer from low durability and lose effectiveness after few washes. The selected antimicrobial agent must be active, must possess antimicrobial activity and show good compatibility with the human skin if these are to be used in wearable apparels or clothing that may come in contact with human skin. Also, the incorporation process should not introduce any harmful substances, detrimental factors or allergens. The concentration of antimicrobial agents must be controlled so that maximum antimicrobial effect is achieved with the use of the least amount of the agent.

Antimicrobial nanoparticles offer a great advantage as the high level activity of nanoparticles means that a lower amount of such nanoparticles can provide effective antimicrobial function. Incorporation of nanoparticles is of great importance here, as the challenge of keeping nanoparticles well dispersed within the textile to ensure that a high level of active surface areas of these particles are still available for antimicrobial activity is not trivial. The dislodgement of these nanoparticles is another important issue. The bonding between the antimicrobial agent and the textile surface is extremely important. While in conventional antimicrobial textiles this is achieved by simply adding a binder, such a solution for binding nanoparticle will work only to the detriment of the antimicrobial activity of these particles as the binder will reduce the effective surface area of the nanoparticles. The key challenge of incorporating antimicrobial nanoparticles into textiles is to ensure this local interfacial bonding between the nanoparticle and the textile over a few millions of square meters and minimize dislodgement. The environmental fate of dislodged nanoparticles is not known currently and there are contradictory data in the literature as regards to the safety of nanoparticles. Many investigations have been conducted on the hazard and toxicity of nanomaterials,[64,65] but the long-term consequences of using nanomaterials are still unknown. Fortunately, this is an active area of research and the standardization of toxicity data has been prioritised by many national, supranational and international research funding bodies and standards organizations.

In respect to the antibacterial effectiveness, it also important to highlight that MRSA is a bacteria that becomes resistant to many conditions, so the bacteriocidal effect of nanomaterials should be determined very carefully.

As described above,[54] in some cases not is only the nanomaterial is likely to display its antimicrobial activity, but ROS can also occur and be responsible for killing bacteria. These factors should be considered when designing nano-particle-loaded textiles for use in MRSA resistance.

14.6   Applications of MRSA-resistant Textiles

One of the most desirable applications of MRSA-resistant textile is in hospitals. The hospital environment is very favourable for the growth of MRSA bacteria that may cause severe infections, affect wounds healing *etc.* Bacteria strains which are obviously present in hospitals and clinics are both Gram-negative (*i.e. E. coli*) and Gram-positive (*i.e.* MRSA). Such infections are usually spread all over the hospitals through ventilation, laboratory equipment and garments. On the other hand, people become affected during a direct patient–patient or staff–patient contact. Areas where such contacts or meetings take place will be where MRSA-resistant textiles can be used.

When biocidal nanomaterial is incorporated into textiles an effective antimicrobial barrier is introduced. The main goal of antibacterial textile is to suppress the growth of harmful microbes. Many social and scientific programs have been implemented to hospitals, with better or worse results.[11] According to the literature,[66] the higher frequency of hand-washing is not sufficient to reduce MRSA contamination. As already mentioned earlier, antibacterial textiles could be used also as curtains in hospitals. Curtains are touched by many people. There is no strict hand-washing rules for people visiting sick people. Even patients do not adhere to the rule of washing hands with antimicrobial substances, which increases the risk of spreading bacteria. MRSA-resistant textiles can be successfully used in a non-surgical environment as an important part of the healthcare system. Antibacterial textile could be used in clothing for medical staff, linens for patients or for general hospital utility garments.

There are many ways of producing textiles with antibacterial activity. Special, antimicrobial sprays or liquids can be used. Choosing appropriate nano-materials and the minimal active concentration that leads to killing microbes is a good direction to follow in obtaining very promising and effective results.

14.7   Conclusions

Nanotechnology is becoming important for the textile industry because of the unique technical and economic advantages that nanomaterials offer. Traditional methods used to impart different characteristics and functionality to textiles often does not lead to the desired results in an efficient way. Fabrics lose their properties during wearing, washing, dry cleaning and hot pressing. Nanotechnology may overcome these limitations of conventional methods and provide durability of textile functions. Nanoparticles that are applied onto the fabric have both – the large surface area-to-volume ratio and high surface energy. Furthermore, nanoparticles that are attached to the fabrics may not significantly affect physical and mechanical properties such as breathability, hand feel, strength, wetting *etc.* Commercially available antimicrobial finishes are resistant to body fluids and processes of sterilization or disinfection. This last property is particularly important in developing MRSA-resistant textiles. We have reviewed works on textiles that offer resistance to MRSA and discussed how antibacterial nanoparticles can provide significant benefits when used in textiles for MRSA resistance in comparison to conventional textiles.

References

1. E. Russell, *Text. Horiz.*, 2002, **9/10**, 7–9.
2. K. F. Anderson and R. A. Sheppard, *Lancet*, 1959, **1**, 514–515.
3. E. McNeil and M. Greenstein, *Proc. Chemical Specialties Manufacturing Association*, 1961, 134–141.
4. E. McNeil and Dev, *Ind. Microbiol.*, 1964, **5**, 30–35.
5. J. C. Wiksell, M. S. Pickett and P. A. Hartman, *Appl. Microbiol.*, 1973, **25**, 431–435.
6. R. Burgess, *J. Appl. Microbiol.*, 1954, **17**, 230–245.
7. M. M. Huycke, D. F. Sahm and M. S. Gilmore, *Emerg. Infect. Dis.*, 1998, **4**, 239–249.
8. A. L. Panlilio, D. H. Culver, R. P. Gaynes, S. Banerjee, T. S. Henderson, J. S. Tolson and W. J. Martone, *Infect. Control Hosp. Epidemiol.*, 1992, **13**, 582–586.
9. S. Pastila, K. T. Sammalkorpi, J. Vuopio-Varkila, S. Kontiainen and M. A. Ristola, *J. Hosp. Infect.*, 2004, **58**, 180–186.
10. J-U. S. Jensen, E. T. Jensen, A. R. Larsen, M. Meyer, L. Junker, T. Rønne, R. Skov, O. B. Jepsen and L. P. Andersen, *J. Hosp. Infect.*, 2006, **63**, 84–92.
11. A. Al-Barrak, J. McLeod, J. Embil, G. Thompson, F. Aoke and L. Nicolle, *Am. J. Infect. Control*, 1998, **26**, 189.
12. G. A. Noskin, V. Stosor, I. Cooper and L. R. Pererson, *Infect. Control Hosp. Epidemiol.*, 1995, **16**, 577–581.
13. S. Oie, S. Suenaga, A. Sawa and A. Kamiya, *Jpn. J. Infect. Dis.*, 2007, **60**, 367–369.
14. R. Porwancher, A. Sheth, S. Remphrey, E. Taylor, C. Hinkle and M. Zervos, *Infect. Control Hosp. Epidemiol.*, 1997, **18**, 771–774.
15. S. Oie, I. Hosokawa and A. Kamiya, *J. Hosp. Infect.*, 2002, **51**, 140–143.
16. H. F. Bonilla, M. J. Zervos and C. A. Kauffman, *Infect. Control Hosp. Epidemiol.*, 1996, **17**, 770–771.
17. S. Oie and A. Kamiya, *J. Hosp. Infect.*, 1996, **34**, 145–149.
18. S. Oie and A. Kamiya, *Biomed. Lett.*, 1996, **57**, 117–119.
19. N. Dickgiesser and C. Ludwig, *Zentbl. Bakteriol. Hyg. Abt. 1 Orig B.*, 1979, **168**, 493–506.
20. F. Pettit and E. J. L. Lowbury, *J. Hyg.*, 1968, **66**, 393–406.

21. C. Wendt, B. Wiesenthal, E. Dietz and H. Ruden, *J. Clin. Microbiol.*, 1998, **36**, 3734–3736.
22. N. Hamza, G. Bazoua, Y. Al-Shajerie, E. Kubiak, P. James and C. Wong, *Ann. R. Coll. Surg. Engl.*, 2007, **89**, 665–667.
23. J. A. Jernigan, A. Pullen, F. S. Nolte, P. Patel and D. Rimland, *Clin. Infect. Dis.*, 1997, **25**, 363.
24. C. Neinhuis and W. Barthlott, *Ann. Bot.*, 1997, **79**, 667–667.
25. W. Barthlott and C. Neinhuis, *Planta*, 1997, **202**, 1–8.
26. H. J. Lee, *J. Mater. Sci.*, 2009, **44**, 4645–4652.
27. T. Onda, S. Shibuichi, N. Satoh and K. Tsuji, *Langmuir*, 1996, **12**, 2125–2127.
28. J. J. Niu, J. N. Wang and Q. F. Xu, *Langmuir*, 2008, **24**, 6918–6923.
29. J. J. Niu and J. N. Wang, *Cryst. Growth Des.*, 2008, **8**, 2793–2798.
30. B. Mahltig and A. Fischer, *J. Polym. Sci., B: Polym. Phys.*, 2010, **48**, 1562–1568.
31. A. D. Rusell, *Int. Biodeterior. Biodegradation*, 1995, **36**, 247–265.
32. R. Jantas and K. Górna, *Fib. Text. East. Eur.*, 2006, **14**, 88–91.
33. T. Ramachandran, K. Rajendrakumar and R. Rajendran, *India J. Text.*, 2004, **84**, 42–47.
34. G. Sun and S. D. Worley, *J. Chem. Educ.*, 2005, **82**, 60–64.
35. L. Song and S. Gang, *Ind. Eng. Chem. Res.*, 2006, **45**, 6477–6482.
36. S. Roessler, R. Zimmermann, D. Scharnweber, C. Werner and H. Worch, *Colloids Surf., B*, 2002, **26**, 387–395.
37. Q. Cheng, C. Li, V. Pavlinek, P. Saha and H. Wang, *Appl. Surf. Sci.*, 2006, **252**, 4154–4160.
38. R. Dastjerdi and M. Montazer, *Colloids Surf., B*, 2010, **79**, 5–18.
39. Q. Li, S. L. Chen and W. C. Jiang, *J. Appl. Polym. Sci.*, 2007, **103**, 412–416.
40. J. H. Li, R. Y. Hong, M. Y. Li, H. Z. Li, Y. Zheng and J. Ding, *Prog. Org. Coat.*, 2009, **64**, 504–509.
41. W. H. Suh, K. S. Suslick, G. D. Stucky and Y. H. Suh, *Prog. Neurobiol.*, 2009, **87**, 133–170.
42. E. Mantovani, P. Zappelli, J. Conde, R. Sitja and F. Periales, Report on Nanotechnology & Textiles. Medical Textiles, Sport/Outdoor Textiles, 2010, http://www.nanotec.it/documenti/FocusReport2010_MedicalTextiles_Sport_ OutdoorTextiles.pdf.
43. J. B. Wright, K. Lam, D. Hansen and R. E. Burrell, *Am. J. Infect. Control*, 1999, **27**, 344–350.
44. J. S. Kim, E. Kuk, K. N. Yu, J. H. Kim, S. J. Park, H. J. Lee, S. H. Kim, Y. K. Park, Y. H. Park and C.Y. Hwang, *Nanomedicine NBM*, 2007, **3**, 95–101.
45. M. J. Wilson, *J. Chem. Ecol.*, 2003, **29**, 1525–1547.
46. Y. Zhang, H. Peng, W. Huang, Y. Zhou and D. Yan, *J. Colloid Interface Sci.*, 2008, **325**, 371–376.
47. A. Nirmala Grace and K. Pandian, *Colloids Surf. A*, 2007, **297**, 63–70.
48. M. Trautmann, A. Pollitt, U. Loh, I. Synowzik, W. Reiter, J. Stecher, M. Rohs, U. May and E. Meyer, *Am. J. of Infect. Control*, 2007, **35**, 643–639.

49. G. Borkow and J. Gabbay, *Med. Hypotheses*, 2008, **70**, 990–994.
50. A. L. Castro, M. R. Nunes, M. D. Carvalho, L. P. Ferreira, J. C. Jumas, F. M. Costa and M. H. Florêncio, *J. Solid State Chem.*, 2009, **182**, 1838–1845.
51. B. Xin, L. Jing, Z. Ren, B. Wang and H. Fu., *J. Phys. Chem.*, 2005, **109**, 2805–2809.
52. S. P. Reddy, A. Venugopal and M. Subrahmanyam, *Water Res.*, 2007, **41**, 379–386.
53. I. Perelstein, G. Applerot, N. Perkas, E. Wehrschuetz-Sigl, A. Hasmann, G. Guebitz and A. Gedanken, *Surf. Coat. Technol.*, 2009, **204**, 54–57.
54. W. Kangwansupamonkon, V. Lauruengtana, S. Surassmo and U. Ruktanonchai, *Nanomedicine NBM*, 2009, **5**, 240–249.
55. http://www.aatcc.org.
56. S. J. O'Hanlon and M. C. Enright, *Int. J. Antimicrob. Agents*, 2009, **33**, 427–431.
57. http://www.seeitsafe.co.uk/download/toray_brochure.pdf
58. http://www.nwupc.ac.uk/docs/Whitakers_Curtain_Cliniweave_Brochure.pdf
59. P. Sarathi and G. Thilagavathi, *J. Text. Apparel*, 2009, **6**, 8.
60. K. Kowal, K. Wysocka-Król, E. Dworniczek, R. Franiczek, M. Kopaczyńska, I. Buzalewicz, M. Wawrzyńska, S. A. M. Tofail and H. Podbielska, ICAR2010, *Int. Conf. Antimicrob. Res.*, 2010, 258.
61. A. Mortensen, *Concise Encyclopaedia of Composite Materials*, Elsevier, Amsterdam, 2nd edn, 2007.
62. Y. W. H. Wong, C. W. M. Yuen, M. Y. S. Leung, S. K. A. Ku and H. L. I. Lam, AUTEX *Res. J.*, 2006, **6**, 1–8.
63. S. Kathirvelu, L. D'Souza and B. Dhurai, *Ind. J. Sci. Tech.*, 2008, **1**, 1–10.
64. R. Landsiedel, M. D. Kapp, M. Schulz, K. Wiench and F. Oesch, *Mutat. Res.*, 2009, **681**, 241–250.
65. V. Stone, B. Nowack, A. Baun, N. van den Brink, F. von der Kammer, M. Dusinska, R. Handy, S. Hankin, M. Hassellöv, E. Joner and T. F. Fernandes, *Sci. Total Environ.*, 2010, **408**, 1745–1754.
66. G. L. French, J. A. Otter, K. P. Shannon, N. M. T. Adams, D. Watling and M. J. Parks, *J. Hosp. Infect.*, 2004, **57**, 31–37.

CHAPTER 15

Inhibition of Encrustation in Urological Devices

S. ROBIN,*[a] T. SOULIMANE[a] AND S. LAVELLE[b]

[a] University of Limerick, Department of Chemical and Environmental Sciences, National Technology Park, Limerick, Ireland; [b] Cook Ireland Ltd., O'Halloran Road, National Technology Park, Limerick, Ireland

15.1 Introduction

Over 100 million urological devices, including urethral catheters, ureteral stents and penile prostheses, are inserted each year in North America alone.[1] Like all abiotic surfaces, they are more sensitive to bacterial adhesion than living tissues due to the lack of the active mechanisms to respond to bacterial infection which, when it happens in the vicinity of the urological device, rapidly leads to biofilm formation.[2] Biofilm-related infections remain the major cause of ureteral stent failure, despite investment of significant resources and several decades of research aimed at its prevention and eradication.[3,4] The risk of infection is directly related to the length of time the device is indwelled. For catheters, this rate has been calculated at 3% per day,[5] which means that all patients undergoing catheterization for >28 days will ultimately develop urinary tract infections (UTIs). Numerous studies revealed that the most implicated bacteria in the development of infection of the urinary tract are *Escherichia coli*, *Pseudomonas aeruginosa*, *Klebsiella pneumoniae*, *Proteus mirabilis*, *Staphylococcus epidermidis*, *Enterococcus faecalis*, *Providencia stuartii*, *Morganella morganii* and *Staphylococcus aureus*.[6,7] The described problem results in

RSC Nanoscience & Nanotechnology No. 21
Biological Interactions with Surface Charge in Biomaterials
Edited by Syed A. M. Tofail
© Royal Society of Chemistry 2012
Published by the Royal Society of Chemistry, www.rsc.org

millions of device-related infections and billions of dollars spent in associated healthcare interventions every year.[1]

Furthermore, these infections can develop into more serious conditions, such as recurrent UTIs, calculi and encrustation and ultimately sepsis.[3] Infection of devices remaining in contact with urine by organisms such as *P. mirabilis* are associated with the formation of crystals of magnesium ammonium phosphate (struvite) and calcium phosphate (hydroxyapatite) which results in mineralization and encrustation of those devices, causing their blockage.[2]

In the absence of microbial infection or urinary devices, the formation of calculi, abnormal mineral concentrations, in the urinary tract of 'stone-former' patients can also naturally be initiated anywhere from the kidney to the urinary bladder. The formation mechanism is complex and its pathogenesis is not completely understood.[8] Under normal circumstances, significant amounts of crystals are excreted in urine without calculus formation and this is because crystallization inhibitors are normally present. Inhibitors include magnesium, citrate, pyrophosphate, peptides and heavy metals.[8] Indwelling of artificial prosthesis in the urinary tract promotes mineral deposition and result in encrustation of the device and this occur in almost all patients with such implants, irrespective of whether they are known 'stone formers' or not. The eventual associated microbial infection results in accelerated encrustation and further complications.

The most devastating consequence of chronic encrustation is obstruction of urine flow which is the main function of the catheter or stent, thus requiring the need for removal. The presence of this hard gangue can additionally result in tissue trauma during the removal of the implants.

This Chapter will describe the complex problem of crystalline biofilm and encrustation in urological devices, the particular the threat that *P. mirabilis* represents, and review the strategies employed to overcome this, which for several decades has mostly been focused on the modification of surfaces of urological devices.

15.2 Crystalline Biofilm Formation and Encrustation of Urological Devices

Most of the nosocomial infections occur in the urinary tract[9] and approximately 80% of those result from implantation of urologic devices, with infections related to indwelling bladder catheterization being the most common and ureteral stent-associated infections becoming more and more existent. It is known that bacterial colonization of stents and the encrustation are closely related.[10] The phenomenon of crystalline biofilm formation and encrustation in urological devices is extremely complex. The physiology of the patient and the nature of biomaterials play a more important role in stent encrustation than in the case of catheters, and studies have been carried out that could not establish a strong association between bacterial colonization of ureteric stent encrustation.[11] Nevertheless, numerous studies showed that 40–90% of retrieved stents have adherent pathogens[12,13] and they are isolated from the stent surface despite the absence of contamination in the collected urine. After infection of the device and bacterial attachment, the biofilm develops to form a three-dimensional structure using extracellular polysaccharidic matrix as cement.[14] In the case of urological devices, eventual biofilm formation will be accompanied by the apparition of inorganic microcrystalline formations.

Human urine is a very complex medium. It is produced as a sterile fluid and contains by-products of the body metabolism that are secreted by the kidneys through the urethra, many of which are rich in nitrogen. The pH of urine is normally acidic, maintained between 5 and 6. When the urinary tract is infected by microorganisms producing urease or urea amidohydrolase (EC 3.5.1.5), this equilibrium is affected and the pH drastically rises. Urease is an enzyme capable of degrading urea, a major component of human urine, and in the presence of this enzyme it is hydrolysed to ammonia and carbonate:

$$H_2NCONH_2 + H_2O \xrightarrow{\text{urease}} 2NH_3^+ + H_2CO_3^{2-} \qquad (1)$$

In aqueous solution, the resulting carbonic acid and two ammonia molecules produced from the above reaction exist in equilibrium with the hydrogenocarbonate and ammonium ions:

$$H_2CO_3 + H_2O \rightleftharpoons H_3O^+ + HCO_3^- \qquad (2)$$

$$2NH_3 + 2H_2O \rightleftharpoons 2OH^- + 2NH_4^+ \qquad (3)$$

The production of ammonium results in an increase in the proportion of hydroxide ions and, hence, the urine becomes more alkaline, reaching a pH value of 8 or 9.[15] Every sample of urine has a characteristic pH, termed the nucleation pH.[16] Once it is reached, it causes supersaturation and crystallization of magnesium and calcium phosphates[2,17] as struvite ($MgNH_4PO_4 \cdot 6H_2O$) and brushite ($CaHPO_4 \cdot 2H_2O$), hydroxyapatite [$Ca_{10}(PO_4)_6 \cdot (OH_2)$] or carbonate-apatite [$Ca_{10}(PO_4)_6(CO_3)$]. Additionally, ammonium and carbonate ions generated by the activity of urease participate in the formation of struvite and carbonate-apatite respectively.[18] Remarkably, the powerful role of urease in virulence has been demonstrated for other pathogenic microorganisms such as *Helicobacter pylori*.[18]

Apatite forms small amorphous crystalline structures, while struvite, in a form of large coffin-shaped crystals, has been shown to lie on the amorphous apatite layer in urine and develop biofilms. Encrustation by struvite and hydroxyapatite is usually reported to occur *in vivo* and *in vitro* for urethral catheters.[19] In case of ureteral stents, crystals of calcium oxalate (CaC_2O_4)[20] are also observed and they can either accompany struvite and hydroxyapatite crystals or exist solely on those devices. Many factors influence the composition and amount of adsorbed mineral material on urological devices, such as urine composition, which itself varies with the physiopathological state of the patient, time of the day and diet.[21] It is not surprising that the nature of the biomaterial also has an effect on encrustation and differences in the apatite/struvite ratio have been reported depending on material tested. Furthermore, this ratio was shown to evolve with the time of indwelling, with the quantity of magnesium crystals increasing gradually after 2 weeks of implantation.[10]

Initial formation of an amorphous apatite layer seeding on the surface of the material constitutes the first step of encrustation. The seeding of such deposit can be triggered by the presence of irregularities on the surface of the device acting as nucleation sites for crystal growth.[22] This early formation of calcium phosphate crystals (Figure 15.1a) has since been elegantly documented for catheters and appears to be not only the initiator of the encrustation but also a support for bacteria, especially *P. mirabilis*, that adhere to it and form biofilms.[23] The aggregates of bacteria can also adhere to inorganic salt crystals in solution. This further facilitates bacterial colonization of the device surface and, once the crystalline material is deposited, the crystalline biofilm is formed (Figure 15.1b). Its subsequent development leads to the blockage of the device.[24]

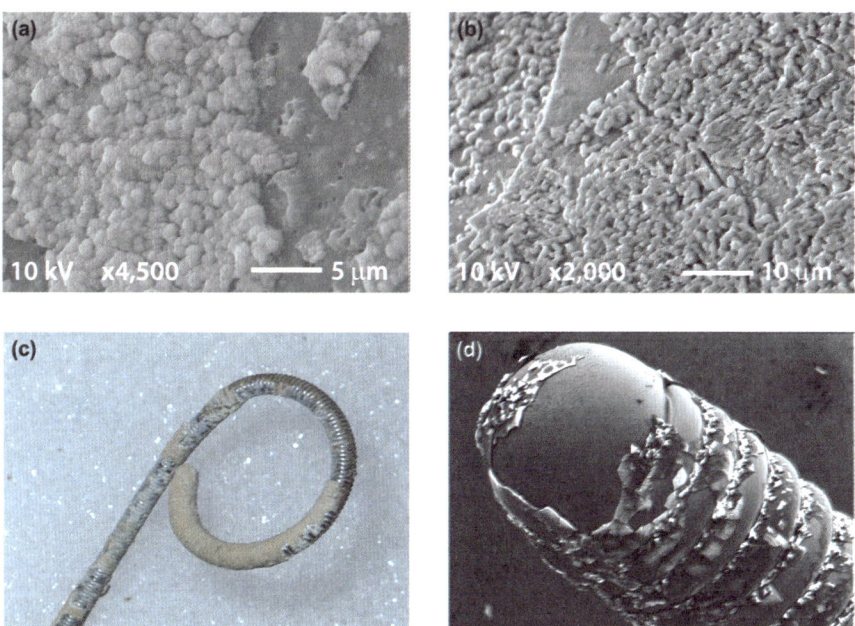

Figure 15.1 Encrustation of urological devices. (a) scanning electron microscopy (SEM) micrograph illustrating the foundation layer of amorphous apatite formed at the surface of a polyurethane-based stent; (b) SEM micrograph showing early crystalline biofilm formed by *P. mirabilis* on the apatite foundation layer; (c) explanted metallic ureteral stent showing encrustation; (d) the distal segment of an explanted metallic stent revealing encrustation deposits by SEM.

Although the importance of bacterial infection in the encrustation phenomenon is obvious, the occurrence of encrustation in sterile urine should not
be neglected.[25] For 'stone former' patients, in absence of infection, calcium
oxalate, which can reach supersaturation in acidic conditions, is the main cause
of encrustation of ureteral devices.[26] Additionally, uric acid has also been
shown to be present in this type of encrustation. Those compounds are found in
renal and bladder calculi.[27] Such a characteristic was demonstrated on full
metallic stents where, in absence of bacterial infection, different levels of
encrustation, comprising mainly of calcium oxalate, could be seen (Figures
15.1c and 15.1d).

15.3   *Proteus mirabilis* and Urological Infections

Most of organisms that cause biofilm-related UTIs exhibit numerous adapta-
tions to human host defences as they naturally come from our own skin, oral
cavity or gastrointestinal tract. *P. mirabilis*, one of the most common UTI-
associated bacterium, is a part of the normal flora of the human gastrointestinal
tract and it can also be found in natural environments such as water and soil.
This Gram-negative bacterium causes 90% of all '*Proteus*' infections in humans
and is most commonly associated with patients suffering from recurrent
catheter encrustation and blockage. There are epidemiological and experimental evidence that *P. mirabilis* is the primary bacterium responsible for the
formation of crystalline biofilms on catheters.[24] Moreover, *Proteus* species
account for 70% of all bacteria isolated from bacteria-induced bladder and
kidney stones.[28]

   P. mirabilis adheres to abiotic surfaces with ease. It can colonize all the biomaterials currently used in the production of urinary devices such as ethylene,
propylene, polystyrene, sulfonated polystyrene, silicone, Teflon[®] and polyurethane.[29–31] After implantation, the surface of catheters or stents is rapidly
covered by proteins and cell debris constituting a conditioning film. *P. mirabilis*
can adhere to this film *via* adhesins that are expressed at its surface and are able
to recognize and bind specific targets on the surface.[1] Nevertheless, this facultative anaerobic rod-shaped bacterium can also bind to inert surface without
specific ligands being present.[32] The recent availability of the *P. mirabilis* genome[33] shed light on the molecular bases that make this bacterium a master of
adhesion, biofilm formation and mobility. The access to the genetic information
allowed the investigation of several of *P. mirabilis* adhesins and revealed the
elaborate tools that this organism utilizes to autoaggregate and form biofilms.[31,34] Once attached to the surface, *P. mirabilis* has the ability to elongate
itself and secrete a polysaccharide, making it extremely motile on items such as
medical equipment. This swarming ability may well have a role in the initiation
of catheter-associated UTIs by facilitating the movement of *P. mirabilis* from
the skin into the bladder through the catheter.[24] Similarly, this extreme mobility
of the microorganism was also reported for ureteral stents[35] and it also mediates

spreading of cells from mature biofilm, participating in the further colonization of a surface.

P. mirabilis produces a very potent urease, much more efficient than in other bacteria, *e.g. M. morganii, P. stuartii, K. pneumoniae, Proteus tettgeri* or *Proteus vulgaris*.[36] As described above, urease hydrolyses urea into ammonia and thus increases the urine alkalinity, which further leads to the formation of crystals of struvite, calcium carbonate and/or apatite. Additionally, the role of lipopolysaccharides, the major components of the outer membrane, in binding cations (Ca^{2+} and Mg^{2+}) and promoting their crystallization by initiating their supersaturation has been demonstrated.[37] Similarly, the crystallization and crystal growth of struvite has been shown to be enhanced by the nature of polysaccharides in the *P. mirabilis* capsule.[38] Also, *P. mirabilis* can be found adhered on the surface of crystals or aggregates of crystals, which participates to the deposition of the bacteria on to the surface, colonization of the device and biofilm formation as described above.

The colonization of surfaces in contaminated urine is a quick phenomenon. It has been shown that after only 4 h pioneer cells can form microcolonies and after 6 h the urine pH rises and crystal formation is observed. As the normal procedure in the case of blocked catheters is to replace it, the new device will be immediately put in contact with contaminated urine already containing crystal aggregates. Hence, the process of encrustation and subsequently formation of mature crystalline biofilm will be rapidly repeated.[17,23]

15.4 Strategies to Reduce Bacterial Adhesion, Biofilm Formation and Encrustation in Endourological Devices

Although urinary implants have proven to be a great tool in physician's armamentarium, their encrustation and associated patient discomfort, tissue trauma and morbidity still present a great challenge to researchers and manufacturers today. For decades, a tremendous effort has also been put in development of urinary devices resistant to biofilm formation and encrustation, with several major avenues of research being of particular interest. However, the plague of ureteral stent encrustation dating back to the 1970s when the 'Double-J' design first emerged is still a prevalent issue today.

Synthetic polymeric compounds, including polyurethane, silicone, Silitek[R] (Surgitek, WI, USA), C-Flex[R] (Cook Medical) and Percuflex[R] (Boston scientific, MA, USA),[39,40] have been extensively used in the development of ureteral stents. Silicone is considered a gold standard with respect to biocompatibility,[41] while polyurethane is the most cost effective material, however, with poor biocompatibility that results in significantly higher urothelial ulceration and erosion.[42] Nevertheless, all of these materials are ultimately susceptible to bacterial adhesion and encrustation. Polymeric materials have been extensively used because they appear to be more inert than metals or other substances. However, polymeric stents exhibit certain restrictions, especially in their ability to stand external compression forces.[43] Thus, more recently, the potential use of metallic stents as ureteral stents has been investigated. Although it has been shown that full metallic stents present lower occurrence of infection related to long-term indwelling,[44] stent encrustation remains a problem and 22% of Resonance® (Cook Medical) stents placed in patients still exhibited encrustation after 8.5 months.[44]

Very early, the effort of the stent industry was put not only in the search for better biomaterials, but also for others substances that could be used to coat the existing materials in order to improve the resistance of stents to bacterial infection and encrustation. Today, stent coatings represent the most significant and promising development in the goal to achieve the ultimate long-lasting device.

15.4.1 Heparin Coatings

Glycosaminoglycans are naturally occurring molecules in normal urine. They have shown to be natural inhibitors of crystallization of urine salt components.[45,46] Heparin, known for its anticoagulant properties, exhibits the strongest inhibitory effect of all glycosaminoglycans.[47] In 2001, Hildebrandt *et al.*[48] presented, for the first time, the use of heparin to decrease the encrustation of urinary stents. Heparin-coated stents showed no encrustation after 1 week of exposure to artificial urine *in vitro*. These promising results were confirmed *in vivo*. The uncoated control stents showed significant encrustation after 120 days of indwelling, with two of the 20 stents tested being completely obstructed. In contrast, the heparin-coated stents did not show any signs of obstruction.[48] The potential of heparin as an effective coating against encrustation was further demonstrated *in vitro* and *in vivo*.[14,49] The use of heparin was further investigated to explore its anti-bacterial adhesion properties. In addition to its strong anti-thrombogenic properties, heparin is highly electronegative, and this is believed to repel both Gram-positive and Gram-negative bacteria that have an overall negative charge at their surfaces.[50] However, while Tenke *et al.*[14] showed that heparin-coated ureteral stents can indeed remain biofilm- and encrustation-free up to 12 months, a more recent study by Cauda *et al.*[20] revealed contradictory results. In the latter *in vivo* study,[20] heparin-coated stents exhibited the presence of biofilm and encrustation after a month of indwelling, although the extent of this encrustation was less than for the uncoated control stents. Surprisingly, the same study showed that stents left in place for 10 months showed no visible encrustation at all. This inconsistency indicates, perhaps, a greater complexity of the stent encrustation and biofilm phenomena that cannot be easily correlated with the presence of heparin and its action. The effect of heparin on bacterial adhesion and biofilm formation in urine has been further put into questions. When uncoated (Endo-Sof®, Cook Medical) or heparin-coated (Radiance®, Cook Medical) commercial stents were challenged in artificial urine *in vitro* with different uropathogenic bacteria, no effect of the coating could be seen.[51] Our *in vitro* study[31] shows that, when exposed to artificial urine containing urease or inoculated with *P. mirabilis,* a foundation layer composed of amorphous calcium phosphates developed rapidly on both uncoated (Endo-Sof®, Cook Medical) or heparin-coated (Radiance®, Cook Medical) stents. The efficacy of heparin coating in decreasing encrustation *in vivo* may therefore reside in its capacity to reduce crystal formation in sterile urine, whereas its role in an environment infected by urease-producing bacteria and biofilm-related encrustation is still uncertain.

15.4.2 Hydrogel Coatings

Hydrogels are hydrophilic polymers that allow the anchoring of water molecules on the stent surface. Hydrogels-coated stents demonstrate improved biocompatibility, hydrophilization and lubrication, which improve deployment of the device and patients' quality of life.[52] Nevertheless, the efficacy in decreasing complications related to stenting of this coating remains controversial, as hydrogel has been shown to both reduce,[53] as demonstrated with poly(vinyl pyrollidone)-based hydrogel,[54] and increase[55] biofilm formation and infection compared with uncoated polyurethane stents. The hydrogel coating can be additionally impregnated with drugs, *e.g.* ciprofloxacin, gentamicin and cefazolin, to create an antimicrobial surface coating[56] and thus it represents an important direction in stent development.

15.4.3 Diamond-like Carbon Coatings

More recently, a new coating based on the anti-adhesive properties and excellent biocompatibility of diamond-like carbon (DLC) material was developed. A preliminary *in vivo* study demonstrated a decrease in friction, encrustation and biofilm formation on the stents where this coating was plasma-deposited.[57] However, its real efficacy in decreasing bacterial adhesion and encrustation *in vivo* or *in vitro* is yet to be proven.[58]

15.4.4 Biomimetic Coatings

Another strategy to make urological devices less sensitive to bacterial adhesion and encrustation involves transforming the surface of the biomaterial to mimic host tissues. Phosphorylcholine (PC) is a natural constituent of the human cell membrane and can be used to mimic lipidic membrane on abiotic surfaces. A clinical trial with a PC-coated stent reported by Stickler *et al.*[59] revealed a real improvement in terms of biofilm formation, with coated stents showing less visible biofilm (36%) than uncoated controls (61%). However, encrustation was not completely inhibited on PC-coated stents. A similar approach was taken by Fu *et al.*[60] and Nakayama *et al.*[61] who recently developed methods for the creation of *in vivo* tissue-engineered autologous tissue-covered stents. This brings forward the concept of 'biocovered' stents that would not be recognized as foreign bodies and may be less sensitive to microbial colonization and associated encrustation.[60,61]

15.4.5 pH Controlling Coatings

As described previously, the rising of pH of urine associated with infection by urease-producing bacteria that subsequently triggers formation of calcium and magnesium phosphates crystals is one of the causes of biomaterial encrustation. Su[62] described an ingenious pH-balancing coating that was demonstrated to maintain neutral pH at the surface of the coated material while the presence of urease causes the pH increase in the surrounding urine. It was shown that this modulated the formation of calcium phosphate deposits on silicone disks incubated *in vitro* in artificial urine.[62] A similar technology is also commercially available (Inlay Optima® stents, Bard). Although it was shown that this technology decreases the accumulation of calcium on the stents in sterile urine, the effect on encrustation related to bacterial infection and biofilm formation *in vivo* is yet to be demonstrated.

15.4.6 Pro-active Coatings

The use of 'pro-active' biomaterials to reduce urological device-associated infections, biofilm formation and encrustation also offers promising opportunities. Examples of this strategy involve the incorporation of antimicrobial agents into urinary device technologies.

Silver based-coating has been in use for 20 years and it is probably the most studied example of pro-active treatment. The broad-spectrum antimicrobial activity of silver has been known since ancient times. This element acts by blocking enzymatic processes *via* interaction with electron donors and through membrane destabilization.[63] Silver alloy, but not silver oxide, coating are effective in reducing catheter-associated UTI rates by up to 45%.[63] Nevertheless, catheters coated with silver (IC® catheters, Bard) could not resist to *P. mirabilis*-generated encrustation *in vivo*. The rapid formation of a foundation layer composed of amorphous apatite shields the surface of the catheter, enabling further attachment of the bacteria and growth of the crystalline biofilm until blockage of the device.[17] Little data are available for silver-coated stents[64] and their efficacy has never been fully demonstrated.

Triclosan is a broad-spectrum antimicrobial agent that has been in use for over 40 years in numerous consumer and medical products.[65] Triclosan-eluting coatings have been investigated recently to prevent device-associated infection and biofilm formation and consequently encrustation. It has been shown to efficiently prevent bacterial growth in an *in vitro* bladder model infected with *P. mirabilis*, subsequently preventing encrustation.[66] In this experiment, the drug-eluting catheters remained clear up to 7 fold longer than the control ones. Recently, the triclosan-eluting stent Triumph® (Boston Scientific Corp. Inc.) demonstrated significant antimicrobial effects during preliminary *in vitro*[67] and *in vivo*[68] studies, as well as anti-inflammatory effects in cell culture.[69] Regrettably, further investigation on triclosan-impregnated stents performed by the same group failed to confirm those promising results *in vivo* and no improvement regarding device colonization or urine cell counts was observed after 24 weeks of indwelling.[70] Importantly, 70% of the triclosan-eluting stents had viable microorganisms and 100% of the patients with a long history of stenting (over 5 years) showed infected stents. Although the authors reported that no resistance developed after long exposure to triclosan in this study,[70] *Enterococci* have already been shown to have acquired moderate resistance to triclosan, while, recently, *P. mirabilis* strains exhibiting extreme resistance to triclosan (over 400 fold) were isolated.[71]

Another kind of pro-active device, with coatings exhibiting biocatalytic activities, has recently emerged and could be of great interest for the development of encrustation-resistant ureteral biomaterials. A very elegant approach was taken by Watterson *et al.*[72] as they established a coating designed to actively degrade oxalate, one of the compounds responsible for stone formation and stents encrustation in the ureter. In this study,[72] silicone disks were coated with the enzyme oxalate decarboxylase from *Oxalobacter formigenes* and, after a month of indwelling in rabbit bladders, due to the breakdown of oxalate by the immobilized enzyme, they showed significantly less encrustation in comparison to uncoated controls. This technology could be incorporated into stents and other urological devices. A second approach employed in this class of devices uses biocatalytic release of antimicrobial compounds. To overcome the limitations of drug-releasing coatings such as short-term or non-uniform release, a lipase is incorporated in a polymeric matrix and this enzyme releases biocatalytically an antimicrobial agent in a controlled manner.[73] In *in vitro* experiments, the coating exhibited antimicrobial activity against planktonic bacteria and showed resistance to microbial colonization.

Possibly the most promising technologies will rely on the combination of various approaches. For instance, Chosa *et al.*[74] reported the development of a hydrophilic erodible surface treatment. The matrix can be loaded with antimicrobial agent such as epigallocatechin-3-gallate (EGCG), an antimicrobial catechin found in green tea extracts.[74] This coating degrades with time, releasing the EGCG near the surface of the biomaterial. Additionally to the antimicrobial effect, the colonization by microbes is impaired as the erodibility of the coating insures detachment, removal or washout of the biofilm.[74]

15.5 Conclusions

The encrustation of urinary devices is a very complex phenomenon. The ultimate anti-encrustation biomaterial would control the adhesion of microbes responsible for alkalinization of urine and prevent the development of biofilms. It should not only reduce the formation of amorphous apatite on its surface, but modulate and prevent deposition and accumulation of other crystalline forms, such as calcium oxalate, and the solution to control one aspect of the problem should not create or promote another one. In the area of microbial adhesion and biofilm control, huge progress has already been made and not only in the scope of endourology. However, the particular crystalline nature of biofilms on urological devices renders the quest even more arduous and even when certain biomaterials are very effective against bacterial adhesion, they usually fail to resist encrustation. The mineralization of abiotic surfaces is a complex mechanism, it is often patient dependent and it can be influenced by many factors such as patient metabolic alterations (hypocitruria, hyperoxaluria or reduced volume) or presence of microorganisms. This complexity makes the development of perfect anti-encrustation urethral and ureteral catheters a real challenge. Although many promising technologies have been developed and are in use today, the ideal device that would be a panacea is still out of reach.

Thought should also not depart from the physical construction and placement techniques of ureteral stents, knowing that it is the implant itself which is the real culprit in the equation. The fact that 'Double-J' ureteral stents maintain the ureteral orifice in an open state gives rise to a conduit for ascending infection in the urinary tract. Smart designs which essentially shield the implant from the outside environment may offer some resistance against encrustation. However, the means of implantation, if still endoscopically guided through the urethra, may introduce the retrograde advancement of bacteria during the surgical procedure. Finally, the difficulty in fully demonstrating the effectiveness of anti-encrustation ureteral stents and comparing strategies resides in the use of test models that are more or less representative of reality. The varying composition of actual urine used poses the greatest challenge. The ideal demonstration is, of course, a bi-lateral placement of an anti-encrustation stent in one ureter and a benchmark stent in the opposite ureter, thus exposing the devices to the same anatomical environment to realize the true end effect across a significant patient sample population.

References

1. S. M. Jacobsen, D. J. Stickler, H. L. Mobley and M. E. Shirtliff, *Clin. Microbiol. Rev.*, 2008, **21**, 26–59.
2. N. S. Morris, D. J. Stickler and R. J. McLean, *World J. Urol.*, 1999, **17**, 345–350.
3. J. D. Denstedt and P. A. Cadieux, *Curr. Opin. Urol.*, 2009, **19**, 205–210.
4. B. H. Chew, M. Duvdevani and J. D. Denstedt, *Expert Rev. Med. Devices*, 2006, **3**, 395–403.
5. R. A. Garibaldi, J. P. Burke, M. L. Dickman and C. B. Smith, *N. Engl. J. Med.*, 1974, **291**, 215–219.
6. D. Stickler, L. Ganderton, J. King, J. Nettleton and C. Winters, *Urol. Res.*, 1993, **21**, 407–411.
7. D. J. Stickler, *Nat. Clin. Pract.*, 2008, **5**, 598–608.
8. J. Sökeland, *Urology: A Pocket Reference*, Thieme, Stuttgart, 2nd edn, 1989.

9. R. W. Haley, D. H. Culver, J. W. White, W. M. Morgan and T. G. Emori, *Am. J. Epidemiol.*, 1985, **121**, 159–167.

10. M. M. Tunney, P. F. Keane, D. S. Jones and S. P. Gorman, *Biomaterials*, 1996, **17**, 1541–1546.

11. G. L. Shaw, S. K. Choong and C. Fry, *Urol. Res.*, 2005, **33**, 17–22.

12. S. H. Paick, H. K. Park, S. J. Oh and H. H. Kim, *Urology*, 2003, **62**, 214–217.

13. G. Reid, J. D. Denstedt, Y. S. Kang, D. Lam and C. Nause, *J. Urol.*, 1992, **148**, 1592–1594.

14. P. Tenke, C. R. Riedl, G. L. Jones, G. J. Williams, D. Stickler and E. Nagy, *Int. J. Antimicrob. Agents*, 2004, **23**(Suppl. 1), S67 S74.

15. N. S. Morris and D. J. Stickler, *Urol. Res.*, 1998, **26**, 275–279.

16. D. J. Stickler and S. D. Morgan, *J. Med. Microbiol.*, 2006, **55**, 489–494.

17. S. D. Morgan, D. Rigby and D. J. Stickler, *Urol. Res.*, 2009, **37**, 89–93.

18. C. Follmer, *J. Clin. Pathol.*, 2010, **63**, 424–430.

19. M. M. Tunney, M. C. Bonner, P. F. Keane and S. P. Gorman, *Biomaterials*, 1996, **17**, 1025–1029.

20. F. Cauda, V. Cauda, C. Fiori, B. Onida and E. Garrone, *J. Endourol.*, 2008, **22**, 465–472.

21. C. Tieszer, G. Reid and J. Denstedt, *J. Biomed. Mater. Res.*, 1998, **43**, 321–330.

22. H. Axelsson, J. Schönebeck and B. Winblad, *Scand. J. Urol. Nephrol.*, 1977, **11**, 283–287.

23. D. J. Stickler and S. D. Morgan, *J. Hosp. Infect.*, 2008, **69**, 350–360.

24. D. J. Stickler and R. C. Feneley, *Spinal Cord*, 2010, **48**, 784–790.

25. F. Grases, O. Sohnel, A. Costa-Bauza, M. Ramis and Z. Wang, *Nephron*, 2001, **88**, 320–328.

26. G. Bithelis, N. Bouropoulos, E. N. Liatsikos, P. Perimenis, P. G. Koutsoukos and G. A. Barbalias, *J. Endourol.*, 2004, **18**, 550–556.

27. S. Bariol, T. Farebrother, S. Ruthven and F. MacNeil, *J. Endourol.*, 2003, **17**, 741–743.

28. C. Coker, C. A. Poore, X. Li and H. L. Mobley, *Microbes Infect.*, 2000, **2**, 1497–1505.

29. L. Hawthorn and G. Reid, *J. Biomed. Mater. Res.*, 1990, **24**, 1325–1332.

30. J. A. Roberts, E. N. Fussell and M. B. Kaack, *J. Urol.*, 1990, **144**, 264–269.

31. S. Robin, T. Soulimane, S. Lavelle and S. Tofail, *unpublished data*.

32. A. Downer, N. Morris, W. J. Feast and D. Stickler, *Proc. Inst. Mech. Eng.*, 2003, **217**, 279–289.

33. M. M. Pearson, M. Sebaihia, C. Churcher, M. A. Quail, A. S. Seshasayee, N. M. Luscombe, Z. Abdellah, C. Arrosmith, B. Atkin, T. Chillingworth, H. Hauser, K. Jagels, S. Moule, K. Mungall, H. Norbertczak, E. Rabbinowitsch, D. Walker, S. Whithead, N. R. Thomson, P. N. Rather, J. Parkhill and H. L. Mobley, *J. Bacteriol.*, 2008, **190**, 4027–4037.

34. P. Alamuri, M. Lower, J. A. Hiss, S. D. Himpsl, G. Schneider and H. L. Mobley, *Infect. Immun.*, 2010.

35. J. D. Watterson, P. A. Cadieux, D. Stickler, G. Reid and J. D. Denstedt, *J. Endourol.*, 2003, **17**, 523–527.
36. P. Tenke, B. Kovacs, M. Jackel and E. Nagy, *World J. Urol.*, 2006, **24** 13–20.
37. A. Torzewska, P. Staczek and A. Rozalski, *J. Med. Microbiol.*, 2003, **52**, 471–477.
38. A. J. Dumanski, H. Hedelin, A. Edin-Liljegren, D. Beauchemin and R. J. McLean, *Infect. Immun.*, 1994, **62**, 2998–3003.
39. F. D. Roemer, *J. Endourol.*, 2000, **14**, 1–4.
40. D. T. Beiko, B. E. Knudsen, J. D. Watterson, P. A. Cadieux, G. Reid and J. D. Denstedt, *J. Urol.*, 2004, **171**, 2438–2444.
41. J. D. Denstedt, T. A. Wollin and G. Reid, *J. Endourol.*, 1998, **12**, 493–500.
42. M. Marx, M. A. Bettmann, S. Bridge, G. Brodsky, L. M. Boxt and J. P. Richie, *J. Urol.*, 1988, **139**, 180–185.
43. A. Al-Aown, I. Kyriazis, P. Kallidonis, P. Kraniotis, C. Rigopoulos, D. Karnabatidis, T. Petsas and E. Liatsikos, *Therapeutic Advances in Urology*, 2010, **2**, 85–92.
44. E. Liatsikos, P. Kallidonis, I. Kyriazis, C. Constantinidis, K. Hendlin, J. U. Stolzenburg, D. Karnabatidis and D. Siablis, *Eur. Urol.*, 2010, **57**, 480–486.
45. A. H. Angell and M. I. Resnick, *J. Urol.*, 1989, **141**, 1255–1258.
46. K. Yoshimura, T. Yoshioka, O. Miyake, M. Honda, S. Yamaguchi, T. Koide and A. Okuyama, *Brit. J. Urol.*, 1997, **80**, 64–68.
47. B. Fellstrom, M. Lindsjo, B. G. Danielson, F. A. Karlsson and S. Ljunghall, *Clin. Chim. Acta*, 1989, **180**, 213–220.
48. P. Hildebrandt, M. Sayyad, A. Rzany, M. Schaldach and H. Seiter, *Biomaterials*, 2001, **22**, 503–507.
49. C. R. Riedl, M. Witkowski, E. Plas and H. Pflueger, *Int. J. Antimicrob. Agents*, 2002, **19**, 507–510.
50. S. F. Rose, S. Okere, G. W. Hanlon, A. W. Lloyd and A. L. Lewis, *J. Mater. Sci. Mater. Med.*, 2005, **16**, 1003–1015.
51. D. Lange, C. N. Elwood, K. Choi, K. Hendlin, M. Monga and B. H. Chew, *J. Urol.*, 2009, **182**, 1194–1200.
52. J. V. Candela and G. C. Bellman, *J. Endourol.*, 1997, **11**, 45–47.
53. S. P. Gorman, M. M. Tunney, P. F. Keane, K. Van Bladel and B. Bley, *J. Biomed. Mater. Res.*, 1998, **39**, 642–649.
54. M. M. Tunney and S. P. Gorman, *Biomaterials*, 2002, **23**, 4601–4608.
55. F. Desgrandchamps, F. Moulinier, M. Daudon, P. Teillac and A. Le Duc, *Brit. J. Urol.*, 1997, **79**, 24–27.
56. T. John, A. Rajpurkar, G. Smith, M. Fairfax and J. Triest, *J. Endourol.*, 2007, **21**, 1211–1216.
57. N. Laube, L. Kleinen, J. Bradenahl and A. Meissner, *J. Urol.*, 2007, **177**, 1923–1927.
58. N. Laube, L. Kleinen, U. Bode, C. Fisang, A. Meissner, J. Bradenahl, I. Syring, H. Busch, W. Pinkowski and S. C. Muller, *Urologe A*, 2007, **46**, 1249–1251.

59. D. J. Stickler, A. Evans, N. Morris and G. Hughes, *Int. J. Antimicrob. Agents*, 2002, **19**, 499–506.
60. W. J. Fu, X. Zhang, B. H. Zhang, P. Zhang, B. F. Hong, J. P. Gao, B. Meng, H. Kun and F. Z. Cui, *BJU Int.*, 2009, **104**, 263–268.
61. Y. Nakayama, Y. M. Zhou and H. Ishibashi-Ueda, *J. Artif. Organs*, 2007, **10**, 171–176.
62. S. H. Su, *Med. Device Technol.*, 2004, **15**, 12–14.
63. R. M. Slawson, M. I. Van Dyke, H. Lee and J. T. Trevors, *Plasmid*, 1992, **27**, 72–79.
64. J. W. C. Leung, G. T. C. Lau, J. J. Y. Sung and J. W. Costerton, *Gastrointest. Endosc.*, 1992, **38**, 338–340.
65. H. N. Bhargava and P. A. Leonard, *Am. J. Infect. Control*, 1996, **24**, 209–218.
66. G. L. Jones, A. D. Russell, Z. Caliskan and D. J. Stickler, *Eur. Urol.*, 2005, **48**, 838–845.
67. B. H. Chew, P. A. Cadieux, G. Reid and J. D. Denstedt, *J. Endourol.*, 2006, **20**, 949–958.
68. P. A. Cadieux, B. H. Chew, B. E. Knudsen, K. Dejong, E. Rowe, G. Reid and J. D. Denstedt, *J. Urol.*, 2006, **175**, 2331–2335.
69. C. N. Elwood, B. H. Chew, S. Seney, J. Jass, J. D. Denstedt and P. A. Cadieux, *J. Endourol.*, 2007, **21**, 1217–1222.
70. P. A. Cadieux, B. H. Chew, L. Nott, S. Seney, C. N. Elwood, G. R. Wignall, L. W. Goneau and J. D. Denstedt, *J. Endourol.*, 2009, **23**, 1187–1194.
71. D. J. Stickler and G. L. Jones, *Antimicrob. Agents Chemother.*, 2008, **52**, 991–994.
72. J. D. Watterson, P. A. Cadieux, D. T. Beiko, A. J. Cook, J. P. Burton, R. R. Harbottle, C. Lee, E. Rowe, H. Sidhu, G. Reid and J. D. Denstedt, *J. Endourol.*, 2003, **17**, 269–274.
73. R. N. Dave, H. M. Joshi and V. P. Venugopalan, *Antimicrob. Agents Chemother.*, 2011, **55**, 845–853.
74. H. Chosa, M. Toda, S. Okubo, Y. Hara and T. Shimamura, *Kansenshogaku Zasshi*, 1992, **66**, 606–611.

CHAPTER 16

The Reduction of Restenosis in Cardiovascular Stents

J. ARKOWSKI,*[a] M. WAWRZYŃSKA,[a] D. BIAŁY,[a]
B. SOBIESZCZAŃSKA[b] AND W. MAZUREK[a]

[a] Wroclaw Medical University, Department of Cardiology, ul. Pasteura 4 50-369, Wrocław, Poland; [b] Wroclaw Medical University, Department of Microbiology, ul. T. Chałubińskiego 4, 50-368 Wrocław, Poland

16.1 Introduction

The first successful coronary stent implantation performed in 1986 intended to resolve all the mechanical problems associated with balloon angioplasty. A stiff metal scaffolding was provided for the vessel wall. The two major causes of suboptimal or even unfavourable effects of angioplasty, namely the elastic recoil of the previously dilated artery and the propagation of the very frequent intimal tear (dissection), were finally prevented.[1] Currently stents are used in as many as 80% of percutaneus coronary procedures.

Nevertheless a new problem occurred. A relatively large quantity of metal was being placed into the small vessel with active endothelial cells on its inner wall. In some patients (every third or every fourth of them) the ischaemic symptoms recurred, usually after few months. Control angiography revealed a new stenosis in the place previously repaired with stent. This time it was not an atherosclerotic plaque but just a thick layer of modified smooth muscle cells called neointima, as they grew in the inner side of the stented vessel. In contrast to the true intima, this neointima formed not a one single cell layer but many of them, thus significantly narrowing the artery lumen.[2]

RSC Nanoscience & Nanotechnology No. 21
Biological Interactions with Surface Charge in Biomaterials
Edited by Syed A. M. Tofail
© Royal Society of Chemistry 2012
Published by the Royal Society of Chemistry, www.rsc.org

Technically, the issue of re-narrowing seems quite easy to resolve. The restenotic tissue is soft, easy to pass with a wire and to dilate with a balloon. No dissection, perforation or thrombus formation can occur as the repeat procedure can be performed in an already stented segment. From an operator's point of view, combating restenosis thus appears to be a quite safe and easy task. From a patient's point of view, it is not necessarily a simple task. Each angioplasty may result in the puncture of the access artery, more radiation exposure, an injection of the contrast medium required for angioplasty, and immobilization and hospitalization of the patient, none of which are pleasant to the patient and most of them poses their own risks.

Coronary angioplasty was always (and is still) a method of treating symptomatic cardiac ischaemia as an alternative to cardiac surgery. The necessity (in some patients) to repeat angioplasty every half a year or so is unacceptable and for many is worse than one surgical procedure with a long-lasting effect. There is also the issue of additional costs of such a procedure each time.

Initial approaches to prevent restenosis were plainly mechanical. As it was noted that restenosis was mainly a reaction to the presence of metal, cardiologists started to carefully choose the optimal stenting strategy. Restenosis was less likely to occur in short and wide stents as they present less amount of metal in the artery. It is also less probable that it occurs in straight artery segments due to the lower amount of tension and movement during each heart contraction, especially among patients who are free of diabetes or kidney failure due to normal endothelial function. Thus for many years the typical approach was to stent short sections of large straight vessels with no calcifications, possibly only the entry of dissection that occurred during angioplasty. The deployment of a stent into narrow, tortuous and calcified vessels of, for example, a diabetic patient was considered to provoke restenosis.

Restenosis was traditionally understood as an incontrollable cell growth, somewhat similar to a neoplasia. Many physicians considered that it can be treated in the same way as tumours – by destroying the cells that causes restenosis. 'Contact' exposure to radiation that penetrates only a few millimeters deep (brachytherapy) was employed to treat restenosis.[3] There were, however, two main issues. First, quite a sophisticated and expensive technology was required to deliver the rays precisely to the stented segment without risking irradiation of any other areas of the patient's body or the medical staff. This was difficult to achieve in practice. Secondly, the radiation had more influence on the artery wall than desired. A new phenomenon of marginal restenosis was observed where vessel segments adjacent to the irradiated section responded with hyperplasia. Additionally, the stent struts uncovered by endothelium were strongly thrombogenic and stent thrombosis was more dangerous than restenosis.

Medical therapy indicated after stent implantation (aspirin and clopidogrel) is designed mainly to prevent thrombosis but possible influences on some pathways leading to vessel cell activation and restenosis cannot be excluded. Steroid anti-inflammatory drugs have been tested in restenosis prevention but showed no clear benefit.[4] Another idea was to administer antioxidant drug (probucol) with or without antioxidant vitamins.[5] It seemed to reduce the restenosis ratio but the regimen required several weeks of drug intake prior to intervention and negative effects on myocardiocyte repolarization and plasma lipoproteins were observed. In a small trial, a cytostatic drug rapamycin was tested but it turned out that most of the patients suffered from its side-effects even before taking this drug in an amount sufficient to influence the restenotic process.[6]

Currently, no systemic medication is believed to act clearly with regards to restenosis prevention. Consequently, all treatment regimens are designated to stop (or even possibly reverse) the development of atherosclerotic disease, prevent the existing atherosclerotic plaques from rupture and reduce blood thrombogenicity as long as the stent struts remain uncovered. While some of these substances may inhibit or delay restenosis, this hypothesis cannot be tested beyond reasonable doubt as all of the medications are considered life-saving and cannot be stopped and/or compared with others in controlled conditions.

16.2   Intravascular Stent Drug and Gene Delivery

Almost from the beginning of the stent era, the concept of stent surface modification attracted great interest. Heparin was the substance of first choice and in the 1990s many commercially available stents had heparin coating. The results of clinical trials showed no clear advantage of such coatings compared to bare metal stents. It was later understood that thrombosis and restenosis are two different phenomena, although activated platelets play an important part in both of them. The amount of heparin that can be immobilized on the stent struts has been found irrelevant to the overall plasma thrombotic or thrombolytic activity.

At the beginning of the 21st century a revolution in interventional cardiology took place. Local release of cytostatic drugs proved to be a very efficient weapon against restenosis.[7] Drug-eluting stents (DESs) were introduced. The first substance used was sirolismus, followed shortly by paclitaxel. Both formulations proved efficient. Over the next few years, several other limus-type drugs were tested, with zotarolimus and everolimus being among the most popular ones. DESs started to be used commonly. At some point, 70% of coronary interventions in the US included DES implantation.

The ratio of the use of DESs to bare metal stent differs significantly between countries mainly, but not only, due to financial reasons. The optimal stent platform, polymer coating and the best cytostatic substance are still subject to scientific debate and commercial competition. Judging by the data available so far, the four most popular DES drugs (sirolimus, paclitaxel, everolimus and zotarolimus) are all efficient in preventing restenosis.[8] There are, it seems, some differences between them, with paclitaxel being probably the most cytostatic and cytotoxic. While each company uses slightly different platform design, polymer formulation, trial design and endpoint criteria, it is hard to tell whether the published differences in numbers of vessel failure, late lumen loss or binary restenosis reflect the actual device efficacy or are just within the variance of the 'normal distribution' of the measured parameters.

It can certainly be said that DESs, while reducing substantially the risk of restenosis, can increase, albeit only slightly, the incidence of thrombosis. The latter, when it occurs, is of course much more dangerous and, in some occasions, even fatal. Most of the cases of stent thrombosis can be prevented by strict adherence to double antiplatelet therapy, but, as with any other medication, in real life no one can guarantee strict compliance by all of the patients undergoing treatment.

DESs do not completely solve the issue of restenosis. They are also more expensive than conventional stents and in many countries they can be used only in the high risk subset of lesions. They require well controlled double anti platelet therapy and therefore are *a priori* excluded in some patients. It should not be forgotten that there is such phenomenon as in-DES restenosis, it is only very rare and usually clinically insignificant.

Drug-eluting balloons (DEBs) were developed a few years after DESs. They are especially useful in cases of in-stent restenosis. A DES implanation into a restenosed stent would double the metal burden. This is because the implantation of a second stent in the same section of the artery results in twice as much metal per artery length (or wall surface unit), thus doubling the unfavourable effects of metal (artery stiffening, thrombogenity).

16.3   Technology Involved in Stents

Typical cardiovascular stents are between 2 and 5 mm in diameter when deployed. They are formed from metal mesh which consists of struts or the so-called 'wires', which are, in reality, more often square or rectangular in cross-section rather than being round. Currently, the best designs are made of wires as thin as 80 μm without sacrificing the radial force. Thinner wires cause less irritation to the vessel wall and consequently give rise to less restenosis. The current research is focused on new materials that are stronger than traditional 316 L steel, but still possess the beneficial properties of stainless steel. Among the most popular are chromium cobalt and chromium platinum alloys.

Another important stent parameter is the 'cell' diameter. The term 'cell' here refers to the space between the struts. The larger the 'cell' size, the less metal is present on a given surface unit. Stents with large cells have typically less radial force and resistance to bending and kinking. There is also a difference between a closed 'cell' and an open 'cell' design. A plain metal mesh (such as the one used in fences) is an example of closed 'cell' design. There is no communication between these closed 'cells'. Open 'cells' communicate with each other and usually there is a narrowing between these 'cells'. As with the 'cell' diameter, there are advantages and disadvantages of each of these constructions. Open 'cells' possess non-jail side branches, conform better to vessel shapes and movements, and cause less irritation. On the other hand, they give less support to the artery wall.

The benchmark of today's stent technology is the DES. The initial difficulties with controlled drug release (drug attached directly to the metal surface) have been overcome.[8] The standard DES stents now consist of a metal scaffold covered with a polymer that releases the active drug formulation. Usually, the polymer is biodegradable and disappears after the dug is released. This avoids any adverse reaction later on. In some stents the drug-containing polymer is placed only on the abluminal (vessel-wall facing) surface of the stent. This design allows the release of cytostatic drugs in higher concentrations directly on to the artery wall cells where it should act, and release in lower concentrations into the circulating blood where the drug is of no use or may even be harmful. Such sophisticated design of drug elution allows a very efficient release of cytostatic drugs and the use of these drugs in much smaller amounts.

16.4 Stent Structure Engineering and Biodegradable Stent Materials

The original idea behind stenting was to deploy stents that were solid, strong and capable of counterbalancing the recoil of the vessel and negative remodelling. Stents are required to do what balloons could not do, *i.e.* to keep the vessel expanded. On the other hand, the disadvantages of vessel stenting originate from the solidness of the stents itself. Stents are made of metal, which is a foreign material. Stents also prevent the vessel from bending and twisting, which is what vessels normally do during every heartbeat. Stents do not shrink and dilate when the vessel wall does as a reaction to different blood flow demands. Finally, stents are placed between blood and endothelium, a very fragile tissue that responds strongly to any foreign stimuli. For example, thrombosis and restenosis are tissue responses to the stent metal.

Stenting thus represents an intrinsic contradiction in the demands of cardiologists in keeping the vessel open. It is, therefore, not surprising that every interventional cardiologist would desire a device that dilates the artery during the intervention, keeps it dilated until the remodelling is completed and then disappears leaving a wide and healthy reactive vessel. The search for such a biodegradable stents has been going on for many years now. At first it was difficult to find a material that is strong and easy to model, does not irritate the vessel wall and dissolves at a desired rate. Several years ago an idea to build a stent from magnesium was quite popular. The stents were even implanted into human arteries.[9] Although bio-absorption was quite good, there were two major difficulties. Firstly, the scaffolding provided was not strong enough as magnesium is a relatively soft metal when compared to steel. Secondly, magnesium stents caused irritation and an inflammatory reaction of the vessel wall before it was dissolved. Currently, these problems are addressed by the use of advanced biopolymer-based stents, which appear to be more promising than magnesium as bio-resorbable stents.[10]

The results of a 3 year follow-up study of biodegradable stent implantation are very promising.[11] The stents platforms are built from poly(L-lactide) and poly(D.L-lactide). They are degraded *in vivo* to lactic acid which is metabolized in the Krebs cycle. In the first group of 30 patients, there were no cases of stent thrombosis and the rates of ischaemia and the occurrences of revascularization were low and comparable to the best that could be achieve through conventional stents. The stents were resorbed completely after three years and many of the stented vessel segments show restored vasomotion, a hallmark of vessel health and normal function. Another larger group of patients who received such bio-resorbable stents is currently being followed.[12]

16.5 Novel Approaches to Stent Surface: Electrical Modifications

A number of electrical phenomena can take place at the blood–vessel wall interface. This is also true for the implant–blood interface. It has been found out that after carotid artery stent implantation, the electrical parameters of erythrocyte cell membranes were changed.[13] The alterations can persist for up to 4 weeks. So far no major clinical effects have been observed, but it can be deduced that the thrombosis/thrombolysis equilibrium may be altered if the erythrocyte surface charge is changed.

Some techniques used in stent (and especially DES) production rely on electrical properties of the stent surface. Electrostatic dry powder deposition technology (EDPDT) has been used on the metal strut followed by fusion to produce uniform, reproducible and accurate coatings of paclitaxel-bearing particles on cardiovascular stents. The coated stents exhibited sustained release profiles for several weeks.[14] This method seems to be an interesting option to current commercial technology.

Another interesting technology is stent electrocoating with C(12)-phenyldiazonium. Electrochemical tests have demonstrated it as efficient and controllable, while the coating showed an increased stability, which makes it possible to release the drug in a controlled fashion.[15]

A coating of nanometals on conventional cells showed enhanced endothelial growth. The cell behaviour in contact with such 'nanometals' (essentially a metal-oxide surface) is currently poorly understood. It may be possible that the proliferation of endothelial cells and smooth muscle cells can be controlled utilizing the surface charge of the stent. The creation of electrostatic domains charge nanodomains may help to modify the behaviour of cells directly adjacent to the stent and help us to understand local reactions of these cells with the stent surface.

16.6 Conclusions

With the first successful coronary balloon angioplasty, it was possible to treat coronary lesions without opening the chest and thereby exposing the patient to the risk of extracorporeal circulation. Coronary stents solved many of the technical problems that angioplasty could not help to eradicate. Coronary stenting has also created several new issues as a consequence of leaving an implant in the vessel lumen. In many cases, DESs solved the problem of restenosis, shifting the balance, however, towards an increased risk of stent thrombosis. The first few trials of bio-resorbable DESs represent a paradigm shift and, if such DESs succeed, they may become the gold standard of treatment for atherosclerotic lesions in coronary arteries. Meanwhile efforts are still being made to optimize the existing metal stents, improving their structure, surface, coating and drug release in order to obtain satisfactory procedural performance and uneventful healing of the lesion site.

References

1. P. W. Serruys, P. de Jaegere, F. Kiemeneij, C. Macaya, W. Rutsch, G. Heyndrickx, H. Emanuelsson, J. Marco, V. Legrand, P. Materne, J. Belardi, U. Sigwart, A. Colombo, J. J. Goy, P. van den Heuvel, J. Delcan and M-a. Morel, Benestent Study Group, *N. Engl. J. Med.*, 1994, **331**(8), 489–495.
2. R. Erbel, M. Haude, H. W. Höpp, D. Franzen, H. J. Rupprecht, B. Heublein, K. Fischer, P. de Jaegere, P. Serruys, W. Rutsch and P. Probst, Restenosis Stent Study Group, *N. Engl. J. Med.*, 1998, **339**(23), 1672–1678.
3. M. B. Leon, P. S. Teirstein, J. W. Moses, P. Tripuraneni, A. J. Lansky, S. Jani, S. C. Wong, D. Fish, S. Ellis, D. R. Holmes, D. Kerieakes and R. E. Kuntz, *N. Engl. J. Med.*, 2001, **344**(4), 250–256.
4. F. Versaci, A. Gaspardone, F. Tomai, F. Ribichini, P. Russo, I. Proietti, A. S. Ghini, V. Ferrero, L. Chiariello, P. A. Gioffrè, F. Romeo and F. Crea., *J. Am. Coll. Cardiol.*, 2002, **40**(11), 1935–1942.
5. J. C. Tardif, G. Côté, J. Lespérance, M. Bourassa, J. Lambert, S. Doucet, L. Bilodeau, S. Nattel and P. de Guise., *N. Engl. J. Med.*, 1997, **337**(6), 365–372.
6. P. S. Brara, M. Moussavian, M. A. Grise, J. P. Reilly, M. Fernandez, R. A. Schatz and P. S. Teirstein, *Circulation*, 2003, **107**(13), 1722–1724.
7. J. E. Sousa, M. A. Costa, A. C. Abizaid, B. J. Rensing, A. S. Abizaid, L. F. Tanajura, K. Kozuma, G. Van Langenhove, A. G. Sousa, R. Falotico, J. Jaeger, J. J. Popma and P. W. Serruys, *Circulation*, 2001, **104**(17), 2007–2011.
8. D. J. Malenka, A. V. Kaplan, F. L. Lucas, S. M. Sharp and J. S. Skinner, *JAMA*, 2008, **299**(24), 2868–2876.
9. R. Erbel, C. Di Mario, J. Bartunek, J. Bonnier, B. de Bruyne, F. R. Eberli, P. Erne, M. Haude, B. Heublein, M. Horrigan, C. Ilsley, D. Böse, J. Koolen, T. F. Lüscher, N. Weissman and R. Waksman, *Lancet*, 2007, **369**(9576), 1869–1875.
10. Y. Onuma, J. Ormiston and P. W. Serruys, *Circ. J.*, 2011, **75**(3), 509–520.
11. Y. Onuma, P. W. Serruys, J. A. Ormiston, E. Regar, M. Webster, L. Thuesen, D. Dudek, S. Veldhof and R. Rapoza, *EuroIntervention*, 2010, **6**(4), 447–453.

12. J. Gomez-Lara, S. Brugaletta, R. Diletti, S. Garg, Y. Onuma, B. D. Gogas, R. J. van Geuns, C. Dorange, S. Veldhof, R. Rapoza, R. Whitbourn, S. Windecker, H. M. Garcia-Garcia, E. Regar and P. W. Serruys, *Eur. Heart. J.*, 2011, **32**(3), 294–304.
13. A. Basoli, C. Cametti, V. Faraglia, T. Gili, L. Rizzo and M. Taurino, *Vasc. Endovascular Surg.*, 2010, **44**(3), 190–197.
14. R. K. Nukala, H. Boyapally, I. J. Slipper, A. P. Mendham and D. Douroumis, *Pharm. Res.*, 2010, **27**(1), 72–81.
15. Y. Levy, N. Tal, G. Tzemach, J. Weinberger, A. J. Domb and D. Mandler, *J. Biomed. Mater. Res. B. Appl. Biomater.*, 2009, **91**(2), 819–830.

CHAPTER 17

Manipulation of Interfaces on Vector Materials

N. HORIUCHI,* M. NAKAMURA, A. NAGAI AND
K. YAMASHITA

Tokyo Medical and Dental University, Institute of Biomaterial and
Bioengineering, 2-3-10 Kanda-Surugadai, Chiyoda-ku, Tokyo 101-0062,
Japan

17.1 Introduction

Biomaterials are being intensively investigated and developed. Tissue or cell engineering using embryonic stem cells, and biomaterial engineering combined with bone morphogenetic proteins (BMPs) and collagens, are the most common methods for development of the biomaterials. Although these methods have high potential, they have some technological and ethical difficulties to overcome. On the other hand, the development of new biomaterials using conventional methods is being intensively performed. Phenomena *in vivo* are dominated by the properties of surface of materials, thus, it is important to understand and manipulate the interaction between biomaterials and its surroundings medium in clinical application of biomaterials.

In the vicinity of interfaces on bone implants, severe inflammation occurs as a defence mechanism of the body on biomaterials during wound healing after implantation (Figure 17.1).[1,2] The intravital reactions, such as inflammation, are heavily affected by the adsorption of inorganic ions, proteins and polysaccharides. Various kinds of bloods cells, including embryonic cells, are attracted to a surface of an implanted material *via* the adsorbed substances.[3] It

RSC Nanoscience & Nanotechnology No. 21
Biological Interactions with Surface Charge in Biomaterials
Edited by Syed A. M. Tofail
© Royal Society of Chemistry 2012
Published by the Royal Society of Chemistry, www.rsc.org

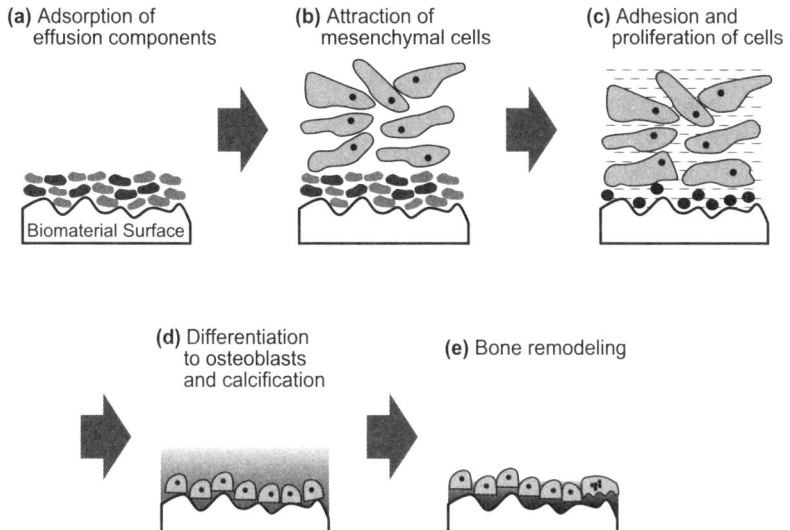

(a) Adsorption of effusion components

(b) Attraction of mesenchymal cells

(c) Adhesion and proliferation of cells

Biomaterial Surface

(d) Differentiation to osteoblasts and calcification

(e) Bone remodeling

Figure 17.1 Schematic illustration of osteoconduction. Woundhealing sequence of bone tissues in the vicinity of biomaterial suface.

is considered that the *in vivo* reactions are determined by these phenomena at an early stage of implantation. Additionally, the size of the substances involved in the reactions at a biointerface is considered as critical issue. The cells have sizes of up to several tens of micrometers; on the other hand, the sizes of ions and ionic groups, proteins and virus range from a few nanometers to submicrons. The scale of the surface roughness of coating or morphology of medical devices ranges from a few millimeter to several tens of centimeters.

As mentioned, we have to focus on the region of nano-sized scale near the biointerfaces. In order to understand these issues concerning biointerfaces, we attempted to use and develop bioactive materials that work at the biointerface.[4] In addition, we tried to manipulate the initial phenomena on a surface of a biomaterial. The vector materials that we propose in this Chapter satisfy these demands.

17.2 Concept of Vector Effects and Ceramics

It is difficult to manipulate a limited area in a quasi-closed system such as diseased area in human body. An external magnetic, sonic or electric field is usually applied to the entire system when the targeted area requires some forces. For example, an external electric field is applied to entire human body when you cure the diseased area (Figure 17.2a). The application of the external field, however, needs a continuous power supply. Additionally, the applied field makes the matters worse, because it produces some secondary competitive effects such as burned skin ($\times$ marks in Figure 17.2a). In this case,

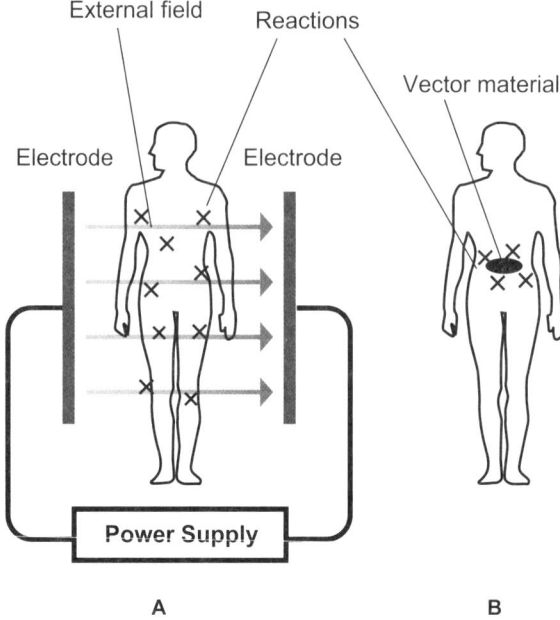

Figure 17.2   Schematic illustration of reactions stimulated by (a) external field and (b) vector material.

independently workable substances, such as magnets, are favourable, because they can bring about effects to a limited and targeted local area (Figure 17.2b). Such materials are expected to have useful affects in some applications for biomedical devices, as well as environmental devices.

An example of a practical use is for application in biomedical usage. In the human body, a local electrostatic force is sometimes required because a typical electrolytic substance and many ionic groups, including proteins and cells, have electrostatic charges. The application of external fields is not always successful, because some destructive reaction or phenomena occur at same time. Alternatively, electrets, which independently form electrostatic fields, can be favourable candidates for such situation. We have found a remarkable result that hydroxyapatite (HAp) can be an excellent electret with a strong biological and favourable effect.[5–13] As mentioned below, HAp can become excellent electrets *via* a polarization procedure,[14–16] and these HAp electrets have an accelerative effect for *in vivo* osteogenesis. It is found that the polarized HAp can manipulate bone-like crystal growth on their own surfaces *via* an *in vitro* experiment using simulated body fluid (SBF),[17,18,19] which includes inorganic ions with almost similar concentrations as those found in human serum.

Here, we employ the term 'electrovector' to refer to a concept exceeding 'bioactivity'. And a 'vector' is used as the upper concept for an 'electrovector' in order to search for new groups of 'vector materials'. The term 'vector' is commonly used in mathematics and biology. The term of vector was originally defined by Webster as follows:

- a noun: (1) (a) a quantity that has magnitude, direction, and sense, (b) a course or compass direction especially of an airplane; or (2) an organism that transmit a pathogen.
- a verb (*vt*): to guide (as an airplane, its pilot, or a missile) in flight by means of a radioed vector.

On the basis of these definitions and usage, we define 'vector', 'vector materials' and 'vector effect' as below:

- vector (*n, vt*): to manipulate the constituents of a living body and environmental system, such as non-living substances of ions, proteins and sugars, and living cells and bacteria, tissues and organs, by biocompatible or the ecomaterials themselves.
- vector materials: a material which has the ability to be a vector.
- vector effect: an effect which a vector material brings forth.

Classified vector materials and their effects are represented in Table 17.1 on the basis of these definitions. Magnets and electrets are thus typical vector materials. Polarized HAp is definitely a member of vector materials. A radioactive material is another candidate to be a vector material because it can individually irradiate and affect the environment near itself. Glass beads that contained radioactive materials have been used in medical applications for cancer therapy. These materials are classified as radioactive vector materials. Bioactive glasses and β-tricalcium phosphate (TCP) are comparatively rapidly dissolved in the body and release chemical constituents or artificial chemicals, resulting in good bone conductivity. These materials are classified as chemicovector materials. Scaffolds can be considered as biovector material, because cells located on the particular area in the scaffold are activated by substances incorporated in the area.

Vector materials can be divided into two categories (Figure 17.3) according to their propagation velocity; magnetic force, electric force and nuclear

Table 17.1 Classified vector materials.

Vector material	Driving force	Possible example
Electrovector	Electrostatic force	Hydroxyapatite
Magnetrovector	Magnetic force	Nd-Fe-B magnet
Chemicovector	Chemical property	Bioactive glass, TCP, TCP-doped YSZ
Biovector	Biological property	Cells with ceramic scaffold
Radioactivevector	Radioactive property	YAS glass
Thermovector	Thermal property	Ferrite deposited glass
Optvector	Optical property	Phosphorescent ceramics: $SrAl_2O_4$:Eu, ZnS:Cu
Mechanovector	Mechanical property	Shape memory ceramics: biphase glass
Chemico-electrovector	Electrostatic force, Chemical property	Bioactive glass

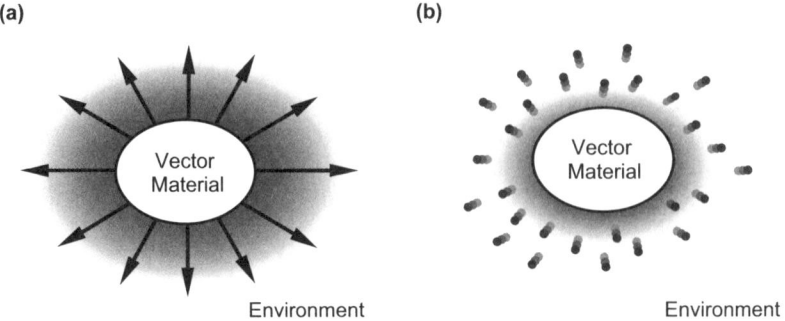

Figure 17.3 Illustrations of vector materials categorized according to propagation type: (a) irradiation and (b) diffusion.

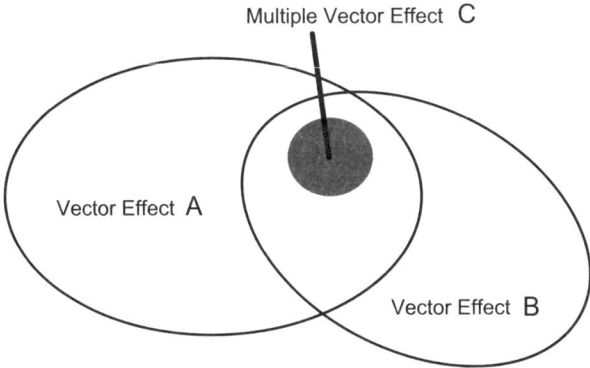

Figure 17.4 Concept of multiple vector effect, in relation to a single vector effect.

radiation propagate quickly (Figure 17.3a), whereas released and diffused ions from the surfaces of vector materials diffuse gradually (Figure 17.3b). The combination of several vector effects may amplify their vector effect (Figure 17.4). An example has been identified as a mixed vector material, *i.e.* polarized 45S5-type bioactive glass as a chemico-electrovector material. Although other vector materials such as mechanovector or optvector materials are rarely under development, a variety of developed functional materials are reasonably expected to have excellent vector effects for biomedical use. The following section will introduce those representative examples of vector ceramics.

17.3 Vector Materials

17.3.1 Electrovector Materials

HAp [$Ca_5(PO_4)_3OH$] has favourable electric properties for biomedical application, in addition to excellent biocompatibility.[14] HAp and its solid solutions have high ion conductivity and are known as the conductors of protons, oxide ions and a mixture of H^+ and O^{2-} ions.[15,16,18–20] HAp can be polarized under a DC voltage at an elevated temperature[5,21,22] and become an electret. HAp electrets are fabricated from HAp ceramics *via* polarization procedures. We have studied the chemical, physical and biological effects of the electrically polarized HAp, or HAp electret. While unpolarized HAp is the most bio-compatible material, we have proved the polarized HAp has more advantages than unpolarized HAp with regards to its osteoconductive property. In this Section, we introduce the biomedical characteristics of HAp electrets and the important electric properties of HAp ceramics.

Electrets form an electric field around the electrets without external con-tinuous power supplies, similar to magnets which form magnetic fields around themselves. However, an important application of electrets has not yet been found. Thus finding of application of electrovector materials (electrets) for biomedical use is expected to be novel method of material development.

Polarization of HAp originates from proton migration in HAp ceramics. The actual polarization of HAp was performed with DC of 1 kV cm^{-1} at 300 °C. The characteristics of the surface of polarized HAp depend on the electric signs of electrodes which are contacted with the surfaces of the HAp when polarization occurs. Therefore, the polarized surfaces are labelled as N- and P-surfaces (Figure 17.5). The surfaces of unpolarized HAp are named 0-surfaces.

The electrovector effects were confirmed by SBF tests. Polarized and unpo-larized HAp ceramics were immersed in 1.5 SBF. After 1 day of immersion, relatively large crystals of 1–4 µm in diameter covered the N-surface of the polarized HAp, whereas small and slight crystals were observed on the unpolarized surfaces (Figure 17.6), Additionally, no crystals were observed on the P-surfaces after 3 days of immersion. The growth rate was increased to 10 µm day^{-1}. The rate is several times larger than the rate on the unpolarized surface. During the early stage of crystal growth, the crystals grown were

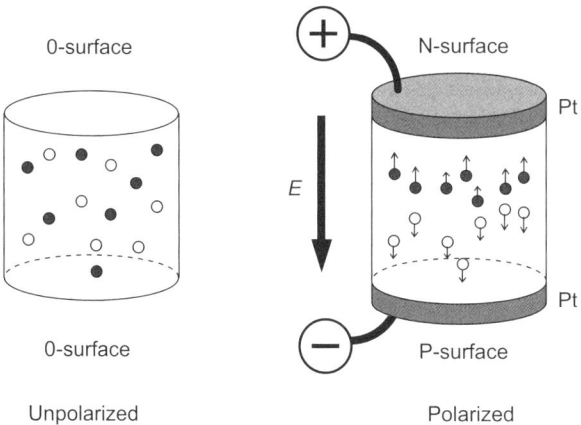

Figure 17.5   Illustration of electric polarization procedure.

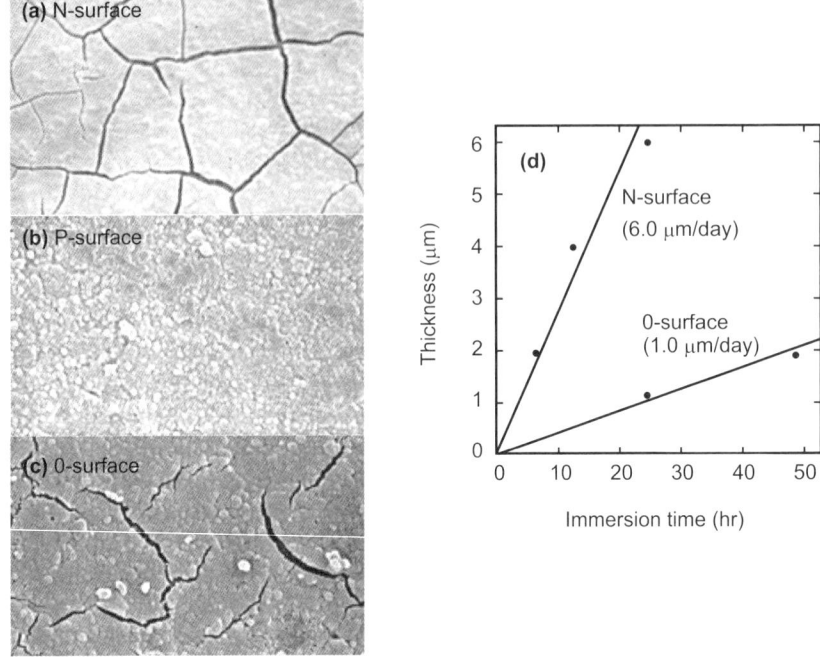

Figure 17.6   (a, b, c) Scanning electron microscopy (SEM) micrograph of bone-
          like apatite layers on electrovector ceramics after 1 day immersion in
          1.5 SBF. (d) Relationship between grown layer thickness and immers-
          ion time.

spherical and their sizes were dependent on the field strength and time of
polarization procedure. The polarization procedure is considered to effect the
nucleation as well as the crystal growth.

The electrovector effect of crystal growth in SBF is understood through
comparison between polarized and unpolarized HAp surfaces. Figure 17.7 is a
schematic illustration of crystal growth of HAp on polarized and unpolarized
HAp ceramics. At an early stage, much more Ca^{2+} ions are predominantly
adsorbed on an N-surface. The amount of adsorption on the N-surface is much
more than that on 0-surface, because the electrostatic attractive force originated from polarized surface [Figure 17.7(1)].

Considering the unusually accelerated growth rate, a suitable amount of
ionic groups of HPO_4^{2-}, HCO_3^- and OH^- are also considered to be gathered
around the Ca^{2+} ions, and nucleation occurs in these condensed ionic groups.
At the next stage, the nuclei grows layer-by-layer [Figure 17.7(2)]. The rate of
N-surface at this stage is larger than that of unpolarized surfaces, indicating the
electrostatic attractive force originated from the electric dipole moment in HAp
remains. On the P-surfaces, chloride ions are mainly adsorbed, which may be
decelerate the crystal growth. The electrovector effect is considered as an
electrostatic attractive and repulsive force between the polarized surfaces and

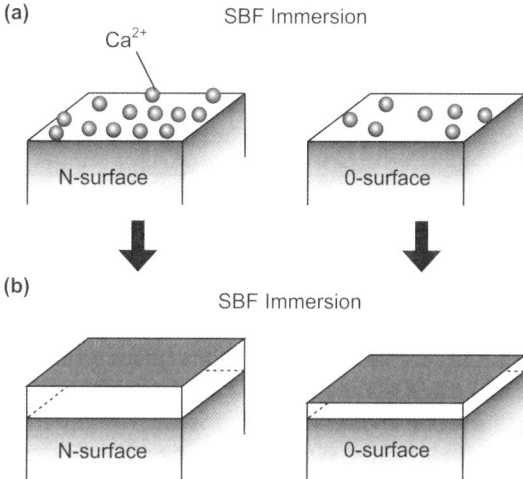

Figure 17.7 Schematic illustration for accelerated bone-like apatite crystal growth on N- and 0-surfaces.

ionic groups, resulting in acceleration and deceleration of nuclei formation and crystal growth.

The *in vivo* electrovector effect has been demonstrated in animal experiment; polarized and unpolarized HA ceramics were implanted into femoral and tibial bones of beagle dogs. The bones were exposed and rectangular holes were bored in the bones. Sterilized HA ceramics were implanted into the holes. The gaps between the faces of the samples and the face of cut cortical bone were fixed at 0.2–0.3 mm. The bones were extracted 7 days after the implantation. In the gap between the negatively charged surfaces (N-surfaces) and the bone, the newly formed bone layers of 0.02 mm thickness were contacted to without any inclusion (Figure 17.8A).

On the other side of the newly formed bone layer, small flat-shaped osteo-blastic cells were structured in a monolayer. The maturing osseous cells were already included in the newly formed bone. The monolayer flattened small osteoblastic cells were also observed on the surface of the cut cortical bone tissues facing the N-surface and partly lengthened towards the N-surface.

On the other hand, in the vicinity of the positively charged surfaces (P-sur-faces), many osteoid-like tissues were surrounded by large square-shaped osteoblastic cells (Figure 17.8B).

Although the newly formed bone surfaces appeared to be directly contacted to the P-surfaces at almost all the interfaces, higher magnification observation revealed fibrous tissues were found at the interface. The newly formed bones directly contacted to the P-surface were osteoid tissues lengthened from the cut cortical bone surfaces towards the ceramic surfaces. The bone formation with layers of the large osteoblastic cells was also observed on the cortical bone surfaces facing the P-surface with a clear line indicating a boundary between the newly formed and old bones. In the gap between the non-polarized HA

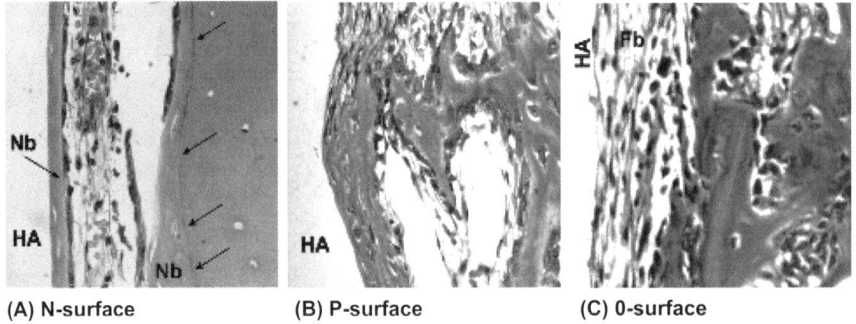

(A) N-surface   **(B) P-surface**   **(C) 0-surface**

Figure 17.8   Femoral tissues of dogs in the vicinity of (A) N-surface, (B) S-surface and
(C) 0-surface.

surfaces (0-surfaces) and the bone, the newly formed bones were isolated from
the surfaces and surrounded with fibrin layers (Figure 17.8C). A large amount
of blood vessels, observed as vacuoles, occupied the space among the osteoid
tissues. Fibrin layers with blood capillaries, including fibroblastic cells with
fusiform nuclei, were observed. The charged surfaces of the HA ceramics
promoted bone reconstruction in wide defects, whereas the processes varied
according to the charge polarity.

17.3.2   Radioactive Vector Materials

Radioactive[23–31] and ferromagnetic[31–41] ceramic microspheres have been used
for intra-arterial therapies of primary and metastatic tumours.[31,32] The
microsphere devices were injected into a target organ through an artery with a
catheter. They formed an embolization and affected the proximate tissue
without damage to the surrounding normal tissues.

The radioactive vector materials for intra-arterial therapies should fulfil the
following requirements: sufficient dose level, appropriate lifetime, effective range
of radioactivity, chemical durability in a biological environment and appropriate
size for embolization. A microspherical glass of $17Y_2O_3 \cdot 19Al_2O_3 \cdot 64SiO_2$
(YAS) was developed by Hyatt and Day[23] in 1987 as a pioneer in intra-arterial
radiation material. The 20–30 μm YAS microspheres activated into β-emitters by
neutron bombarding were administered to the hepatic artery using a catheter.
The YAS microspheres formed an embolization in the hepatic artery and blocked
the supply of nutrition to tumours (Figure 17.9).

Consequently, the YAS microspheres were predominantly found in the
tumours and selectively led to malnutrition of the tumours. The effective range
of the β-ray from activated ^{90}Y is approximately 2.5 mm, thus radionercrosis of
the normal tissues does not occur. The high chemical durability of YAS prevented dissolution of the ^{90}Y and its diffusion to normal tissues. The ^{90}Y
radioactivity, with a 64.1 h half-life period, decreased to an undetectable level
21 days later. After such successful preliminary results, the YAS microspheres

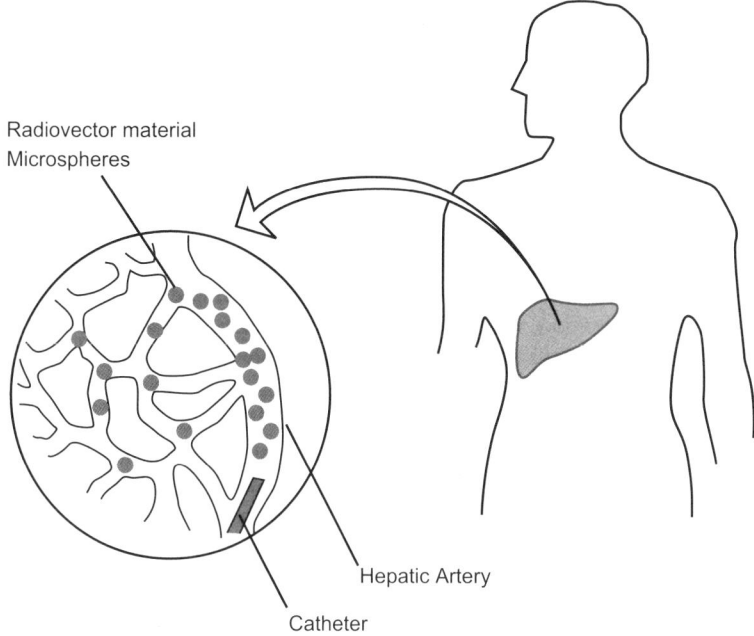

Radiovector material
Microspheres

Hepatic Artery

Catheter

Figure 17.9 Schematic illustration of hepatic arterial embolization in oncotherapy.

have been made commercially available as TheraSphere^R (MDS Nordion, Canada) and have been clinically applied to hepatic tumors in Canada, Hong Kong and USA. Treatment with the YAS glass microspheres enabled the patients to be treated in the hospital and return home in 1 day. However, the YAS glass microspheres required more than 2 weeks' neutron bombarding for activation because of the low concentration of Y_2O_3. Moreover, the activated YAS glass microspheres could not be preserved due to the shortness of the half-life period.

Rhenium glass microspheres composed of metallic rhenium particles dispersed in MgO-Al_2O_3-B_2O_3 glass matrix are also a candidate for radioactive vector materials.[27]

The rhenium glass microspheres were neutron-activated to a therapeutic level within several hours, whereas the YAS glass microspheres required more than 2 weeks for activation. That is because the thermal neutron cross sections of ^{186}Re and ^{188}Re (112×10^{-28} m^2 and 74.6×10^{-28} m^2, respectively) were approximately 100 times larger than that of ^{90}Y (1.28×10^{-28} m^2). The half-life periods of 90.6 h for ^{186}Re and 17.0 h for ^{188}Re were almost equal to that of ^{90}Y. The β-emissions of the ^{186}Re and ^{188}Re accompanied by γ-emissions had an advantage in determination of the injection position using a γ-camera, while the γ-emissions of the ^{90}Y were too weak to detect. Although the rhenium glass microspheres had a relatively high solubility, based on *in vitro* tests, the total body dose from the dissolution was estimated as approximately 1 mGy in the case of a 100 Gy dose to a cancerous liver.

Crystalline Y_2O_3 microspheres were synthesized using a radio-frequency induction thermal plasma-melting method.[29] The obtained microspheres had a 20–30 μm diameter and noticeably higher chemical durability than the YSA glass, even under the acidic condition of pH 6 as a simulated condition of cancerous tissues. Moreover, the required time of neutron activation was remarkably shortened, because the Y_2O_3 amount in the crystalline Y_2O_3 microspheres was approximately 5 times higher than that of the YAS glass.

Neutron-activated phosphorus-containing microspheres are also under development for radio-embolization therapy. The ^{32}P of the neutron-activated phosphorus was a β-emitter without γ-emission. The half-life period of 14.3 days for ^{32}P was significantly longer than that of ^{90}Y. Phosphorus-containing SiO_2 glass microspheres were successfully prepared by P^+ ion implantation at more than 200 keV.[30] Crystalline YPO_4 microspheres were also synthesized using a radio-frequency induction thermal plasma-melting method.[29]

17.3.3   Thermovector Materials

Ferromagnetic materials are heated by applied AC magnetic field. Therefore, the region contacting to the materials can locally heated by thermal diffusion. The local heating has been used for carcinoma therapy because the heat induces thermal cytotoxicity in cancerous cells. Moreover, the reduced bloodstream supplied to cancerous tissues resulted in a higher temperature increase due to the lack of coolant.

Magnetically mediated hyperthermia (MMH), which is a method of cancer therapy involving local heating of cancerous tissues using ferromagnetic materials, was introduced in 1954.[32] MMH had fewer side effects than chemotherapy and radiotherapy. Ceramic ferromagnetic materials have a relatively high chemical durability and have been applied to arterial embolization hyperthermia. Lithium ferrite ($LiFe_5O_8$) deposited Al_2O_3-P_2O_5-SiO_2 glass ceramics were developed as hyperthermia media. The glass ceramics were administered to the mammary carcinoma-transplanted femora of mice with an AC magnetic field of 10 kHz and 500 Oe.[33] The animal experiments revealed that the hyperthermia induced using the glass ceramics selectively provoked apoptosis of cancerous cells and was also effective for the prevention of repullulation, while the glass ceramics had an insufficient size, chemical durability and heat quantity for intra-arterial therapies.

Various attempts to develop ferromagnetic glass ceramics have been carried out, *e.g.* magnetite (Fe_3O_4) containing CaO-SiO_2-B_2O_3-P_2O_5,[34] and α-Fe containing CaO-SiO_2.[35] It was difficult to produce ferromagnetic ceramic microspheres with chemical durability, a thermal conversion efficiency and a diameter for hyperthermic embolization therapy. One of the prime candidates is crystalline magnetite microspheres synthesized by the post-heating of β-FeOOH particles prepared from SiO_2 microspheres and an Fe_3O_4 component containing a hydrofluoric solution.[36] The crystalline magnetite microspheres consisted of 50 nm magnetite domains and indicated an approximately 4 times higher heat quantity than the magnetite-containing glass ceramics in an AC magnetic field of 100 kHz and 300 0e, while further improvement has been required for clinical application. Another candidate is a colloidal sol of magnetite nanoparticles with a much higher specific absorption rate than ferrite microspheres in an AC magnetic field.[37–41] The magnetite–dextran[37–39,41] and magnetite–silan[41] complexes with embolic materials successfully showed arterial embolization and selective heating using animals.

Applications of radioactive vector and thermovector materials, as well as the combination of both materials, have far fewer restrictive side effects than chemotherapy and would confer remarkable advantages in cancer therapy.

17.3.4 Chemicovector Materials

Materials are dissolved, and the constituent ions and molecules are released from the surfaces of the materials. The released substances influence surrounding environments of the material. The effect is called the chemicovector effect, and materials which have chemicovector effects are chemicovector materials. In the biomedical field, the chemicovector effects have been applied to bone grafts. The representative bone graft Bioglass[R] of Na_{10}-CaO-SiO_2-P_2O_5 glass was developed by Hench in 1972.[42–44] The Bioglass[R] released ions and formed thin HA layers on the glass surfaces, and formed HA layers bonded to the bone tissues.[45] Consequently, the Bioglass[R] conducted new bone formation on the glass surfaces without any inclusions such as fibrous tissues, namely, the bioglass has osteoconductivity. The osteoconductive materials such as HA showed similar behaviour in bone tissues and were also categorized as bioactive materials.[46] Moreover, almost all of the osteoconductive materials exhibited apatite formabilities in the immersion test using SBF, but there were a few exceptions.[47–52] The SBF test was supported by the similarity of the interfacial reactions in biological and pseudobiological environments and widely accepted as an osteoconductivity evaluation.[53] The essentially required ions for osteoconduction were reported to be calcium ions.[54] On the basis of the SBF test, the glasses containing CaO and SiO_2, without P_2O_5, were recognized to be osteoconductive.[55] Of course body fluids contain supersaturated calcium and phosphorus ions, however, inhibitory substances (*e.g.* magnesium ions, carboxylic acids and certain proteins) prevented the precipitation of HA. The family of the CaO and SiO_2 glasses and the effects of the additives, such as Al_2O_3,[56] BaO,[57] MgO,[57] Na_2O,[58] Nb_2O_5,[59] Ta_2O_5[59] and TiO_2,[60] have been studied by many researchers.

In the prestigious works of Kokubo and colleagues,[46,47,61–64] A-W including crystalline oxyfluoroapatite [$Ca_{10}(PO_4)_6(O, F_2)$] and b-wollastonite ($CaSiO_3$) with MgO-CaO-SiO_2 residual glass released calcium ions into the surrounding body fluid and had favourable sites for apatite nucleation, while the glass ceramics showed a higher mechanical strength and bioactivity than those of dense HA ceramics and have been applied to bone grafts as Cerabone[R] A-W. Ceravital glass ceramics with a composition of CaO-SiO_2-P_2O_5-Na_2O-MgO-K_2O were developed[65] and have clinically applied to the replacement of the ossicular chain in the middle ear.[66] Bioverit I® and Bioverit II® are mica-apatite and mica-cordierite glass ceramics with chemical compositions from the base glass systems of $CaO-SiO_2-P_2O_5-Al_2O_3-MgO-Na_2O-K_2O-F$ and $SiO_2-Al_2O_3-Na_2O-MgO-K_2O-F$, respectively.[67,68] Bioverit III® prepared from ZrO_2 or $FeO/Fe2O3$-doped phosphate invert glass of $CaO-P_2O_5-Al_2O_3-MgO-Na_2O-K_2O-F$ contained microcrystals of apatite, berlinite ($AlPO_4$) and other complex phosphates.[69] The inverted glass ceramics with a composition of $CaO-P_2O_5-Na_2O-TiO_2$ without SiO_2 contained microcrystalline TCP [$Ca_3(PO_4)_2$] and indicated an osteo-conductivity as a chemicovector effect due to the release of calcium ions and apatite nucleation sites of β-TCP.[70,71]

Although β-TCP[72] did not have an apatite-forming ability in SBF due to its higher solubility,[47,49–52,73] β-TCP was proved to be osteoconductive in the bone tissues and has been clinically applied to bone grafts. The β-TCP single phase[74] and the β-TCP–HA biphase ceramics[75–77] were absorbed and substituted by newly formed bones in a shorter time compared to HA ceramics.

As mentioned above, the release of calcium ions and the appropriate nucleation sites of chemicovector ceramics directly affected the apatite formabilities of their surfaces and caused osteoconductivity. Therefore, the osteoconductivity mechanisms were not concerned with the biological process of bone formation but were medical modifications of the interface of the ceramics and the surrounding biological environment. Zinc-doped β-TCP and zinc-doped β-TCP–HA composite ceramics, designed by Ito *et al.*,[78,79] released zinc ions into the bone tissues. According to studies on bone metabolism, zinc activated the osteoblasts[80] but suppressed the osteoclasts.[81] The zinc-doped calcium phosphate ceramics promoted proliferation and differentiation of cultured osteoblastic cells derived from rats.[82] An implant of the zinc-doped calcium phosphate ceramics significantly raised the cortical bone apposition rates in the femora of a rat model for osteoporosis.[83,84] The zinc-doped calcium phosphate ceramics were not absorbed due to the suppressive effects of the released zinc ions to osteoclasts in the area directly contacting the bong tissues.[85] The ceramics producing osteoconductivity, which has been recognized a kind of chemicovector effect, have been clinically applied to bone grafts, artificial joints and tooth roots.

Another example of a chemicovector ceramic is TCP-added yttria-stabilized zirconia (YSZ) composites. Due to the excellent fracture strength and toughness, YSZ ceramics are used in orthopedic applications such as hip and knee joints. However, as YSZ ceramics are bioinert, they do not directly bond to natural bone in hip–joint replacements. The addition of a small amount of calcium phosphate to YSZ was developed in order to produce a composite biomaterial with a high bioactivity for a heavy load. The calcium phosphate employed here was TCP because of its solubility in aqueous solutions. An SBF immersion test proved the chemicovector effect of TCP added-YSZ; the surface of the undoped YSZ showed no change after immersion in SBF, whereas heavy coverage of bone-like crystals on the surface of the TCP-doped YSZ was observed (Figure 17.10).[86] The effect was attributed to the release of calcium and phosphate ions due to the dissolution of TCP particles.

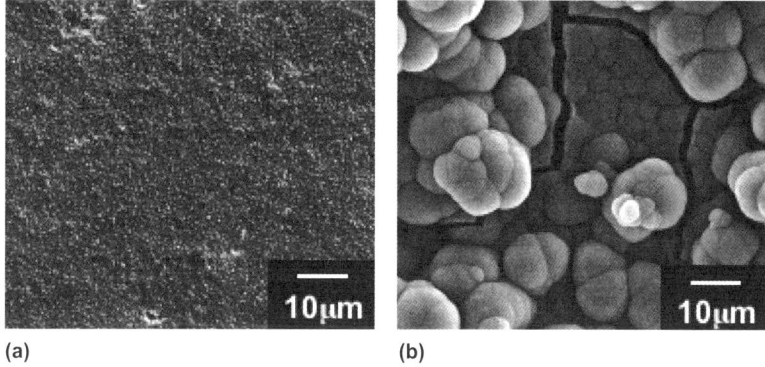

(a)   (b)

Figure 17.10 Scanning electron microscopy (SEM) micrograph of (a) undoped YSZ and (b) TCP–YSZ surfaces after 5 days of immersion in SBF.

17.3.5 Chemico-electrovector Materials

As mentioned above, a mixed vector effect is not expected to be easy to realize. We can present one example as a mixed vector ceramic; a chemicovector ceramic 45S5-type bioactive glass has been proved to produce an electrovector effect by polarization.[87–89] As a result, the polarized 45S5-type bioactive glass is a chemico-electrovector material. The bioglass contains 24.5 mass % sodium ions in the structure which generally reduces the rigidity of the SiO_4^{4-} and PO_4^{3-} tetrahedral networks. Taking this into consideration, we experimentally revealed the rather high ionic mobility or conductivity by complex impedance measurements, and established the polarization method for the 45S5-type bioactive glasses on the basis of the ionic conductivity results. The 45S5-type bioactive glass showed conductivity on the order of 10^{-4} cm^{-2} at 400–500 °C, and it can be polarized under a DC field of 50 V cm^{-1} at 500 °C. The polarized state of 45S5-type bioactive glass is proved by the thermally stimulated depolarized current (TSDC) method; depolarization peaks were definitely observed in the TSDC spectrum.

To demonstrate the electrovector effect of the polarized 45S5-type bioactive glass, SBF tests were employed.[87] The effect was seen in the morphologic difference among the 0-, N- and P-surfaces after immersion in SBF for the shorter period of 2 h (Figure 17.11). Although granular deposits with a diameter of approximately 100 pm covered the entire surface regardless of the polarization, the differences among the surfaces were denoted by the size of the deposited granules. On the 0-surface, the deposits were dense and in a thin micaceous, while the size of the granules on the P-surface was smaller than those of the 0- and N-surfaces. As no difference was seen among the three kinds of surfaces after a 24 h of immersion in SBF, the electrovector is effective during the early stage of nucleation and growth in the 45S5-type bioactive glass. In a sense, as shown above, the chemicovector effect surpassed the electrovector effect after the middle stage.

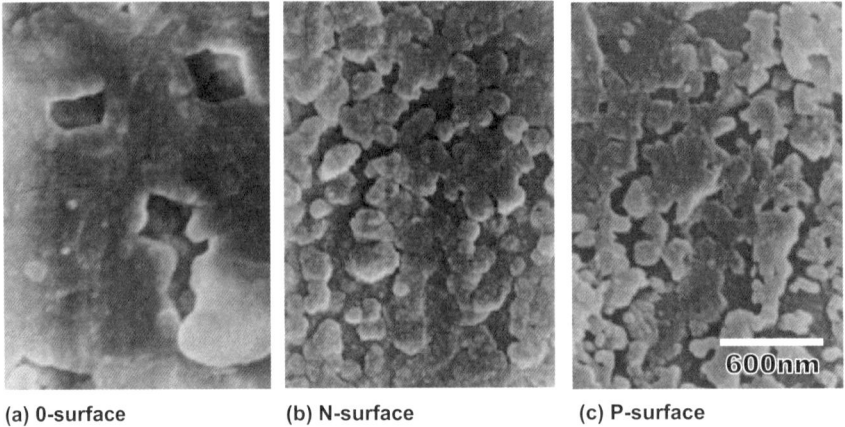

(a) 0-surface   **(b) N-surface**   **(c) P-surface**

Figure 17.11   Scanning electron microscopy (SEM) micrograph of 45S5-type bioactive glass surfaces after 2 h of immersion in SBF.

17.4   Conclusions

The concept of vector ceramics is expected to open a new way for the innovation or creation of ceramic science. Our first finding of electrovector effects in
polarized HA ceramics has lead to the discovery of new bone formation
mechanisms. The chemicovector effects of the 45S5type-bioactive glass have
inspired us to develop new bioglasses. Radioactive vector ceramics have been
commercialized and used for clinical use. A family of vector ceramics is
expected to be developed not only for material science, but also for novel
devices in the near future.

References

1. D. A. Puleo and A. Nanci, *Biomaterials*, 1999, **20**, 2311–2321.
2. Z. Schwartz and B. D. Boyan, *J. Cell. Biochem.*, 1994, **56**, 340–347.
3. E. Ruoslahti and B. Obrink, *Exp. Cell Res.*, 1996, **227**, 1–11.
4. K. Yamashita, *Phosphorus Sulfur*, 2002, **177**, 1889–1892.
5. K. Yamashita, N. Oikawa and T. Umegaki, *Chem. Mater.*, 1996, **8**, 2697–
   2700.
6. T. Kobayashi, S. Nakamura and K. Yamashita, *J. Biomed. Mater. Res.*,
   2001, **57**, 477–484.
7. M. Ueshima, S. Tanaka, S. Nakamura, T. Kobayashi and K. Yamashita,
   J. Biomed. Mater. Res., 2002, **60**, 578–584.
8. S. Nakamura, T. Kobayashi and K. Yamashita, *J. Biomed Mater. Res.*,
   2002, **61**, 593–599.
9. S. Nakamura, T. Kobayashi and K. Yamashita, *J. Biomed. Mater. Res. A*,
   2004, **68**, 90–94.

10. M. Ohgaki, T. Kizuki, M. Katsura and K. Yamashita, *J. Biomed. Mater. Res. A*, 2001, **57**, 366–373.
11. H. Takeda, Y. Seki, S. Nakamura and K. Yamashita, *J. Mater. Chem.*, 2002, **12**, 2490–2495.
12. M. Ueshima, S. Nakamura, M. Ohgaki and K. Yamashita, *Solid State Ionics*, 2002, **151**, 29–34.
13. Y. Toyama, M. Ohgaki, S. Nakamura, K. Katayama and K. Yamashita, *Solid State Ionics*, 2002, **151**, 159–163.
14. K. Yamashita, K. Kitagaki and T. Umegaki, *J. Am. Ceram. Soc.*, 1995, **78**, 1191–1197.
15. K. Yamashita, H. Owada, H. Nakagawa, T. Umegaki and T. Kanazawa, *J. Am. Ceram. Soc.*, 1986, **69**, 590–594.
16. H. Owada, K. Yamashita, T. Urnegaki, T. Kanazawa and M. Nagai, *J. Ceram. Soc. Jpn*, 1986, **94**, 837–841.
17. T. Kokubo, H. Kushitani, S. Sakka, T. Kitsugi and T. Yamarnuro, *J. Biomed. Mater. Res.*, 1990, **24**, 721–734.
18. K. Yamashita, H. Owada, T. Limegaki, T. Kanazawa and T. Futagami, *Solid State Ionics*, 1988, **28–30**, 660–663.
19. H. Owada, K. Yamashita, T. Umegaki, T. Kanazawa and M. Nagai, *Solid State Ionics*, 1989, **35**, 401–404.
20. K. Yamashita, H. Owada, T. Umegaki, T. Kanazawa and K. Katayama, *Solid State Ionics*, 1990, **40–41**, 918–921.
21. S. Nakamura, H. Takeda and K. Yamashita, *J. Appl. Phys.*, 2001, **89**, 5386–5392.
22. M. Ueshima, S. Nakamura and K. Yamashita, *Adv. Mater.*, 2002, **14**, 591–595.
23. M. J. Hyatt and D. E. Day, *J. Am. Ceram. Soc.*, 1987, **70**, C283–C287.
24. E. M. Erbe and D. E. Day, *J. Biomed. Mater. Res.*, 1987, **27**, 1301–1308.
25. G. J. Ehrhardt and D. E. Day, *Nucl. Med. Biol.*, 1987, **14**, 233–242.
26. D. E. Day and T. E. Day, in *Introduction to Bioceramics*, ed. L. L. Hench and J. Wilson, World Science, Singapore, 1993, p. 305.
27. S. D. ConZon, U. O. Häfeli and D. E. Day, *J. Biomed. Mater. Res.*, 1998, **42**, 617–625.
28. U. O. Häfeli, S. Casillas, D. W. Dietz, G. J. Pauer, L. A. Rybicki, S. D. Conzone and D. E. Day, *Int. J. Radiat. Oncol. Biol. Phys.*, 1999, **44**, 189–199.
29. M. Kawashita, R. Shineha, H. M. Kim, T. Kokubo, Y. Inoue, N. Araki, Y. Nagata, M. Hiraoka and Y. Sawada, *Biomaterials*, 2003, **24**, 2955–2963.
30. M. Kawashita, F. Miyaji, T. Kokubo, G. H. Takaoka, I. Yamada, Y. Suzuki and K. Kajiyama, *J. Am. Ceram. Soc.*, 1999, **82**, 683–688.
31. J. Overgaard, D. G. Gonzalez, M. C. C. M. Hulshof, G. Arcangcli, O. Dahl, O. Melia and S. M. Bentzen, *Lancet*, 1995, **345**, 540–543.
32. C. Breedis and G. Young, *Am. J. Path.*, 1954, **30**, 969–977.
33. A. A. Luderer, N. F. Borrelli, J. N. Panzarino, G. R. Mansfield, D. M. Hess, J. L. Brown, E. H. Barnett and E. W. Hahn, *Radiol. Res.*, 1983, **94**, 190–198.

34. Y. Ebisawa, Y. Sugimoto, T. Hayashi, T. Kokubo, K. Ohkura and T. Yamamuro, *J. Ceram. Soc. Jpn.*, 1991, **99**, 7–13.
35. H. Konaka, F. Miyaji and T. Kokubo, *J. Ceram. Soc. Jpn*, 1997, **105**, 833–836.
36. M. Kawashita, M. Tanaka, T. Kokubo, T. Yao, S. Hamada and T. Shinjo, in *Bioceramics 14: Key Engineering Materials*, ed. S. Brown, I. Clarke and P. Williams, TransTech Publications, Switzerland, 2002, p. 645.
37. M. Mitsunciori, M. Hiraoka, T. Shibata, Y. Okuno, S. Masunaga, M. Koishi, K. Okajima, Y. Nagata, Y. Nishimura and M. Abe, *Int. J. Hyperthermia*, 1994, **10**, 785–793.
38. M. Mitsunciori, M. Hiraoka, T. Shibata, Y. Okuno, Y. Nagata, Y. Nishimura, M. Abe, M. Hasegawa, H. Nagae and Y. Ebisawa, *Hepatogastroenterology*, 1996, **43**, 1431–1437.
39. A. Jordan, R. Scholz, P. Wust, H. Fahling, J. Krause, W. Wlodarczyk, B. Sander, T. Vogl and R. Felix, *Int. J. Hyperthermia*, 1997, **13** 587–605.
40. A. Jordan, R. Scholz, P. Wust, T. Schiestel, H. Schmidt and R. Felix, *J. Magn. Mangn. Mater.*, 1999, **194**, 185–196.
41. P. Moroz, S. K. Jones and B. N. Gray, *Int. J. Hyperthermia*, 2002, **18**, 267–284.
42. L. L. Hench, R. J. Splinger, W. C. Allen, T. K. Greenlee and Jr , *J. Biomed. Mater. Res.*, 1972, **2**, 117–141.
43. L. L. Hench and J. W. Wilson, *Science*, 1984, **226**, 630–636.
44. L. L. Hench, *J. Am. Ceram. Soc.*, 1998, **81**, 1705–1728.
45. T. Kokubo, *Biomaterials*, 1991, **12**, 155–163.
46. T. Kokubo, S. Ito, Z. T. Huang, T. Hayashi and S. Sakka, *J. Biomed. Mater. Res.*, 1990, **24**, 331–343.
47. C. Ohtsuki, T. Kokubo, M. Neo, S. Kotani, T. Yamamuro, T. Nakamura and Y. Banda, *Phosphorus Res. Bull.*, 1991, **1**, 191–196.
48. Y. Fajita, T. Yamamuro, T. Nakamura, S. Kotani, C. Ohtsuki and T. Kokubo, *J. Biomed. Mater. Res.*, 1991, **25**, 991–1003.
49. S. Kotani, Y. Fujita, T. Kitsugi, T. Nakamura, T. Yamamuro, C. Ohtsuki and T. Kokubo, *J. Biomed. Muter. Res.*, 1991, **25**, 1303–1315.
50. M. Neo, S. Kotani, Y. Fujita, T. Nakamura, T. Yamamuro, Y. Bando, C. Ohtsuki and T. Kokubo, *J. Biomed. Mater. Res.*, 1992, **26**, 255–267.
51. M. Neo, S. Kotani, T. Nakamura, T. Yanaamuro, C. Ohtsuki, T. Kokubo and Y. Bando, *J. Biomed. Mater. Res.*, 1992, **26**, 1419–1432.
52. M. Neo, T. Nakamura, C. Ohtsuki, T. Kokubo and T. Yamamuro, *J. Biomed. Mater*, 1993, **27**, 999–1006.
53. A. Oyane, H. M. Kim, T. Furuya, T. Kokubo, T. Miyazaki and T. Nakamura, *J. Biomed. Mater. Res.*, 2003, **65A**, 188–195.
54. C. Ohtsuki, T. Kokubo, K. Takatsuka and T. Yamamuro, *J. Ceram. Soc. Jpn.*, 1991, **99**, 1–6.
55. C. Ohtsuki, T. Kokubo and T. Yamamuro, *J. Non-Cryst. Solids*, 1992, **143**, 84–92.

56. C. Ohtsuki, T. Kokubo and T. Yamamuro, *J. Mater. Sci. Mater. Med.*, 1992, **3**, 119–125.
57. K. Tsuru, C. Ohtsuki and A. Osaka, *Proc. XII International Congress Glass*, 1995, **5**, 85–90.
58. H. M. Kim, F. Miyaji, T. Kokubo, C. Ohtsuki, T. Nakamu and T. Yarnamuro, *J. Am. Ceram. Soc.*, 1995, **78**, 2405–2411.
59. N. Imayoshi, S. Hayakawa, C. Ohtsuki and A. Osaka, *Mem. Fac. Eng., Okayama Univ.*, 1997, **31**, 39–44.
60. C. Otsuki, A. Osaka and T. Kokubo, in *Bioceramics 7*, ed. O. H. Anderson and A. Yli-Urpo, Butterworth–Heinemann, 1994, p. 73.
61. T. Kokubo, M. Shigematsu, Y. Nagashima, M. Tashiro, T. Yamamuro and S. Higashi, *Bull. Inst. Chem. Res., Kyoto Univ.*, 1982, **60**, 260–268.
62. T. Kokubo, S. Ito, S. Sakka and T. Yamamuro, *J. Mater. Sci.*, 1986, **21**, 536–540.
63. T. Kitsugi, T. Yamamuro, T. Nakamura, T. Kakutani, Y. Hayashi, S. Ito, T. Kokubo and T. Shibuya, *J. Biol. Mater. Res.*, 1987, **21**, 467–484.
64. K. Ono, T. Yamamuro, T. Nakamura and T. Kokubo, *Biomaterials*, 1990, **11**, 265–271.
65. H. Brömer, E. Pfeil and H. Käs, *German Patent 2*, 1973, **326**, 100.
66. R. Reck, *Laryngoscope*, 1984, **94**(Suppl. 33), 1.
67. W. Holand, W. Vogel, K. Naumann and J. Gurnmel, *J. Biomed. Mater. Res.*, 1985, **19**, 303–312.
68. W. Holand, P. Wange, K. Naumann, J. Vogel, G. Carl, C. Jana and W. Gotz, *J. Non-Cryst. Solids*, 1991, **129**, 152–162.
69. P. Wang, W. Vogel, L. Horn, W. Höland and W. Vogel, *Silic. Incl.*, 1990, **7–8**, 231–236.
70. T. Kasuga, T. Fujimoto and M. Nogami, *J. Ceram. Soc. Jpn.*, 2003, **111**, 633–635.
71. H. Maeda, T. Kasuga and M. Nogami, *J. Eur. Ceram. Soc.*, 2004, **24**, 2125–2130.
72. T. D. DriskelI, M. J. O'Hara, H. D. Sheets, Jr, G. W. Greene, Jr, J. R. Natiella and Armitage, *J. Biomed. Mater. Res.*, 1972, **6**, 345–361.
73. M. Neo, T. Nakamura, C. Ohtsuki, T. Kokubo and T. Yamamuro, *J. Biomed. Mater. Res.*, 1993, **27**, 999–1006.
74. M. Ozawa, *Seitaizairyou*, 1995, **13**, 167–175.
75. Y. Abe, T. Kasuga, H. Hosono and K. de Groot, *J. Am. Ceram. Soc.*, 1984, **67**, C142–C144.
76. C. P. Klein, Y. Abe, H. Hosono and K. de Groot, *Biomaterials*, 1984, **5**, 362–364.
77. C. P. Klein, Y. Abe, H. Hosono and K. de Groot, *Biomaterials*, 1987, **8**, 234–236.
78. A. Ito, K. Ojirna, H. Naito, N. Ichinose and T. Tateishi, *J. Biomed. Mater. Res.*, 2000, **50**, 178–183.
79. A. Ito, H. Kawamura, M. Otsuka, M. Ikeuchi, H. Ohgushi, K. Ishikawa, K. Onuma, N. Kanzaki, Y. Sogo and N. Ichinose, *Mater. Sci. Eng., C*, 2002, **22**, 21–25.

80. M. Yamaguchi, H. Oishi and Y. Suketa, *Biochem. Pharmacol.*, 1987, **22**, 4007–4012.

81. B. S. Moonga and D. W. Dempster, *J. Bone Min. Res.*, 1995, **10**, 453–457.

82. M. Ikeuchi, A. Ito, Y. Dohi, H. Ohgushi, H. Shimaoka, K. Yonernasu and T. Tatcishi, *J. Biomed. Mater. Res.*, 2003, **67A**, 1115–1122.

83. H. Kawamura, A. Ito, S. Miyakawa, P. Layrolle, K. Qiima, N. Ichinose and T. Tateishi, *J. Biomed. Mater. Res.*, 2000, **50**, 184–190.

84. H. Kawamura, A. Ito, T. Murarnatsu, S. Miyakawa, N. Ochiai and T. Tateishi, *J. Biomed. Mater. Res.*, 2003, **65A**, 468–474.

85. A. Ito, H. Kawamura, S. Miyakawa, P. Layrolle, N. Kanzaki, G. Treboux, K. Onuma and S. Tsutsumi, *J. Biomed. Mater. Res.*, 2002, **60**, 224–231.

86. M. Inuzuka, S. Nakamura, S. Kishi, K. Yoshida, K. Hashimoto, Y. Toda and K. Yamashita, *Solid State Ionics*, 2004, **172**, 509–513.

87. A. Obata, S. Nakamura, Y. Moriyoshi and K. Yamashita, *J. Biomed. Mater. Res.*, 2003, **67A**, 413–420.

88. A. Obata, S. Nakamura and K. Yamashita, *Biomaterials*, 2004, **25**, 5163–5169.

89. A. Ohata, S. Nakamura, Y. Sekijima and K. Yamashita, *J. Ceram. Soc. Jpn.*, 2004, **112** (Suppl.), S822–S825.

Subject Index

Note: Figures are indicated by *italic page numbers*, Tables by **emboldened numbers**